Biophysics of DNA

Surveying the last 60 years of research, this book describes the physical properties of DNA in the context of its biological functioning. It is designed to enable both students and researchers of molecular biology, biochemistry and physics to better understand the biophysics of DNA, addressing key questions and facilitating further research.

The chapters integrate theoretical and experimental approaches, emphasizing throughout the importance of a quantitative knowledge of physical properties in building and analyzing models of DNA functioning. For example, the book shows how the relationship between DNA mechanical properties and the sequence specificity of DNA–protein binding can be analyzed quantitatively by using our current knowledge of the physical and structural properties of DNA. Theoretical models and experimental methods in the field are critically considered to enable the reader to engage effectively with the current scientific literature on the physical properties of DNA.

Alexander Vologodskii began his research career as a theorist, pioneering theoretical studies of knots and links in circular DNA and designing statistical mechanical models and computational methods to predict the appearance of alternative structures in supercoiled DNA. He later ran a research laboratory at New York University, where he made important contributions to topics related to DNA topology and supercoiling, DNA topoisomerases and the physics of DNA bending.

Biophysics of DNA

ALEXANDER VOLOGODSKII

Formerly Research Professor at New York University

CAMBRIDGE
UNIVERSITY PRESS

CAMBRIDGE
UNIVERSITY PRESS

University Printing House, Cambridge CB2 8BS, United Kingdom

One Liberty Plaza, 20th Floor, New York, NY 10006, USA

477 Williamstown Road, Port Melbourne, VIC 3207, Australia

314-321, 3rd Floor, Plot 3, Splendor Forum, Jasola District Centre, New Delhi - 110025, India

79 Anson Road, #06-04/06, Singapore 079906

Cambridge University Press is part of the University of Cambridge.

It furthers the University's mission by disseminating knowledge in the pursuit of education, learning and research at the highest international levels of excellence.

www.cambridge.org
Information on this title: www.cambridge.org/9781009045810

First published 2015
First paperback edition 2021

A catalogue record for this publication is available from the British Library

ISBN 978-1-107-03493-8 Hardback
ISBN 978-1-009-04581-0 Paperback

Cambridge University Press has no responsibility for the persistence or accuracy of URLs for external or third-party internet websites referred to in this publication, and does not guarantee that any content on such websites is, or will remain, accurate or appropriate.

Contents

Color plates are between pages 152 and 153.

Preface

This book presents a comprehensive survey of DNA physical properties. Over the last 60 years physical properties of DNA molecules have been studied in detail, and the author believes that it is the right time to review them. The book is intended for molecular biologists, biochemists and physicists who want to know more about DNA biophysics. In addition, the book briefly considers physical aspects of DNA interaction with proteins that bind DNA specifically and nonspecifically. The level of presentation assumes some knowledge of molecular biology and physical chemistry, mainly at the level of introductory courses.

In accordance with the author's own research experience, the book combines theoretical and experimental approaches to the properties. Wherever it is possible, the chapters start from the theoretical descriptions of the phenomena followed by the application of these descriptions to experimental studies of particular problems.

There are no descriptions of methods that are used in DNA studies, except short remarks on the application of the major biophysical methods in this field. These major methods have been described in detail in other books on molecular biophysics (van Holde *et al.* 1998, Bloomfield *et al.* 1999). Excellent descriptions of general principles of the physical chemistry, often used in this book, as well as descriptions of many methods can be found in the book by Tinoco *et al.* (1995). However, we analyze the methods that were developed specifically for DNA studies.

I am grateful to my colleagues, who made very valuable comments on the chapters of this book, Maxim Frank-Kamenetskii, Neville Kallenbach, Ioulia Rouzina, and to my wife, Maria Vologodskaia, for her help during the book preparation.

References

Bloomfield, V. A., Crothers, D. M. & Tinoco, I., Jr. (1999). *Nucleic Acids: Structures, Properties, and Functions*. Sausalito, CA: University Science Books.

Tinoco, I., Jr., Sauer, K. & Wang, J. C. (1995). *Physical Chemistry*. Upper Saddle River, NJ: Prentice-Hall.

van Holde, K. E., Johnson, W. C. & Ho, P. S. (1998). *Principles of Physical Biochemistry*. Upper Saddle River, NJ: Prentice Hall.

1 DNA structures

1.1 Chemical structure and conformational flexibility of single-stranded DNA

Single-stranded DNA (ssDNA) is the building base for the double helix and other DNA structures. All these structures are formed due to noncovalent interactions between the components of ssDNA. ssDNA consists of a backbone of repeating units and bases that are attached to each unit as side chains (Fig. 1.1).

An isolated part of the repeating unit that consists of a base and the sugar is called a *nucleoside*. If a phosphate group is added to the nucleoside, it becomes a 3'- or 5'- *nucleotide*, depending on where the phosphate group is bound. Each repeating unit of the backbone consists of sugar and phosphate and has six skeletal bonds. The backbone has clear directionality, and the method of numbering of carbon atoms of the sugar, shown in Fig. 1.1, identifies 3'–5' or 5'–3' directions. It is common to assume a 5'–3' direction of the polynucleotide chain when presenting a sequence of bases.

There is an important degree of freedom in isolated nucleosides that is related to rotation around the bond connecting the sugar and a base, the β-glycosidic bond. The rotation angle, χ, is measured with reference to the orientation of $O^{1'}$–$C^{1'}$ and N^9–C^8 bonds for purines and to the orientation of $O^{1'}$–$C^{1'}$ and N^1–C^6 bonds for pyrimidines. Although many different values of χ are sterically allowed, two major rotational isomers, called *anti* and *syn*, are particularly important. For the *anti* conformation χ is close to 0°, and for *syn* χ is around 210°. The conformations are diagramed in Fig. 1.2.

The bond lengths and the angles between the adjacent bonds do not change notably. The remarkable conformational flexibility of ssDNA is due to six rotation angles in each repeating unit of the backbone. Of course, the rotation angles cannot accept just any values, since there are many potential steric clashes between chemical groups of the unit.

The deoxyribose ring is also able to adopt different conformations. Four of the five atoms of the ring tend to be in the plane, while the fifth is out of the plane. Depending on which atom is out of the plane, $C^{2'}$ or $C^{3'}$, the ring can be in one of four conformations. In the $C^{2'}$-*endo* conformation $C^{2'}$ lies above the plane, on the same side as the base and 5'-carbon (Fig. 1.3). The $C^{3'}$-*exo* conformation is similar to $C^{2'}$-*endo*, although the $C^{3'}$ atom deviates from the plane more than $C^{2'}$. Correspondingly, in the $C^{3'}$-*endo* conformation $C^{3'}$ is located above the ring, and the $C^{2'}$-*exo* conformation is similar to $C^{3'}$-*endo*.

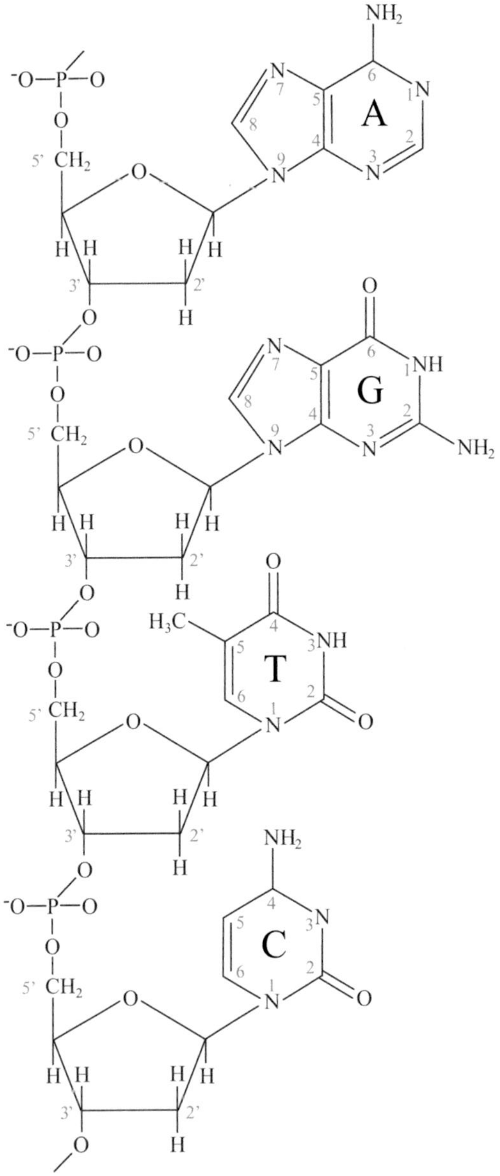

Figure 1.1 Chemical structure of ssDNA. Each repeating unit of the molecule backbone consists of 2′-deoxyribose and a negatively charged phosphate residue. The side chains are represented by one of four bases: adenine (A), guanine (G), thymine (T) or cytosine (C). Adenine and guanine belong to the group of purines, while the smaller thymine and cytosine are pyrimidines. The numbering systems of carbon and nitrogen atoms in the backbone and the bases are shown by gray digits.

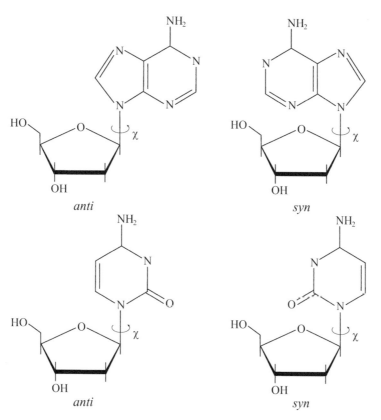

Figure 1.2 *Anti* and *syn* conformations of purines and pyrimidines. The actual conformations differ by 210° in angle χ and are only approximately represented in the figure.

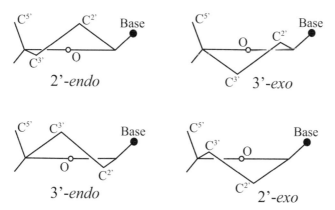

Figure 1.3 Major conformations of the deoxyribose ring. The plane containing O, $C^{1'}$ and $C^{4'}$ is perpendicular to the page.

The nucleosides are uncharged at pH values between 3 and 9, while the phosphate residues are negatively charged. Thus, each nucleotide of ssDNA has a single negative charge, with the exception of the end of the polynucleotide chain, where the phosphate group has a double negative charge at pH 7.

1.2   Double helices

The most important structural form of DNA is called the B form. It is this form of DNA, discovered by Watson and Crick in 1953 (Watson & Crick 1953), that started modern biology. It became clear later that DNA can adopt other double-stranded forms as well. All double-stranded forms share the most important structural feature, the base pairing. The base pairs are formed by bases from opposite strands. There are two canonical base pairs, AT and GC. Thus, a base in one strand determines the base at the corresponding position in the other strand. This means that the sequence of bases in one strand completely determines the sequence in the other strand. This is a key property for DNA biological functioning.

There are two hydrogen bonds in an AT base pair and three in a GC pair (Fig. 1.4). This difference in the number of hydrogen bonds is responsible for a higher thermal stability of GC base pairs (although base pairing per se does not stabilize the double helix (Yakovchuk *et al.* 2006)). The base pairs do not just have very close dimensions, but also their external geometries related to the backbones are nearly identical and have an important pseudosymmetry. The bonds between the nitrogens of the bases and $C^{1'}$ atoms of the sugar rings form the same angle of 51.5° with the $C^{1'}$–$C^{1'}$ line for all four bases of AT and GC base pairs (see Fig. 1.4). Base-pair geometry specifies the distance between $C^{1'}$ atoms across the pair, and this distance is exactly the same for both AT and GC pairs. Therefore, in either of two orientations (AT or TA, and GC or CG) the base pairs can be very well incorporated in a uniform helical structure of the backbones. It is this pseudosymmetry that makes Watson–Crick base pairing so unique. Many other possible patterns of base pairing do not have this symmetry and cannot be incorporated into a uniform structure of the backbones.

1.2.1   B-DNA

B-DNA, a form that the double helix has in aqueous solutions, is a right-handed helix with a helical period close to 10.5 bp per turn (Wang 1979, Peck & Wang 1981, Goulet *et al.* 1987). Its external diameter is approximately 2.0 nm (Dickerson & Ng 2001). The complementary strands have antiparallel orientations. The base pairs are located inside the helix, while the backbones are at the helix exterior (Fig. 1.5). The helix has two dyad symmetry axes that are perpendicular to the helix axis (assuming that the helix ends are extended to infinity). One of them passes through the plane of the bases (see Fig. 1.4). The other one passes between two base pairs. These are true symmetry axes for the backbone and pseudosymmetry axes for the base pairs. For

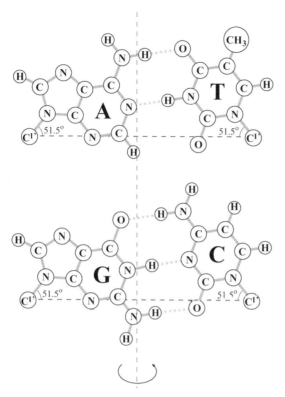

Figure 1.4 Complementary base pairs of DNA. The bonds directed to $C^{1'}$ atoms of the deoxyribose from the nitrogens of the bases form the same angles with the $C^{1'}$–$C^{1'}$ line. The dyad symmetry axis is shown by the dashed line. The distance between $C^{1'}$ atoms is equal to 1.085 nm for both base pairs.

the double helix formed by self-complementary polynucleotides poly(AT)poly(AT) and poly(GC)poly(GC), however, the latter axis is a true symmetry axis for the entire helix.

The average planes of adjacent base pairs are nearly perpendicular to the helix axis, although the base pairs are not flat. In B-DNA they form a propeller with 16° between the planes of the bases. The planes are separated by 0.34 nm. There is a strong interaction between the bases of adjacent base pairs, called the stacking interaction. This is the most important interaction stabilizing the double helix.

The helix has two grooves, the minor and major ones. The bases are more exposed in the major groove (see Fig. 5.5). The major groove is approximately 2.2 nm in width, while the width of the minor groove is close to 1.2 nm (Wing *et al.* 1980). The glycosyl angles in B-DNA correspond to *anti* conformations. The sugar ring has $C^{2'}$-*endo* conformation.

Structural parameters of B-DNA weakly depend on the sequence, although precise information on this dependence is limited. There are an enormous number of X-ray data on structures of oligonucleotide duplexes (Berman *et al.* 1992), but these structural data are affected by the forces of crystal packing. Clear indications for this follow, for example, from comparison between the average helical repeat of B-DNA, γ, in the

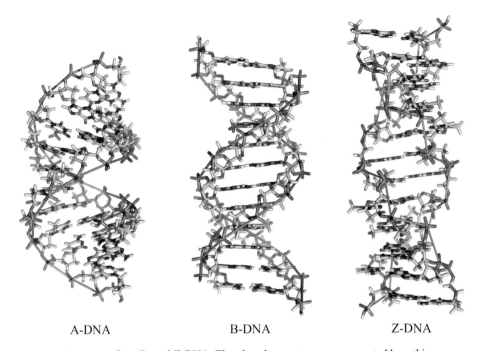

A-DNA   B-DNA   Z-DNA

Figure 1.5 Structure of A-, B- and Z-DNA. The phosphorus atoms are connected by a thin orange line to emphasize the geometry of the backbones. Image made by Richard Wheeler, published with permission. A black-and-white version of this figure will appear in some formats. For the color version, please refer to the plate section.

crystal structures and in solutions. While the X-ray data show that the average value of γ is equal to 10.0 bp/turn, it is close to 10.5 bp/turn in solution (see Section 3.2 for details). On the other hand, solution methods have not given sufficient information on the sequence variation of the conformational parameters. In general, thermal fluctuations of the helix parameters exceed their sequence variations (see Section 3.2).

1.2.2   A-DNA

A-DNA is also a right-handed double helix with the same rules of base pairing. B-DNA converts into the A form of the double helix in solutions with low water activity, such as concentrated solutions of ethanol. It is also formed in DNA fibers at humidities of less than 75%. A-DNA has also been found in cocrystals of some DNA–protein complexes. Since RNA duplexes cannot adopt the B conformation, the A form is their normal conformation in solution. A-DNA has 11 bp/turn (Krylov *et al.* 1990, Dickerson & Ng 2001). The base pairs are displaced from the helix axis, and this results in a larger helix diameter, 2.3 nm. The base pairs are tilted by 20° with respect to the helix axis (see Fig. 1.5). In A-DNA they form a propeller with even larger angle than in B-DNA, 19°, between the planes of the bases. It has smaller rise/turn, the distance between base pairs projected to the helix axis, 0.25 nm versus 0.34 nm in B-DNA. The sugar has

$C^{3'}$-*endo* conformation in this DNA form, and the glycosyl angles correspond to *anti* conformations.

1.2.3 Z-DNA

DNA conformation in the Z form is very different from those in the A and B forms (Rich *et al.* 1984). Z-DNA is a left-handed double helix rather than a right-handed one (Fig. 1.5). Its repeating unit consists of two base pairs rather than one. The line connecting the phosphorus atoms does not form a smooth curve, as it does in A- and B-DNA; it zigzags (initiating the name "Z form"). The glycosyl angles in each strand of Z-DNA alternate between *syn* and *anti* conformations, and in each base pair one base is in *anti* and the other is in *syn* conformation. The winding angle (the angle of the helix rotation between adjacent base pairs) also alternates between $-9°$ and $-51°$, for the *anti–syn* and *syn–anti* contacts, respectively. Thus, there are 12 bp/turn in Z-DNA. The sugar conformations also alternate along each polynucleotide chain: $C^{2'}$-*endo* and $C^{3'}$-*endo* correspond to *anti* and *syn* conformations of the nucleotide.

Each base pair has two sides. In Z-DNA the base pairs are inverted relative to the direction of the sugar–phosphate chains, as compared with A- and B-DNA. This inversion changes the conformation of the glycosyl angle for purines. In the case of pyrimidines the sugar is rotated together with the base, so they remain in *anti* conformations. Clearly, the necessity to rotate base pairs makes the transition between B and Z forms more difficult.

A linear double helix can adopt the Z form only at very high concentration of sodium ions, or in the presence of some multivalent ions (Rich *et al.* 1984). Even in these conditions Z-DNA can be formed nearly exclusively in duplexes with alternating purine–pyrimidine sequences. This very strong sequence requirement results from the very high energy of the *syn* conformation of pyrimidines. In DNA segments with alternating purine–pyrimidine sequences, all pyrimidines are in *anti* conformations while all purines are in *syn*. The most favorable sequence for Z-DNA is $d(GC)_n \cdot d(GC)_n$. Still, the sequence requirements are not strict, and under sufficient negative supercoiling many segments of natural DNA adopt the Z form (see Sections 2.4 and 6.5).

1.2.4 B'-DNA

B'-DNA is a conformation that is adopted by poly(dA)poly(dT) in a wide range of solution conditions. Although the structure is close to the B form, it has a narrower minor groove and a large positive propeller twist of the base pairs (Alexeev *et al.* 1987). It has been established that any tract with the sequence $d(A_m T_{n-m})$ adopts this conformation if $n \geq 4$ (see Hagerman (1990) for a review). B'-DNA is able to introduce substantial intrinsic curvature into DNA molecules. The bends appear, mainly, at the stacks between B and B' helices (Hagerman 1986, Koo *et al.* 1986). For the track of six As in a row the total bend is close to $19°$ (Koo *et al.* 1990, MacDonald *et al.* 2001). If these bends are repeated in phase with DNA helical periodicity, they can introduce a substantial intrinsic bend in the double helix. This intrinsic curvature was first discovered

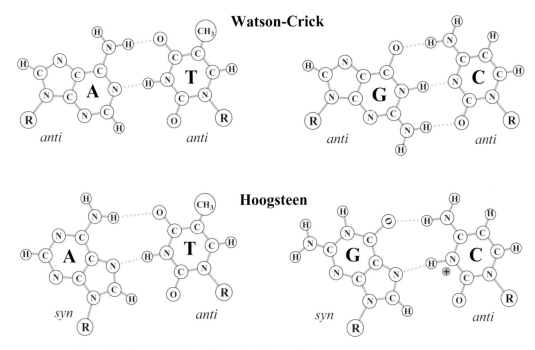

Figure 1.6 Watson–Crick and Hoogsteen base pairing.

in the fragments isolated from kinetoplast DNA that contains phased A-tracts of five to eight nucleotides (Marini *et al.* 1982). It was surprising that X-ray studies of A-tracts did not detect DNA bends observed in solution. The puzzle was solved when it was found that in solutions of 2-methyl-2,4-pentanediol, the dehydrating agent commonly used in crystallization of oligonucleotides, the intrinsic curvature associated with the A-tracts nearly disappears (Sprous *et al.* 1995). Probably this happens due to B′–B transition in the A-tracts.

In addition to the canonical base pairs, many other patterns of base pairing are possible, and some of them have been observed experimentally (Egli & Saenger 1984, Sinden 1994). Some patterns were found in pairs formed by modified bases or in DNA–protein complexes. Probably the most important among these are Hoogsteen and reverse Hoogsteen base pairs (Fig. 1.6), since they were found in triple helices.

1.3   Triple helices

A third polynucleotide chain can be incorporated in the major groove of B-DNA. Although the double helix participating in the triplex formation changes its conformation to some extent, it maintains the major features of B-DNA. In particular, the glycosyl angles correspond to *anti* conformations and the sugars are in C2′-*endo* conformations. The helical repeat of the triple helix corresponds to 12 triads per helix turn (Raghunathan

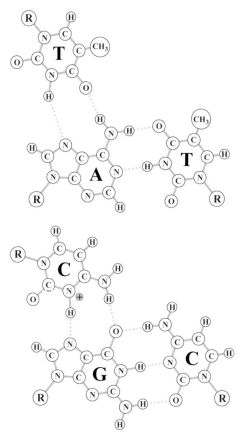

Figure 1.7 Base triads in the Hoogsteen triplexes. The cytosine of the third strand is protonated in the N3 position.

et al. 1993). Over all the varieties of the triple helices the third strand has to form hydrogen bonds with purines of the double helix (Frank-Kamenetskii & Mirkin 1995). This requirement, due to structural restrictions, reduces the types of sequence of B-DNA able to form the triple-stranded helices to homopurine–homopyrimidine segments. The patterns of hydrogen bonding between purines of the B-like DNA and bases of the third strand correspond to Hoogsteen and reverse Hoogsteen base pairs (where the planes of the bases have the same orientations as in Watson–Crick base pairs). According to these patterns, the triple helices can be divided into two groups.

Triplexes with Hoogsteen base pairing. The third strand consists of all pyrimidines and forms Hoogsteen base pairs TA and CG. The two strands connected by Hoogsteen base pairs have to be parallel to each other. Thus, the third strand has to be antiparallel to the pyrimidine strand of the Watson–Crick duplex, and its sequence has to be the reverse of the sequence of that strand. The patterns of hydrogen bonds in the triads of Hoogsteen triplexes are shown in Fig. 1.7. It is important that cytosine participating in the Hoogsteen pairing is protonated in the N3 position.

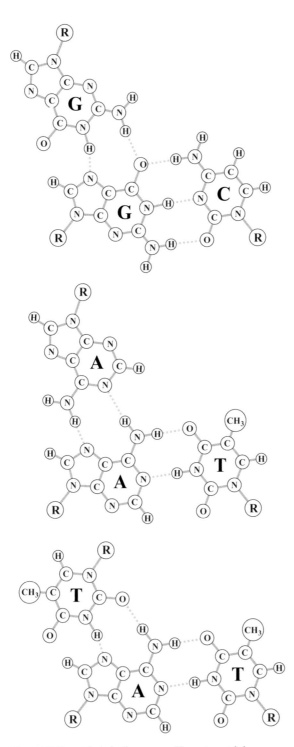

Figure 1.8 Base triads in the reverse Hoogsteen triplexes.

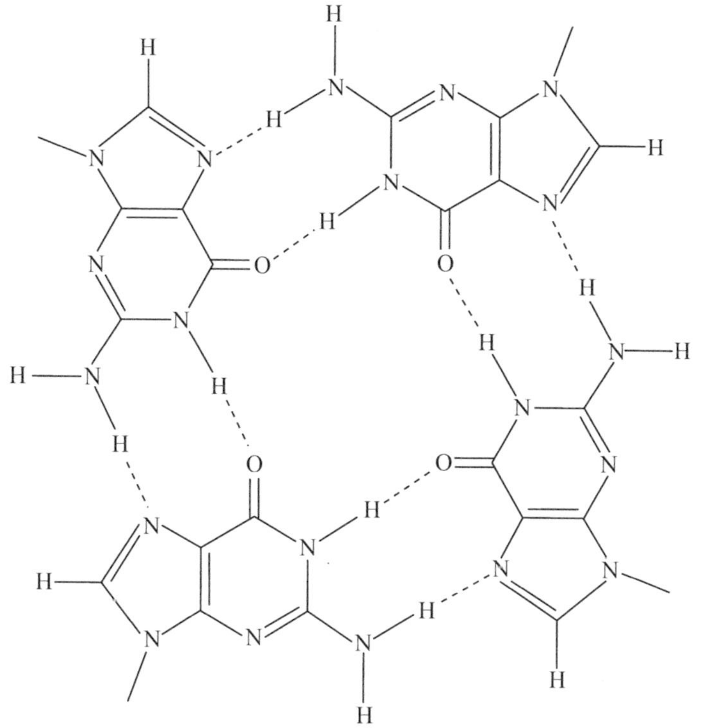

Figure 1.9 The pattern of hydrogen bonds in the G-quartet. The guanines serve here as both donors and acceptors of hydrogen bonds. The quartet is stabilized by K^+ or, to a lesser extent, by Na^+ that is coordinated at the center and interacts with electronegative carbonyl oxygens.

Triplexes with reverse Hoogsteen base pairing. It was assumed initially that in this case the third strand has to consist of purines to form triads A∗AT and G∗GC. It was found later, however, that triad T∗AT can be incorporated into the triplex and even increases its thermal stability (Beal & Dervan 1991). The patterns of hydrogen bonds in the triads of the reverse Hoogsteen triplexes are shown in Fig. 1.8. The two DNA strands connected by the reverse Hoogsteen pairing are antiparallel to each other.

Of course, any deviation of the sequence requirements for the triplexes results in a substantial reduction of their stability.

1.4   G-quadruplexes

It has been known for a long time that a concentrated solution of guanylic acid (GMP) forms a gel (Bang 1910). In 1962, Gellert and co-authors (Gellert *et al.* 1962) investigated the phenomenon in detail and concluded that the guanines form planar quartets where each base is connected with two others by two hydrogen bonds (Fig. 1.9). ssDNA molecules with G-repeats can form four-stranded structures based on the G-quartets.

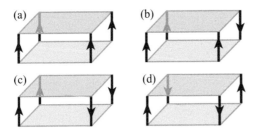

Figure 1.10 Possible orientations of DNA strands in the G-quadruplexes.

Interest in the structures involving G-quartets jumped when multiple guanine-rich repeats were found at telomeres, termini of eukaryotic chromosomes. In human telomeres, for example, the GGGTTA sequence is repeated thousands of times. In particular, the repeats constitute the single-stranded region at the telomere end, which can reach 300 nucleotides in length. It is well established that ssDNA molecules with the corresponding sequences form four-stranded structures in solution at physiological ionic conditions and temperature.

The four-stranded structures based on G-quartets represent a stack of G-quartets arranged in a helical manner. Fiber diffraction of poly(G) suggested a quadruple right-handed helix structure with a rise of 0.34 nm and 30° helical twist, with all of the glycosidic angles in the *anti* conformation and all four strands having parallel orientations (Zimmerman *et al.* 1975). This structure is also formed by oligonucleotides d(TTAGGG) (Wang & Patel 1992). Thus, in the absence of loop constraints this structure has the lowest free energy among G-quadruplex structures. However, further studies showed that numerous four-stranded structures based on G-quartets are possible (reviewed by Williamson (1994), Lane *et al.* (2008) and Phan (2010)). The main variable feature of the structures is the polarity of the strands, which has been observed in all possible combinations (Fig. 1.10). The majority of the studied structures were intramolecular, formed by oligonucleotides with telomere-type sequences. In such intramolecular structures a DNA strand forms loops, whose geometry depends on the polarity of the strand segments in a particular structure. Although the necessity of the loop formation influences the structural preferences, numerous kinds of loop were found in the structures based on G-quartets. Oligonucleotide d(GTGGTGGGTGGGTGGGT), which inhibits HIV integrase (Wyatt *et al.* 1994) and for this reason was studied in detail, presents a surprising example. The oligonucleotide forms a parallel-stranded quadruplex structure with single-nucleotide loops (Li *et al.* 2010, Mukundan *et al.* 2011). The possibility of such small loops differentiates Q-quadruplexes from the hairpins with B-form stems, where the loops have to contain at least three nucleotides (see Section 2.2).

There is no simple rule that would allow us to predict what structure should be formed by a particular oligonucleotide with G-repeats. Although there are 26 possible arrangements of the loops in intramolecular G-quadruplexes, only some of them have been found experimentally, by X-ray or NMR analysis (Phan 2010). It is interesting that 2G-quartets are sufficient to form a stable quadruplex structure (Macaya *et al.* 1993). The glycosyl angles of guanines in G-quartets can correspond to either *anti* or *syn* conformations, correlating with the strand directionality (Williamson 1994, Phan 2010).

The folding of the intramolecular quadruplexes is associated with a relatively small gain in an oligonucleotide free energy that hardly exceeds 10 kcal/mol at 37 °C, when the structure contains three G-quartets (Lane *et al.* 2008). This is much lower than the free energy gain due to formation of the duplexes by two complementary oligonucleotides of the same length (considering that the length is 24 nucleotides), which should be in the range of 30 kcal/mol (see Table 2.1). This means that a G-quadruplex cannot be formed with a notable probability inside regular double-stranded DNA (dsDNA) in a segment containing G-repeats, since its formation has to be associated with the duplex unwinding. Formation of G-quadruplexes should be facilitated by negative supercoiling, since this would result in great reduction of the supercoiling free energy due to unwinding of the corresponding DNA segment. The estimation shows, however, that the reduction of the supercoiling free energy should make formation of quadruplexes favorable only at a superhelix density of approximately −0.06 (see Section 6.5). The torsional stress corresponding to this superhelix density is hardly achievable in vivo, and, due to the competition with other alternative DNA structures, may be difficult to obtain in vitro. Therefore, the negative supercoiling per se is probably insufficient to induce formation of the quadruplexes by unwinding a corresponding segment of the dsDNA. The formation of the quadruplexes can be additionally facilitated by proteins that specifically bind this DNA form. So far, however, formation of G-quadruplexes by unwinding segments of dsDNA has not been observed even in vitro.

Single-stranded overhangs at the ends of eukaryotic chromosomes are much more favorable candidates for quadruplex formation. In these cases the conformation is thermodynamically favorable and can only be blocked by proteins that bind a different conformation of the overhangs. The existence of G-quadruplexes at the overhangs remains to be proven, however.

The characteristic time of the quadruplex formation depends on reaction order. For the intramolecular folding of the quadruplexes this time is in the range of tens of milliseconds, while for the structures formed by four oligonucleotides the characteristic time can constitute many days (reviewed by Lane *et al.* (2008)).

1.5   Stacking in single-stranded DNA

A single DNA strand can form an ordered structure even without forming hydrogen bonds with other single-stranded segments. Stacking interaction between adjacent bases of poly(dA) and poly(dC) results in formation of a helical structure. This structure gradually disappears when temperature increases from negative values to nearly 100 °C (Dewey & Turner 1979). This melting of regular structure can be monitored by increase of light absorbance in the range of 230–280 nm or by changes in the circular dichroism spectrum. It was shown that this stacking interaction increases the size of the polynucleotide coils (Stannard & Felsenfeld 1975). This effect is large only in the vicinity of 0 °C, due to shortening of the length of the polynucleotide segments with continued stacking at higher temperatures (Eisenberg & Felsenfeld 1967). In general, disruption of stacking in single-stranded polynucleotides is a noncooperative process, so there is very little correlation between the stacking at adjacent dinucleotide steps. Studies of single-stranded poly- and oligonucleotides have been performed mainly for RNA molecules, although the results for poly(rA) and poly(dA), and poly(rC) and poly(dC), are close. It was shown that there is no notable stacking interaction in poly(rU) (Inners & Felsenfeld 1970), and we can suggest that the conclusion is valid for poly(dT) as well. Formation of stacking in the single-stranded polynucleotides is a fast process, with rate constants around 100 ns (Pörschke 1978, Dewey & Turner 1979).

The same stacking interaction takes place in ssDNA molecules with natural sequences. In such molecules, however, segments with complementary sequences can form hairpins with double-stranded stems. What fraction of nucleotides is involved in the hairpin formation depends on the DNA sequence. If hairpin loops are sufficiently small, interaction between the bases in these loops can selectively stabilize particular conformations of the loops. A clear illustration of such an effect came from the studies of interaction between complementary anticodons of tRNA. It was found that the equilibrium constant of association of complementary anticodons is three orders of magnitude larger than the association constant for the anticodon and complementary codon (Eisinger 1971). This finding is a strong indication that the preferable conformation of the anticodon in the loop is close to its conformation in the double helix, so the unfavorable change of entropy associated with formation of the double helix is dramatically reduced in this case. There are no similar data for DNA hairpin loops, but there are reports on unusually stable hairpin loops in DNA oligonucleotites (Yoshizawa *et al.* 1997, Moody & Bevilacqua 2004). Probably stacking interactions between the loop bases are involved there as well.

1.6   Structure of mismatches and bulge loops

Isolated mismatches incorporated into DNA duplexes disturb the structure only for adjacent base pair steps (Patel *et al.* 1984a, 1984b, Brown *et al.* 1985, Hunter *et al.* 1986a, 1986b). All mismatched bases form pairs with hydrogen bonds. Two examples of such base pairs are shown in Fig. 1.11. Although the external parameters of these base pairs do not exactly match the parameters of the canonical base pairs, they are relatively close to the latter (compare Fig. 1.4 and Fig. 1.11). A remarkable flexibility of the sugar–phosphate backbone allows easy accommodation of these base pairs into B, A or Z forms of the double helix. Still, mismatched base pairs destabilize the double helix to some extent (see Section 3.2).

Of course, the mismatched base pairs have to be recognized by DNA-repairing enzymes. The recognition is definitely possible due to the variety of functional groups exposed in the helix grooves and able to interact with the enzymes. It is not clear if a different conformation of the DNA backbone surrounding the mismatches contributes to the recognition.

The structure of the single-nucleotide bulge loops brings another manifestation of the backbone flexibility. It has been shown in the NMR studies that an extra base in one strand can be incorporated into the double helix without pairing with a base in the opposite strand (Hare *et al.* 1986, Woodson & Crothers 1988b, Woodson & Crothers 1989). Although the helix is slightly unwound at the adjacent steps of the duplex and

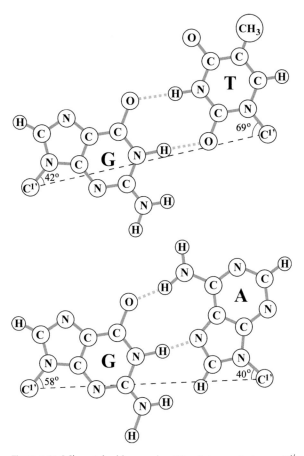

Figure 1.11 Mismatched base pairs. The distances between $C^{1'}$ atoms for these base pairs are equal to 1.03 and 1.07 nm, correspondingly, versus 1.085 nm in AT and GC base pairs.

kinked by 20°, the bases in the strand that lacks a nucleotide are separated by a much larger distance than in regular B-DNA, and the structure proves that this is possible as well. The disturbance of the double-helix conformation surrounding the unpaired base is mainly spread to the nearest base pairs, and nearly undetectable 3–4 bp away from the base. It should be noted that in some cases the unpaired base is looped out of the double helix, with the flanking base pairs stacked on each other (Kalnik *et al.* 1989, Morden & Maskos 1993).

1.7   Experimental and theoretical studies of DNA structure

1.7.1   X-ray analysis of DNA crystals and fibers

X-ray analysis of DNA fibers and crystals made the main contribution to our current knowledge of DNA structures. Fibers were used in the first few decades of DNA studies, but from 1979 the analysis of crystallized oligonucleotide duplexes became the method of choice in DNA structural studies. This change was mainly due to the breakthrough in oligonucleotide synthesis that made it possible to obtain DNA oligonucleotides with desired sequences at relatively low cost. While the analysis of fibers allows only determination of the sequence-averaged parameters, the crystal-based X-ray analysis allows one to investigate how these parameters depend on DNA sequence. It also allows one to study structures of duplexes with various chemical modifications and irregularities. It is interesting that the first structure of a DNA crystal decoded by Rich and co-workers resulted in the discovery of left-handed Z-DNA (Wang *et al.* 1979). The great majority of DNA duplexes decoded later were found, of course, in B form. The first such structure of a dodecamer duplex, obtained by Dickerson and co-workers (Wing *et al.* 1980), attracted a lot of attention in the biophysical community. The opportunity to study the sequence dependence of DNA structure looked very attractive at the beginning, and structures of many duplexes with different sequences were obtained at that time.

Pretty soon it became clear, however, that the crystal packing strongly affects the structures (Dickerson *et al.* 1987, DiGabriele *et al.* 1989). Substantially different structures were obtained for the same duplexes depending on the crystallization conditions (Berman 1997). Even the average parameters of B-DNA obtained from X-ray analysis of DNA duplexes were notably different from the parameters in solution. While the averaged helical repeat of B-DNA in solution is close to 10.5 bp per helix turn (Wang 1979, Peck & Wang 1981, Strauss *et al.* 1981), it is equal to 10.0 bp per turn in the crystallized duplexes that were found in B form (Gorin *et al.* 1995). Therefore, it is hardly possible to obtain the sequence variation of the helix parameters from the X-ray analysis of oligonucleotide duplexes, although some interesting conclusions have been derived from these data (Young *et al.* 1995). The priorities of X-ray studies of DNA duplexes shifted from the sequence dependence of DNA structure to the structures with various mismatches and chemical modifications. Although crystal packing should affect these structures as well, the packing effects are usually smaller than the disturbances caused by the chemical irregularities.

Decoded structures of oligonucleotides are deposited into the Nucleic Acid Database (Berman *et al.* 1992). At the beginning of 2014 the database contained more than 900 structures obtained by X-ray analysis of DNA duplexes. The database also contains nearly 3000 structures of DNA–protein complexes obtained by X-ray analysis of the crystals. X-ray analysis definitely continues to be the main method for structural studies of DNA-related problems.

1.7.2   NMR of small DNA molecules

The great advantage of NMR as a structural method is that it allows structure determination in solutions, where intermolecular interaction does not disturb the structures. However, NMR does not allow direct determination of DNA and protein structures as does X-ray analysis of crystals. The method only allows us to obtain some parameters of the structures, such as distances between close atoms, provided by the nuclear Overhauser effect (NOE). Molecular modeling is needed to reconstruct structures from this kind of data. Until the late 1990s it was nearly impossible to reconstruct global features of the structures, such as orientation of protein subunits and curvature of oligonucleotide duplexes, because NMR data only provided local structural constraints. The situation changed radically when the methods of measuring the residual dipolar couplings (RDCs) were developed (Tjandra & Bax 1997). RDCs allow us to obtain restrictions on orientation of internuclear vectors relative to a fixed vector that specifies the preferable orientation of the molecules in solution. This information is complementary to the internuclear distances provided by NOEs, and together these data allow us to obtain good reconstruction of both local and global structural features (see Vermeulen *et al.* (2000), Wu *et al.* (2003) for analysis of the procedural accuracy). This approach was first used to determine structures of oligonucleotide duplexes containing A-tracts (Vermeulen *et al.* 2000, MacDonald *et al.* 2001, Barbic *et al.* 2003, Wu *et al.* 2003). It is interesting that more and more accurate structure determination results in more and more regular duplexes (Wu *et al.* 2003). Even before the use of RDC in the structure determination, NMR studies showed that in normal solution conditions DNA duplexes always have B-conformation (Van de Ven & Hilbers 1988). Earlier X-ray studies had found many duplexes in A-form which was clearly a consequence of the packing forces and/or the crystallization conditions.

In general, application of RDC allows high-resolution structures of oligonucleotide duplexes in solution to be obtained. Although the method remains more laborious than X-ray analysis of the same DNA duplexes, the number of structures in the Nucleic Acid Database obtained by NMR is growing fast. The method is also used in structural studies of relatively small DNA–protein complexes. At the beginning of 2014 the database contained nearly 600 structures of DNA duplexes, mainly with various chemical modifications, and 120 DNA–protein complexes obtained by NMR.

1.7.3   Molecular dynamics simulation

If we know all forces acting between atoms of a macromolecule, molecules of the solvent, and ions that are present in the solution, we could try to explicitly model the dynamic behavior of the system by solving Newton's equations of motion. If such modeling is sufficiently accurate and can be extended to sufficiently long time intervals, it would solve the great majority of biophysical problems. Such modeling, molecular dynamics (MD) simulation, has been under intensive development since the late 1950s. Of course, the increasing power of computers allows us to consider larger and larger molecular systems and to extend the simulations to larger time intervals. Although impressive progress in this field has been achieved over the last two decades (see Lavery *et al.* (2010) for a review), two major internal problems of the method limit applications of MD simulations in DNA studies.

Properties of DNA molecules strongly depend on solution conditions, so water molecules and ions have to be included in the simulated system. As a result, even for very short DNA molecules the simulated systems include tens or hundreds of thousands of atoms. Therefore, even modern computers do not allow MD simulation of the system dynamics for time spans longer than a few hundred nanoseconds. This constitutes the first major problem of the method. Although time spans of 100 ns seem to be sufficient

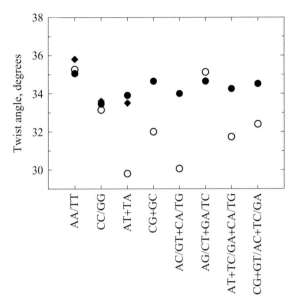

Figure 1.12 Comparison of the twist angles obtained in the extended MD simulations with the experimental data. To make the comparison possible the angle magnitudes for individual base-pair steps, obtained in the simulations, were reduced to the average values over eight linear combinations that were only available from the experiments. The combinations of the corresponding base-pair steps are shown along the horizontal axis. The simulation results (o) were taken from Lavery *et al.* (2010); the shown experimental data are based on the *j*-factors of short DNA fragments (●) (Geggier & Vologodskii 2010) and on the analysis of the topoisomer distribution in circular DNAs (♦) (Peck & Wang 1981, Strauss *et al.* 1981, Goulet *et al.* 1987).

to address conformational properties of the most populated states of DNA duplexes (Lavery *et al.* 2010), much longer simulation runs are needed to analyze transient states and conformational transitions. A special computer for MD simulations was designed and built recently that allows us to simulate dynamic trajectories 100–1000 times longer than universal computers, and this has brought very interesting results in the field of protein folding (Shaw *et al.* 2010). This is definitely a promising direction. However, there is another, probably even more serious, problem of the simulation method.

The force fields used in MD simulations do not provide precise energies of interactions between various atoms. These force fields are permanently improved, however. Clearly, only careful comparison with experimental data can allow us to judge if a particular force field is sufficiently accurate for a particular problem. An example of this kind of comparison is shown below. It is based on the results of the recent large project intended to determine the sequence dependence of DNA conformational properties by MD simulations. The simulations used the best force field that has been designed for DNA (Perez *et al.* 2007). Many research groups participated in the project, and together they were able to obtain statistically reliable results for conformational properties of 4 bp DNA segments with all possible sequences (Lavery *et al.* 2010). Figure 1.12 shows a comparison of the sequence dependence of the twist angle obtained in the simulations with available experimental data. The experimental data were obtained by different methods and tested internally, so they are quite reliable.

The comparison shows a substantial difference between the simulation results and the experimental data. One has to conclude that the force field used in this project is still not good enough to address the sequence dependence of DNA conformational parameters. Maybe this force field is sufficiently good to analyze other interesting problems related to DNA conformational properties, but they require longer simulation runs and careful testing of the computational method. It seems, however, that MD simulation has the potential to become a very powerful and unique research instrument.

References

Alexeev, D. G., Lipanov, A. A. & Skuratovskii, I. Y. (1987). Poly(dA).poly(dT) is a B-type double helix with a distinctively narrow minor groove. *Nature* 325, 821–3.

Bang, I. (1910). Examination or the guanyle acid. *Biochem. Z.* 26, 293–311.

Barbic, A., Zimmer, D. P. & Crothers, D. M. (2003). Structural origins of adenine-tract bending. *Proc. Natl. Acad. Sci. U. S. A.* 100, 2369–73.

Beal, P. A. & Dervan, P. B. (1991). 2nd structural motif for recognition of DNA by oligonucleotide-directed triple-helix formation. *Science* 251, 1360–3.

Berman, H. M. (1997). Crystal studies of B-DNA: the answers and the questions. *Biopolymers* 44, 23–44.

Berman, H. M., Olson, W. K., Beveridge, D. L., Westbrook, J., Gelbin, A., Demeny, T., Hsieh, S. H., Srinivasan, A. R. & Schneider, B. (1992). The nucleic acid database. A comprehensive relational database of three-dimensional structures of nucleic acids. *Biophys. J.* 63, 751–9.

Brown, T., Kennard, O., Kneale, G. & Rabinovich, D. (1985). High-resolution structure of a DNA helix containing mismatched base pairs. *Nature* 315, 604–6.

Dewey, T. G. & Turner, D. H. (1979). Laser temperature-jump study of stacking in adenylic acid polymers. *Biochemistry* 18, 5757–62.

Dickerson, R. E., Goodsell, D. S., Kopka, M. L. & Pjura, P. E. (1987). The effect of crystal packing on oligonucleotide double helix structure. *J. Biomol. Struct. Dyn.* 5, 557–79.

Dickerson, R. E. & Ng, H.-L. (2001). DNA structure from A to B. Proc. Natl. *Acad. Sci. U. S. A.* 98, 6986–8.

DiGabriele, A. D., Sanderson, M. R. & Steitz, T. A. (1989). Crystal lattice packing is important in determining the bend of a DNA dodecamer containing an adenine tract. *Proc. Natl. Acad. Sci. U. S. A.* 86, 1816–20.

Egli, M. & Saenger, W. (1984). *Principles of Nucleic Acid Structure*. New York: Springer.

Eisenberg, H. & Felsenfeld, G. (1967). Studies of the temperature-dependent conformation and phase separation of polyriboadenylic acid solutions at neutral pH. *J. Mol. Biol.* 30, 17–37.

Eisinger, J. (1971). Complex formation between transfer RNA's with complementary anticodons. *Biochem. Biophys. Res. Commun.* 43, 854–61.

Frank-Kamenetskii, M. D. & Mirkin, S. M. (1995). Triplex DNA structures. *Annu. Rev. Biochem.* 64, 65–95.

Gellert, M., Lipsett, M. N. & Davies, D. R. (1962). Helix formation by guanylic acid. *Proc. Natl. Acad. Sci. U. S. A.* 48, 2013–18.

Gorin, A. A., Zhurkin, V. B. & Olson, W. K. (1995). B-DNA twisting correlates with base-pair morphology. *J. Mol. Biol.* 247, 34–48.

Goulet, I., Zivanovic, Y. & Prunell, A. (1987). Helical repeat of DNA in solution. The V curve method. *Nucleic Acids Res.* 15, 2803–21.

Hagerman, P. J. (1986). Sequence-directed curvature of DNA. *Nature* 321, 449–50.

    (1990). Sequence-directed curvature of DNA. *Annu. Rev. Biochem.* 59, 755–81.

Hare, D., Shapiro, L. & Patel, D. J. (1986). Extrahelical adenosine stacks into right-handed DNA: solution conformation of the d(C-G-C-A-G-A-G-C-T-C-G-C-G) duplex deduced from distance geometry analysis of nuclear Overhauser effect spectra. *Biochemistry* 25, 7456–64.

Hunter, W. N., Brown, T., Anand, N. N. & Kennard, O. (1986a). Structure of an adenine–cytosine base pair in DNA and its implications for mismatch repair. *Nature* 320, 552–5.

Hunter, W. N., Kneale, G., Brown, T., Rabinovich, D. & Kennard, O. (1986b). Refined crystal structure of an octanucleotide duplex with G · T mismatched base-pairs. *J. Mol. Biol.* 190, 605–18.

Inners, L. D. & Felsenfeld, G. (1970). Conformation of polyribouridylic acid in solution. *J. Mol. Biol.* 50, 373–89.

Kalnik, M. W., Norman, D. G., Zagorski, M. G., Swann, P. F. & Patel, D. J. (1989). Conformational transitions in cytidine bulge-containing deoxytridecanucleotide duplexes: extra cytidine equilibrates between looped out (low temperature) and stacked (elevated temperature) conformations in solution. *Biochemistry* 28, 294–303.

Koo, H. S., Drak, J., Rice, J. A. & Crothers, D. M. (1990). Determination of the extent of DNA bending by an adenine–thymine tract. *Biochemistry* 29, 4227–34.

Koo, H. S., Wu, H. M. & Crothers, D. M. (1986). DNA bending at adenine–thymine tracts. *Nature* 320, 501–6.

Krylov, D. Y., Makarov, V. L. & Ivanov, V. I. (1990). The B-A transition in superhelical DNA. *Nucleic Acids Res.* 18, 759–61.

Lane, A. N., Chaires, J. B., Gray, R. D. & Trent, J. O. (2008). Stability and kinetics of G-quadruplex structures. *Nucleic Acids Res.* 36, 5482–515.

Lavery, R., Zakrzewska, K., Beveridge, D., Bishop, T. C., Case, D. A., Cheatham, T., 3rd, Dixit, S., Jayaram, B., Lankas, F., Laughton, C., Maddocks, J. H., Michon, A., Osman, R., Orozco, M., Perez, A., Singh, T., Spackova, N. & Sponer, J. (2010). A systematic molecular dynamics study of nearest-neighbor effects on base pair and base pair step conformations and fluctuations in B-DNA. *Nucleic Acids Res.* 38, 299–313.

Li, M. H., Zhou, Y. H., Luo, Q. & Li, Z. S. (2010). The 3D structures of G-quadruplexes of HIV-1 integrase inhibitors: molecular dynamics simulations in aqueous solution and in the gas phase. *J. Mol. Model.* 16, 645–57.

Macaya, R. F., Schultze, P., Smith, F. W., Roe, J. A. & Feigon, J. (1993). Thrombin-binding DNA aptamer forms a unimolecular quadruplex structure in solution. *Proc. Natl. Acad. Sci. U. S. A.* 90, 3745–9.

MacDonald, D., Herbert, K., Zhang, X., Polgruto, T. & Lu, P. (2001). Solution structure of an A-tract DNA bend. *J. Mol. Biol.* 306, 1081–98.

Marini, J. C., Levene, S. D., Crothers, D. M. & Englund, P. T. (1982). Bent helical structure in kinetoplast DNA. *Proc. Natl. Acad. Sci. U. S. A.* 79, 7664–8.

Moody, E. M. & Bevilacqua, P. C. (2004). Structural and energetic consequences of expanding a highly cooperative stable DNA hairpin loop. *J. Am. Chem. Soc.* 126, 9570–7.

Morden, K. M. & Maskos, K. (1993). NMR studies of an extrahelical cytosine in an A. T rich region of a deoxyribodecanucleotide. *Biopolymers* 33, 27–36.

Mukundan, V. T., Do, N. Q. & Phan, A. T. (2011). HIV-1 integrase inhibitor T30177 forms a stacked dimeric G-quadruplex structure containing bulges. *Nucleic Acids Res.* 39, 8984–91.

Patel, D. J., Kozlowski, S. A., Ikuta, S. & Itakura, K. (1984a). Deoxyadenosine–deoxycytidine pairing in the D(C-G-C-G-a-a-T-T-C-a-C-G) duplex – conformation and dynamics at and adjacent to the dA.dC mismatch site. *Biochemistry* 23, 3218–26.

(1984b). Deoxyguanosine–deoxyadenosine pairing in the d(C-G-A-G-A-A-T-T-C-G-C-G) duplex: conformation and dynamics at and adjacent to the dG × dA mismatch site. *Biochemistry* 23, 3207–17.

Peck, L. J. & Wang, J. C. (1981). Sequence dependence of the helical repeat of DNA in solution. *Nature* 292, 375–8.

Perez, A., Marchan, I., Svozil, D., Sponer, J., Cheatham, T. E., 3rd, Laughton, C. A. & Orozco, M. (2007). Refinement of the AMBER force field for nucleic acids: improving the description of alpha/gamma conformers. *Biophys. J.* 92, 3817–29.

Phan, A. T. (2010). Human telomeric G-quadruplex: structures of DNA and RNA sequences. *FEBS J.* 277, 1107–17.

Pörschke, D. (1978). Molecular states in single-stranded adenylate chains byrelaxation analysis. *Biopolymers* 17, 315–23.

Raghunathan, G., Miles, H. T. & Sasisekharan, V. (1993). Symmetry and molecular structure of a DNA triple helix: $d(T)_n \cdot d(A)_n \cdot d(T)_n$. *Biochemistry* 32, 455–62.

Rich, A., Nordheim, A. & Wang, A. H.-J. (1984). The chemistry and biology of left-handed Z-DNA. *Annu. Rev. Biochem.* 53, 791–846.

Shaw, D. E., Maragakis, P., Lindorff-Larsen, K., Piana, S., Dror, R. O., Eastwood, M. P., Bank, J. A., Jumper, J. M., Salmon, J. K., Shan, Y. & Wriggers, W. (2010). Atomic-level characterization of the structural dynamics of proteins. *Science* 330, 341–6.

Sinden, R. R. (1994). *DNA Structure and Function.* San Diego, CA: Academic.

Sprous, D., Zacharias, W., Wood, Z. A. & Harvey, S. C. (1995). Dehydrating agents sharply reduce curvature in DNAs containing A-tracts. *Nucleic Acids Res.* 23, 1816–21.

Stannard, B. S. & Felsenfeld, G. (1975). The conformation of polyriboadenylic acid at low temperature and neutral pH. A single-stranded rodlike structure. *Biopolymers* 14, 299–307.

Strauss, F., Gaillard, C. & Prunell, A. (1981). Helical periodicity of DNA, poly(dA).poly(dT) and poly(dA-dT). poly(dA-dT) in solution. *Eur. J. Biochem.* 118, 215–22.

Tjandra, N. & Bax, A. (1997). Direct measurement of distances and angles in biomolecules by NMR in a dilute liquid crystalline medium. *Science* 278, 1111–14.

Van de Ven, F. J. & Hilbers, C. W. (1988). Nucleic acids and nuclear magnetic resonance. *Eur. J. Biochem.* 178, 1–38.

Vermeulen, A., Zhou, H. & Pardi, A. (2000). Determination DNA global structure and DNA bending by application of NMR residual dipolar coupling. *J. Amer. Chem. Soc.* 122, 9638–9647.

Wang, A. H., Quigley, G. J., Kolpak, F. J., Crawford, J. L., van Boom, J. H., van der Marel, G. & Rich, A. (1979). Molecular structure of a left-handed double helical DNA fragment at atomic resolution. *Nature* 282, 680–6.

Wang, J. C. (1979). Helical repeat of DNA in solution. *Proc. Natl. Acad. Sci. U. S. A.* 76, 200–203.

Wang, Y. & Patel, D. J. (1992). Guanine residues in $d(T_2AG_3)$ and $d(T_2G_4)$ form parallel-stranded potassium cation stabilized G-quadruplexes with anti glycosidic torsion angles in solution. *Biochemistry* 31, 8112–19.

Watson, J. D. & Crick, F. H. C. (1953). The structure of DNA. *Nature* 171, 123–31.

Williamson, J. R. (1994). G-quartet structures in telomeric DNA. *Annu. Rev. Biophys. Biomol. Struct.* 23, 703–30.

Wing, R., Drew, H., Takano, T., Broka, C., Tanaka, S., Itakura, K. & Dickerson, R. E. (1980). Crystal structure analysis of a complete turn of B-DNA. *Nature* 287, 755–8.

Woodson, S. A. & Crothers, D. M. (1988). Structural model for an oligonucleotide containing a bulged guanosine by NMR and energy minimization *Biochemistry* 27, 3130–41.

  (1989). Conformation of a bulge-containing oligomer from a hot-spot sequence by NMR and energy minimization. *Biopolymers* 28, 1149–77.

Wu, Z., Delaglio, F., Tjandra, N., Zhurkin, V. B. & Bax, A. (2003). Overall structure and sugar dynamics of a DNA dodecamer from homo- and heteronuclear dipolar couplings and ^{31}P chemical shift anisotropy. *J. Biomol. NMR* 26, 297–315.

Wyatt, J. R., Vickers, T. A., Roberson, J. L., Buckheit, R. W., Jr., Klimkait, T., DeBaets, E., Davis, P. W., Rayner, B., Imbach, J. L. & Ecker, D. J. (1994). Combinatorially selected guanosine-quartet structure is a potent inhibitor of human immunodeficiency virus envelope-mediated cell fusion. *Proc. Natl. Acad. Sci. U. S. A.* 91, 1356–60.

Yakovchuk, P., Protozanova, E. & Frank-Kamenetskii, M. D. (2006). Base-stacking and base-pairing contributions into thermal stability of the DNA double helix. *Nucleic Acids Res.* 34, 564–74.

Yoshizawa, S., Kawai, G., Watanabe, K., Miura, K. & Hirao, I. (1997). GNA trinucleotide loop sequences producing extraordinarily stable DNA minihairpins. *Biochemistry* 36, 4761–7.

Young, M. A., Ravishanker, G., Beveridge, D. L. & Berman, H. M. (1995). Analysis of local helix bending in crystal structures of DNA oligonucleotides and DNA–protein complexes. *Biophys. J.* 68, 2454–68.

Zimmerman, S. B., Cohen, G. H. & Davies, D. R. (1975). X-fay fiber diffraction and model-building study of polyguanylic acid and polyinosinic acid. *J. Mol. Biol.* 92, 181–92.

2 Conformational transitions

The main form of the double helix, the B form, is stabilized by only weak hydrogen bonds and van der Waals interactions. If we also take into account the remarkable flexibility of the backbone of ssDNA, it is not surprising that depending on the solution conditions DNA can be found in various alternative forms. Conformational flexibility of DNA is needed for its functioning, since it facilitates specific DNA-DNA, DNA-RNA and DNA–protein interactions inside the cell. When we are changing solution conditions gradually, DNA can undergo transitions from one form to another. Studying these transitions has brought a lot of important information about DNA conformational flexibility and the stability of the various forms. This is why the conformational transitions have been a subject of biophysical investigation for decades. In this chapter we start from general theoretical analysis of the transitions, and then consider individual transitions: DNA melting or the helix–coil transition, B–A and B–Z transitions. We mainly consider only equilibrium properties of the transitions; the corresponding dynamic properties will be the subject of Chapter 4.

2.1   Theoretical analysis of conformational transitions in DNA

2.1.1   Preliminary remarks

A key concept of the theoretical description of conformational transitions that will be used in this chapter is a concept of a macrostate. It seems that Zimm and Bragg were the first to apply this approach to the analysis of the helix–coil transition in polypeptides (Zimm & Bragg 1959), although a few groups were moving in the same direction at that time. In this approach all microscopic states of a base pair (or nucleotides that can form the base pair) are divided into two groups, which correspond to the two DNA forms under consideration. The exact numbers of microstates in the macrostates and their corresponding energies are not specified in this approach. To apply it we only need to know the ratio of the statistical weights of the macrostates. The approach is similar to the description of a chemical reaction, where the equilibrium constant, K, also specifies the ratio of statistical weights of two macrostates of a system. It would be very difficult to calculate the value of K by analyzing microstates of the reaction components, but K can be measured experimentally. It is common to measure both enthalpy and entropy of the reaction, which gives us the temperature dependence of K. The same is true for the transitions in a DNA molecule: the ratios of statistical weights are considered as the model parameters and are determined from experimental data. The difference between applying this approach to a chemical reaction and to a conformational transition in DNA is that the entire DNA molecule has 2^N rather than two macrostates (N is the number of base pairs in the molecule). As a result, analysis of the model is more complex than for a chemical reaction. Also, more than one equilibrium constant is usually needed to model a conformational transition in DNA.

Sometimes the macrostate-based model is also called the "two-state model," since only two macrostates of each base pair are accounted for. This is not a very important feature of the model, however. Indeed, in general we can introduce in the same way as many macrostates for each base pair as necessary. For example, in the description of the B–Z transition we have to introduce three macrostates for each base pair (see below). The key feature of this approach is accounting only for macrostates rather than for microstates. Correspondingly, the energy of microstates is replaced by the free energy of macrostates. Of course, the macrostates used in the approach have to be distinguishable experimentally.

The main limitation of this approach is its low resolution. For example, we can estimate the probability that a single base pair will be open in a helical region of DNA, but we can say almost nothing about typical conformations of the open base pair. Also, by its nature this is a thermodynamic approach and therefore it cannot give any information about DNA dynamic properties.

Over the years there have been attempts to develop more traditional physical models of DNA, operating with microstates. The most direct model of this kind is the all-atom model used in MD simulations. Despite all their promise, MD simulations have their own problems and limitations (see Section 1.7). In particular, the MD simulations are hardly practical to study conformational transitions in DNA molecules hundreds and thousands of base pairs in length. Therefore, simpler models based on reduced sets of microstates received attention. In these models DNA base pairs are replaced by relatively small sets of pseudo-atoms with artificial potentials (see Peyrard & Bishop (1989), for example). The potentials are parameters of the models, which should be chosen such that the models will mimic properties of a real DNA molecule. It is definitely possible to choose the model parameters to obtain a reasonable agreement with experimental data for a particular conformational property. However, there is no reason to assume that this kind of model will be able to predict a broader spectrum of DNA conformational properties that were not used for the adjustment of their parameters. Indeed, DNA conformational space and its interaction with water molecules and ions is so complex that it seems hardly possible to accumulate them in relatively simple artificial potentials (see Frank-Kamenetskii & Prakash (2014) for a review). On the other hand, these models are more complex than the macrostate-based models. For these reasons DNA models based on reduced sets of microstates will not be analyzed here.

2.1.2   General consideration

A DNA base pair or a nucleotide of ssDNA can be in one of many different conformations. In this chapter we do not distinguish the conformations that belong to a particular

BBBBBAAAAAAAAABBBBBBBBBAAAAAAAAAABBB

Figure 2.1 A diagram of a DNA molecule undergoing the transition between A and B forms. The state of each base pair is shown by a single letter, A or B.

form of DNA. Thus, only one variable, the DNA form, will specify a macrostate of each base pair or nucleotide. Each of these macrostates includes many elementary microstates. The detailed distribution of base-pair conformations in a particular form is beyond the scope of this analysis. This method of description is illustrated in Fig. 2.1.

The main feature of conformational transitions in DNA molecules is their cooperativity. The cooperativity means that any boundary between two forms increases the free energy of the molecule. The physical origin of the cooperativity could be different and will be analyzed when we consider particular transitions. The main consequence of the cooperativity is that a typical state of DNA molecule in the transition interval represents a succession of extended alternating stretches of each form. This is illustrated in Fig. 2.1 for the DNA B–A transition.

A statistical-mechanical model that accounts for the additional energy of boundaries between different forms is called the Ising model. In this model the energy (or the free energy when we operate in terms of macrostates) associated with a unit depends not only on the state of this unit but also on the state of adjacent units. Thus, the model accounts for the nearest-neighbor interaction between the units. This model provides a good theoretical description of conformational transitions in biopolymers, nucleic acids and polypeptides. The model was first applied for analysis of the α-helix–coil transition in polypeptides by Zimm and Bragg, and sometimes it is called the Zimm–Bragg model (Zimm & Bragg 1959). Describing the model we will continue to call the two states of each base pair A and B, although with some modifications the model can be applied to other transitions as well. In this general consideration we will consider the state where all base pairs are in B form as a reference state in the free-energy calculations. There are two major parameters in the model, the free energy of A-form elongation, ΔG, and the free energy of A-form nucleation, ΔG_{nucl}. The value of ΔG specifies the free-energy change for extending a region of base pairs in the A form by one base pair. This can be diagramed as

$$AAABBB \xrightarrow{\Delta G} AAAABB.$$

During this extension the number of base pairs in the A form increases by one, but the number of boundaries does not change. Formation of a new segment of A form, one base pair in length, can be diagramed as

$$BBBBBAA \xrightarrow{\Delta G+\Delta G_{nucl}} BBABBAA.$$

This corresponds to the free-energy change of $\Delta G + \Delta G_{nucl}$. Thus, if a state i has n base pairs in m regions of the A form, its free energy, $\Delta G(i)$, is given by

$$\Delta G(i) = n\Delta G + m\Delta G_{nucl}. \tag{2.1}$$

Equation (2.1) assumes that the base pairs at the ends of a DNA molecule are in the B form; if the first or/and last base pair of the DNA are in the A form, we will assume that the pair does not form a border between the two forms. The free energy specifies the probability of macrostate i, $P(i)$, similarly to how energy specifies the probability of a microstate:

$$P(i) = A \exp(-\Delta G(i)/RT), \tag{2.2}$$

where R is the gas constant, T is the absolute temperature and A is a coefficient. The coefficient has to be chosen so that the sum of $P(i)$ over all macrostates will be equal to unity. From this condition we conclude that

$$1/A = \sum_j \exp(-\Delta G(j)/RT). \tag{2.3}$$

In statistical physics the sum in Eq. (2.3) is called the *partition function*; we will use letter Q for it. It is important to emphasize that the summation in Eq. (2.3) has to be taken over all possible macrostates of the molecule. If the DNA molecule has N base pairs, the number of different macrostates is 2^N (assuming that only A and B forms appear with a notable probability under the considered solution conditions).

It is convenient to rewrite Eqs. (2.2) and (2.3) in terms of the corresponding equilibrium constants,

$$s = \exp(-\Delta G/RT) \tag{2.4a}$$

$$\sigma = \exp(-\Delta G_{\text{nucl}}/RT). \tag{2.4b}$$

When expressed over these constants, the probability of state i is equal to

$$P(i) = s^n \sigma^m \Bigg/ \left(\sum_j s^k \sigma^l \right). \tag{2.5}$$

It is also convenient to use the concept of *statistical weight* of state i, W_i. The statistical weight is proportional to $P(i)$. We can assume that the statistical weight of state i is equal to $s^n \sigma^m$. The entire set of statistical weights can be multiplied by a constant: this simply corresponds to the change of the reference level for the free energy. Such renormalization of W_i does not change observable values, such as $P(i)$. Therefore, Eq. (2.5) can be also written as

$$P(i) = \frac{W_i}{\sum_j W_j}. \tag{2.6}$$

Let us now discuss major properties of the model when it is applied to DNA (or polypeptide). First, the value of δG_{nucl} is always positive and large compared with RT, so σ is much less than unity. On the other hand, the value of s is of the order of unity and changes with solution conditions. The value of σ also changes with solution conditions, but remains small and usually these changes are difficult to detect, so σ is usually considered to be a constant. A small value of σ means that the statistical weights of

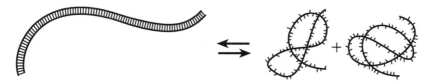

Figure 2.2 A diagram of the helix–coil transition. Separated single-stranded molecules are much more flexible than the double helix, so the melting is accompanied by an increase of the system entropy.

states with a large number of A–B boundaries are small. Therefore, the DNA molecule has a tendency to have rather extended regions of A and B forms at solution conditions that correspond to the transition interval. In an extreme case, when the DNA molecule consists of only a few base pairs, in a good approximation we can account for only two states, when all base pairs are in state A and all base pairs are in state B (see below). However, in a sufficiently long molecule the number of states with many A–B boundaries is so large that the total contribution of these states becomes important, despite the small statistical weight of the individual states.

The DNA double helix consists of two different base pairs, AT and GC, and the constant s depends on the type of base pair. It can also depend on the type and orientation of the adjacent base pairs. We will consider this issue in detail when we analyze specific conformational transitions.

2.2 DNA melting

2.2.1 Experimental registration of DNA melting

DNA melting (also called the helix–coil transition or denaturation) is the most important conformational transition in the molecule. During this transition the complementary strands of the double helix are separated and form much smaller coils, since ssDNA is much more flexible than the double helix (Fig. 2.2).

The study of DNA melting was pioneered by Doty and co-workers, who unambiguously clarified the nature of the transition by 1957 (Rice & Doty 1957). The transition is accompanied by changes of many DNA physical properties: a huge reduction of DNA viscosity, changes in optical activity and light scattering, and an increase of UV absorption (Rice & Doty 1957). The increase of DNA optical density soon became the method of choice for monitoring the transition, as the most simple and accurate (Fig. 2.3). The increase of UV absorption on melting is due to the disruption of the stacking interaction between DNA bases during the transition (Devoe & Tinoco 1962). It is important that the increase of absorption is proportional to the fraction of melted base pairs, $\theta(t)$, so the fraction can be calculated as

$$\theta(t) = \frac{A(t) - A(20)}{A(95) - A(20)}, \tag{2.7}$$

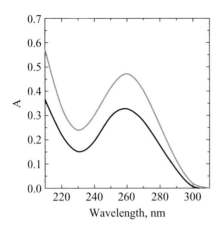

Figure 2.3 Absorption spectrum of dsDNA at room temperature (black line) and denatured DNA at 95 °C (gray line). DNA concentration is equal to 16 µg/ml. The absorption at the wavelength of 260 nm is usually used to monitor the helix–coil transition.

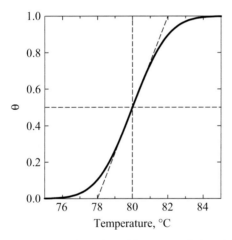

Figure 2.4 A schematic melting curve of DNA. The melting temperature, t_m, which corresponds to a θ of 0.5, is equal to 80 °C for this example. The width of the melting interval, Δt_m, is defined by the equation $\Delta t_m = 1/\frac{d\theta}{dt}\big|_{t=t_m}$. The value of Δt_m has a simple geometrical interpretation: it is specified by the intersections of the tangent line in the middle of the melting curve with lines $\theta = 0$ and $\theta = 1$. In the example shown Δt_m is equal to 4 °C.

where $A(t)$ is the absorbance at temperature t (measured in °C). Equation (2.7) assumes that DNA is completely melted at 95 °C, which is not necessarily the case at $[Na^+] \geq 0.2$ M, so a higher temperature should be used in the equation. Although Eq. (2.7) is not precise for DNA duplexes of a few base pairs in length, it is very accurate for long DNA molecules. The function $\theta(t)$ is called the melting curve. A schematic DNA melting curve is shown in Fig. 2.4. The melting temperature, t_m, is defined as the temperature at which 50% of base pairs are melted ($\theta(t_m) = 0.5$).

Early studies established that the transition occurs in a relatively narrow range of temperatures that depends on DNA sequence and solution conditions. The melting temperature is relatively insensitive to pH in the range between 5 and 9. The value of t_m decreases for pH beyond this interval, and at pH below 3 DNA can be denatured at room temperature (Rice & Doty 1957).

2.2.2 Salt and sequence dependence of DNA melting

Systematic investigation of the helix–coil transition by Marmur and Doty showed that DNA melting temperature, t_m, is proportional to the fraction of GC base pairs in DNA molecule, x_{GC}:

$$t_m = t_{AT} + x_{GC} t_{GC}, \tag{2.8}$$

where t_{AT} and t_{GC} are the melting temperatures of the molecules consisting of only AT or only GC base pairs (Marmur & Doty 1959). The values of t_{AT} and t_{GC} depend on the concentration of ions in solution. At ionic conditions close to physiological ones (0.2 M NaCl and 10 mM MgCl$_2$), the difference between t_{GC} and t_{AT} is close to 40 °C. If sodium ions are the only cations in solution, the following empirical equation specifies the melting temperature (Owen *et al.* 1969, Frank-Kamenetskii 1971):

$$t_m = 176.0 - (2.60 - x_{GC})(36.0 - 7.04 \log[Na^+]), \tag{2.9}$$

where $[Na^+]$ is measured in mol/l. The effect of divalent ions is much more complicated: depending on the ion type they can either increase or decrease t_m (Eickbush & Moudrianakis 1978). The most biologically important divalent ions, Mg^{2+}, increase DNA melting temperature up to ion concentrations of 10^{-2} M (Blagoi *et al.* 1978, Owczarzy *et al.* 2008). It was found that in a solution of 3 M tetramethylammonium chloride or 2.4 M tetraethylammonium chloride, the value of $t_{GC} - t_{AT}$ becomes equal to zero (Melchior & Von Hippel 1973). Of course, under these conditions all DNA molecules melt at the same temperature regardless of their GC-content and the melting curves become very narrow (Melchior & Von Hippel 1973, Voskoboinik *et al.* 1975).

2.2.3 Theoretical description of the melting transition

The helix–coil transition in DNA usually involves formation of one duplex, D, from two complementary single-stranded chains, S and S_c, so it has to be considered as a bimolecular reaction. Although accounting for this factor is not essential for melting long DNA molecules until the very end of the process, it is very important for the helix–coil transition in duplexes up two a few dozens base pairs in length. In general, we can diagram the process as

$$S + S_c \xleftrightarrow{K} D, \tag{2.10}$$

where the equilibrium constant K accounts, in particular, for the entropy loss due to the association of two chains in the duplex. This constant K can be calculated as a sum over

Figure 2.5 A diagram of a DNA state in the middle of the melting interval. There are many single-stranded loops in long DNA molecules. Their locations depend on the DNA sequence. In the middle of the transition a typical loop includes hundreds of nucleotides. The fact that internal single-stranded regions form closed loops diminishes their entropy.

all possible arrangements of the base pairing between the complementary strands in line with the Ising model described above. Thus, K should be considered as the partition function over all states that include at least one base pair.

In the case of DNA melting it is common to define constants s as the equilibrium constants for joining one base pair to an already existing helical region (this means that the melted DNA is considered as a reference state). Since GC base pairs are more stable than AT base pairs, the values of s have to be different for the two types of base pair. Formation of a first base pair between the complementary strands can be written as βs, where β has units of reverse concentration. Thus, the contribution to K from the configuration with a single helical region starting from base pair i and ending by base pair j should be written as

$$\beta \prod_{k=i}^{j} s_k.$$

Formation of the second and all subsequent helical regions have to be treated differently, however. First of all, they have to be associated with the nucleation constant σ, accounting for the formation of a new helical region. Then, it has to be taken into account that the single-stranded loops are formed between each pair of adjacent helical regions (Fig. 2.5).

These loops restrict the conformational freedom of the single-stranded region relative to the corresponding region with free ends, so the entropy of a single-stranded region in a closed loop is smaller than the entropy of a single-stranded region of the same size with free ends. This reduction is related to the probability of spontaneous loop closing, P_c, during thermal motion of the single-stranded segment,

$$P_c = p(0)\delta v, \tag{2.11}$$

where $p(0)$ is the distribution of the end-to-end distance for the corresponding linear single-stranded segment, and δv is a small volume where the segment end should appear for the covalent closing of the loop. It is known that for large loops of length L the latter probability is proportional to $L^{-\alpha}$, where α is close to $3/2$ (see Section 3.1). It has not been studied in detail how this probability depends on the loop size for relatively short loops, although the available data suggest that α is larger than two (Goddard et al. 2000). However, accurate parameterization of small loops is less important for the DNA melting since the states with small loops have very low probability of appearance in any case. Indeed, each loop is associated with small constant σ and only large number of melted base pairs can compensate it. The volume element δv is not known, although careful analysis of the loop energetics suggests that its value is very small and makes a major contribution to the cooperativity factor of the helix–coil transition (Krueger et al. 2006).

Thus, the statistical weight, W_i, of the state with two helical regions, (i_1, j_1) and (i_2, j_2), has to be written as

$$W_i = \beta \left(\prod_{k=i_1}^{j_1} s_k \right) \sigma (i_2 - j_1)^{-\alpha} \left(\prod_{k=i_2}^{j_2} s_k \right). \tag{2.12}$$

Generalization of the last expression for the case of a few helical regions is straightforward. Assuming that the complementary strands consist of N nucleotides each, there are $2^N - 1$ states that contribute to the partition function K (the state with no base pairs is not included in K). It should be noted that the appearance of the loop factor means a deviation from the Ising model. Indeed, now the free energy associated with formation of a new base pair can depend not only on the nearest neighbors but also on the distance to another helical region. This more complex model maintains, however, all major properties of the Ising model, including first of all the cooperativity of transitions between two DNA forms.

Of course, constants s_{AT} and s_{GC}, for AT and GC base pairs correspondingly, depend on temperature,

$$s_{AT} = \exp[-(\Delta H_{AT} - T\Delta S)/RT] \tag{2.13a}$$

$$s_{GC} = \exp[-(\Delta H_{GC} - T\Delta S)/RT] \tag{2.13b}$$

where T is the temperature measured in kelvin. The values of ΔH_{AT} and ΔH_{GC} account for energy gain due to adding the stacking interaction and the hydrogen bonds associated with the base-pair formation. Thus, ΔH_{AT} and ΔH_{GC} are negative. The value of ΔS is also negative, reflecting the reduction of the conformational entropy when flexible single-stranded chains form a very rigid double helix. The values of ΔH_{AT}, ΔH_{GC} and ΔS were determined in the microcalorimetric experiments (Shiao & Sturtevant 1973, Gruenwedel 1974, Delcourt & Blake 1991), and in the experiments on DNA melting (see below). The value of ΔS in a very good approximation is equal to 25 cal/(mol · degree) for both AT and GC base pairs, and hardly depends on ionic conditions (Gruenwedel 1974, Delcourt & Blake 1991). The values of ΔH_{AT} and ΔH_{GC} depend on ionic conditions and this dependence is well described by Eq. (2.9), if we take into account that $T_{AT} = \Delta H_{AT}/\Delta S$ and $T_{GC} = \Delta H_{GC}/\Delta S$.

The value of ΔG_{nucl} accounts for two features of the transition. First, there are only $(n-1)$ stacking interactions in a helical region of n base pairs, so formation of the first base pair of a helical region does not involve energetically favorable stacking interaction. The reduction of the conformational entropy due to the closed single-stranded loop also makes a contribution to the value of σ. Part of this reduction is in the multiplier $L^{-\alpha}$, while a small element of volume, δv, contributes to σ. Both of these contributions increase ΔG_{nucl} and, correspondingly decrease the value of σ. For near-physiological ionic conditions (≈ 0.2 M NaCl) the value of σ is equal to $(2.5–5) \times 10^{-5}$ (Amirikyan et al. 1981, Kozyavkin et al. 1987). It reduces by three orders of magnitude when the concentration of sodium ions decreases from 1 M to 0.01 M (Kozyavkin et al. 1987). The value of β should have a similar dependence on ionic conditions, although experimental data are available only for a solution of 0.2 M NaCl, where β is equal to 10^{-3} 1/M (Craig et al. 1971). For a solution of 1 M NaCl the value of β can be calculated from the set of parameters shown in Table 2.1.

To calculate the value of K we have to perform summation of all $2^N - 1$ terms for a given DNA sequence. There are efficient computational algorithms that allow rigorous calculation for the Ising model, but the problem is more difficult for the model of DNA that accounts for the entropy reduction in the DNA loops. A very accurate approximate solution of the problem was found by Fixman and Freire (1977). Their algorithm allows very fast computations of melting curves for DNA molecules many thousands of base pairs in length. However, problems related to melting of short DNA duplexes do not require rigorous consideration, and in these cases the theoretical treatment of the model is very simple.

2.2.4 Short DNA duplexes

Melting of short DNA duplexes has become the most important application of the helix–coil transition due to its role in polymerase chain reaction (PCR) and various diagnostic methods. If the length of DNA, N, does not exceed a few dozen base pairs, the probability of states with more than one helical region is very low (except for duplexes with very special sequences). Thus, with very good accuracy K can be calculated as

$$K = \sum_{i=1}^{N} \sum_{j=i}^{N} \beta \prod_{k=i}^{j} s_k. \qquad (2.14)$$

The double sum in Eq. (2.14) contains only $N(N+1)/2$ terms, so the calculation of K is very simple. It is convenient for further analysis to introduce the total concentration of strands, C_T, and the fraction of paired DNA strands, x:

$$x = 2[D]/(2[D] + [S] + [S_c]). \qquad (2.15)$$

Then, assuming that $[S] = [S_c]$, we obtain that $[D] = xC_T/2$, $[S] = (1-x)C_T/2$ and

$$K = 2x/[C_T(1-x)^2]. \qquad (2.16)$$

Solving Eq. (2.16) gives the value of x as a function of K,

$$x = \frac{1 + KC_{\mathrm{T}} - \sqrt{1 + 2KC_{\mathrm{T}}}}{KC_{\mathrm{T}}}. \tag{2.17}$$

The average fraction of paired nucleotides under particular conditions, $1 - \theta$, can be calculated as

$$1 - \theta = \frac{x}{N} \sum_{k} P(k)n_k, \tag{2.18}$$

where $P(k)$ is the probability of state k (under the condition that the duplex exists) and n_k is the number of base pairs in state k of the duplex. The value of $P(k)$ has to be calculated according to the rules of statistical mechanics as a ratio of its statistical weight W_k to the partition function,

$$P(k) = \frac{1}{K} \beta \prod_{k=i}^{j} s_k. \tag{2.19}$$

Equations (2.14) and (2.17)–(2.19) allow one to calculate the melting curve of oligonucleotide duplexes with any sequence of base pairs.

Let us analyze the above equations in some detail. The melting temperature (the temperature where $x = 0.5$) corresponds to the condition

$$KC_{\mathrm{T}} = 4. \tag{2.20}$$

It follows from Eq. (2.20) that since constant β is of the order of 10^{-3} $1/\mathrm{M}$ and C_{T} is in the range of $\mu\mathrm{M}$, the average values of s_k at T_{m} of oligonucleotide duplexes 6–20 bp in length are substantially larger than unity. Therefore, at $T = T_{\mathrm{m}}$ all partially melted states of the duplexes have relatively low probability compared with the state with the maximum number of base pairs (see Eq. (2.19)) and the completely melted state (with the exception of such sequences as $\mathrm{dA}_m\mathrm{G}_n \cdot \mathrm{dC}_n\mathrm{T}_m$ that can melt in two steps if n and m are larger than 15). Thus, we can approximate the double sum in Eq. (2.14) by the single most important term,

$$K \cong \beta \prod_{k=1}^{N} s_k. \tag{2.21}$$

Equation (2.21) corresponds to the two-state approximation for the oligonucleotide melting when only fully melted and fully paired states are taken into account. The approximation is quite good for this range of duplex length, better for shorter molecules. Two examples of the melting curves calculated with the use of the two-state approximation and by accounting for all states with a single helical region are shown in Fig. 2.6.

Some oligonucleotides have self-complementary (palindromic) sequences, so they can form duplexes by pairing with themselves. In this case we have to consider the equilibrium between the duplex and one single-stranded olgonucleotide,

$$\mathrm{S} + \mathrm{S} \xleftrightarrow{K} \mathrm{D}. \tag{2.22}$$

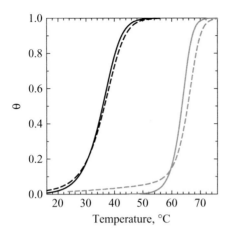

Figure 2.6 Theoretical melting curves of two DNA duplexes calculated in the two-state approximation (solid lines) and by accounting for all states with a single helical region (dashed lines). The black curves correspond to the melting of a 10 bp duplex, d(ATTATGGGGC) · d(GCCCCATAAT), and the gray curves are for a 20 bp duplex, d(AACGCGTGAATTCTGGCCAA) · d(TTGGCCAGAATTCACGCGTT). Clearly, the two-state approximation is more accurate for shorter duplexes.

Correspondingly, Eq. (2.16) is replaced by

$$2K = x/[C_T(1 - x)^2],\qquad(2.23)$$

and in Eq. (2.17) K has to be replaced by $4K$.

Short DNA duplexes can melt at rather low temperatures. At these temperatures substantial stacking interaction between nucleotides is maintained in the single strands, reducing the increase of UV absorption during the melting. If for long DNA molecules the absorption increase is close to 40% at the wavelength of 260 nm, for short duplexes it can be as small as 10%. The stacking interaction in the single strands disappears gradually over a broad temperature interval (see Cantor & Schimmel (1980) for details), and the absorption gradually increases with increasing temperature. This definitely complicates precise measurement of the melting curves. Another complication arises from the fact that melting curves become broader when the duplex length is reduced. It can be shown, by using the two-state model, that the width of the melting interval is specified by the following equation:

$$\Delta T_m = 6RT_m^2/\Delta H,\qquad(2.24)$$

where ΔH is the total enthalpy of the duplex melting. It follows from Eq. (2.24) that, for example, for a duplex of 8 bp in length $\Delta T_m \approx 15\,°C$. These factors make it difficult to measure the T_m of short duplexes precisely.

How well can we predict T_m by using the theoretical model? The theoretical description of the oligonucleotide melting presented above gives the melting temperature of the oligonucleotide duplexes with an average accuracy of $\pm 4\,°C$. Clearly, it is desirable to get a better predictive power of the theory, and attempts to improve it started in the first

Table 2.1 The nearest-neighbor thermodynamic parameters for Watson–Crick base pair formation in 1 M NaCl (SantaLucia & Hicks 2004). $t = 37\,°C$.

Base pair step	ΔH, kcal/mol	ΔS, cal/mol/K	ΔG_k, kcal/mol
AA/TT	−7.6	−21.3	−1.00
AT/AT	−7.2	−20.4	−0.88
TA/TA	−7.2	−21.3	−0.58
CA/TG	−8.5	−22.7	−1.45
GT/AC	−8.4	−22.4	−1.44
CT/AG	−7.8	−21.0	−1.28
GA/TC	−8.2	−22.2	−1.30
CG/CG	−10.6	−27.2	−2.17
GC/GC	−9.8	−24.4	−2.24
GG/CC	−8.0	−19.9	−1.84
Initiation	+0.2	−5.7	+1.96
Terminal AT penalty	+2.2	+6.9	+0.05

half of the 1980s. All this development followed essentially the same theoretical model as described above. The only modification of the model was the assumption that the helix propagation constants, s_i, depend on two adjacent base pairs rather than on the type of a single base pair that extends a helical region. The assumption means that the stacking interaction between the adjacent pairs should depend on the types of contacting base. There are ten different dinucleotide steps in the double helix, AA/TT, AC/GT, AG/CT, AT/AT, CA/TG, CC/GG, CG/CG, GA/TC, GC/GC, and TA/TA. The new constants are specified by ten pairs of enthalpies and entropies, which were determined experimentally from the melting data for large sets of DNA duplexes (Breslauer *et al.* 1986, Doktycz *et al.* 1992, Sugimoto *et al.* 1996, Allawi & SantaLucia 1997). The set of parameters based on the data from SantaLucia *et al.* (1996) and Sugimoto *et al.* (1996) seems to give the best prediction for the melting temperatures of DNA duplexes (SantaLucia & Hicks 2004). The set of these parameters for a solution of 1 M NaCl is shown in Table 2.1 (SantaLucia & Hicks 2004). Using Table 2.1, the free energy of the duplex formation, ΔG, which corresponds to the two-state approximation for constant K, should be calculated as

$$\Delta G = \Delta G_{\text{init}} + \sum_{k=1}^{10} n_k \Delta G_k, \qquad (2.25)$$

where ΔG_{init} is the free energy of helix initiation, n_k is the number of base-pair steps of type k, and ΔG_k is the free energy associated with a step of type k. The values of ΔG_{init} in the table assume that C_T is measured in mol/l.

This set is capable of predicting the melting temperature of oligonucleotide duplexes within $\pm 2\,°C$, on average. The relation between the two sets of the propagation constants, one that depends only on the type of the base pair and the other that depends on the nearest neighbors, will be discussed later in this chapter.

Thus, the dinucleotide model gives a better approximation of DNA stability. Still, its accuracy is definitely limited. According to the model two duplexes with identical dinucleotide compositions have to have the same melting temperature, but in reality their melting temperatures can differ from one another. For example, the t_m of duplexes (GCCGGC)$_2$ and (GGCGCC)$_2$ are equal to 67.2 °C and 65.2 °C, respectively (Freier et al. 1986).

Although the parameters shown in Table 2.1 correspond to a solution of 1 M NaCl, it was also investigated in detail how the values of ΔH and ΔS depend on the concentration of monovalent ions and magnesium ions (Owczarzy et al. 2004, Owczarzy et al. 2008). The cited and many earlier studies showed that magnesium ions have much larger effect on DNA melting than sodium ions, affecting the melting temperature even when their concentration is much lower than the concentration of monovalent ions. The two most important monovalent ions, Na$^+$ and K$^+$, influence DNA melting temperature in nearly the same way (Owczarzy et al. 2008). A few websites offer t_m calculation for chosen oligonucleotide duplexes and ionic conditions (see Integrated DNA Technologies, for example).

2.2.5   Mismatches and bulge loops

Different nucleotide mismatches and small loops can appear in DNA duplexes, both in laboratory manipulations and inside the cells. It is important to know how these defects influence the stability of DNA duplexes. Over the years many data have been accumulated on these issues. Different structural and thermodynamic methods showed that the mismatches introduce relatively small disturbances in regular DNA structure and their energetic cost is relatively low – the free energy of mismatches is much smaller than the free energy of a loop formed due to the opening of a single base pair inside a helical region. Thus, mismatched nucleotides surrounded by regular base pairs do not form loops. Instead, they only introduce some irregularity in the structure of DNA duplexes, and the stacking interaction between regular and mismatched base pairs is essentially preserved (Patel et al. 1982, Patel et al. 1984, Arnold et al. 1987). This is due to the remarkable flexibility of the ssDNA backbone, which can accommodate many distortions of the regular structure over a single base-pair step.

Of course, the distortion depends on the type of the mismatch. An internal GT pair introduces a very small destabilization in the double helix or can even stabilize it. This is what one would expect since the pair forms a wobble hydrogen-bonded structure (Allawi & SantaLucia 1998c). It was less expected that GG pair can, depending on the sequence context, have an even smaller destabilizing effect. The contributions of all eight possible mismatches to the duplex stability were carefully investigated, depending on the sequence context (Aboul-ela et al. 1985, Allawi & SantaLucia 1997, Allawi & SantaLucia 1998a, Allawi & SantaLucia 1998b, Allawi & SantaLucia 1998d, Peyret et al. 1999). In a brief form the data are accumulated in Table 2.2. A more detailed set of parameters that accounts for the nearest-neighbors of each type of mismatch can be found in a comprehensive review on the subject by SantaLucia and Hicks (2004).

DNA duplexes can also contain "bulge loops," where one or more consecutive bases of one strand have no bases to pair with in the other strand. Structural studies showed that single-base bulge loops can assume two types of conformation: a "stacked" structure,

Table 2.2 The contributions to the duplex free energy from single internal mismatches at 37 °C, in 1 M NaCl, pH 7. The range of the free energy shown in the table corresponds to different sequence context for a particular mismatched pair.

The data were adopted from Allawi and SantaLucia (1997), Allawi and SantaLucia (1998a), Allawi and SantaLucia (1998b), Allawi and SantaLucia (1998d) and Peyret *et al.* (1999), which contain detailed information on the sequence dependence of free energies for all eight mismatched base pairs.

Base pair	Contribution to the duplex free energy, kcal/mol
AA	(0.34)–(1.38)
AC	(1.22)–(2.25)
AG	(−0.78)–(1.16)
CC	(1.4)–(2.66)
CT	(1.02)–(1.95)
GG	(−2.22)–(0.88)
GT	(−1.05)–(1.05)
TT	(−0.24)–(1.38)

Figure 2.7 Diagram of two conformations of a single-base bulge loop. The unpaired base can be stacked between neighboring bases (left) or looped out of the duplex (right).

in which the unpaired base is stacked between neighboring bases, and a "looped-out" structure, in which the unpaired base is pushed away from the helix axis (Fig. 2.7) (Woodson & Crothers 1988a, Woodson & Crothers 1988b, Woodson & Crothers 1989, Kalnik *et al.* 1989, Kalnik *et al.* 1990, Morden & Maskos 1993). In general, single-base bulge loops destabilize DNA duplexes, increasing their free energy by 2.5–4 kcal/mol in solution containing 1 M NaCl at 37 °C (Zhu & Wartell 1999, Tanaka *et al.* 2004). To a good approximation this free energy is specified by the bulged base and its surrounding base pairs (Zhu & Wartell 1999, Tanaka *et al.* 2004), although deviations from the nearest-neighbor model were documented as well (Conceição *et al.* 2010).

Overall, the nearest-neighbor model and corresponding sets of parameters for Watson–Crick base pairs, mismatched pairs and bulge loops allow reasonably good estimation of thermal stability of DNA duplexes (SantaLucia & Hicks 2004).

2.2.6   Hairpins

Single-stranded oligonucleotides can form hairpin-like structures that include a double-stranded stem and a hairpin loop (Fig. 2.8). Formation of the hairpin is an intramolecular

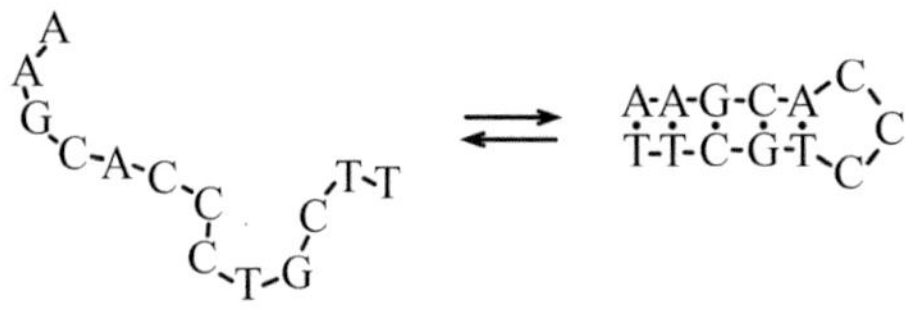

Figure 2.8 Formation of hairpin structure by a single-stranded oligonucleotide. A stem of 5 bp can be formed by this oligonucleotide, with a loop of three nucleotides.

process. Therefore, the equilibrium in the process does not depend on the concentration of oligonucleotide S and can be diagramed as

$$S \xrightarrow{K} H.$$

In a typical hairpin there are only a few base pairs in the stem, although in principle the stem length is specified by DNA sequence and can be very large. The size of the hairpin loop cannot be smaller than three nucleotides since the loop contour length has to exceed 2 nm. The free energy of the loop formation depends, first of all, on the loop size. This free energy is always positive, since the loop has reduced entropy relative to the corresponding DNA segment with free ends. It has been found empirically that the entropy change upon formation of an n ($n > 3$) nucleotide loop, $\Delta S_{\text{loop}}(n)$, can be approximated as

$$\Delta S_{\text{loop}}(n) = \Delta S^0_{\text{loop}} - \alpha \ln(n - 3), \tag{2.26}$$

where $-T\Delta S^0_{\text{loop}} = 3.5$ kcal/mol at 37 °C for a loop of four nucleotides, and α is a coefficient (Doktycz *et al.* 1990, SantaLucia & Hicks 2004). The best fit to the experimental data was found for an α of 2.44 rather than 1.5–1.75, the value for large loops formed during the melting of long DNA molecules (see above). The majority of the experimental data show that the loop entropy is nearly independent on the concentration of sodium ions. It should be noted that some loops of three and four nucleotides with specific sequences have lower free energy due to interaction between the bases (Senior *et al.* 1988, Blommers *et al.* 1989, Antao *et al.* 1991, Varani 1995, Nakano *et al.* 2002). In a good approximation the enthalpy of the hairpin loops is equal to zero, so Eq. (2.26) specifies the free energy of the loop, which includes formation of the first base pair (SantaLucia & Hicks 2004). The free energies of other base pairs of the stem are calculated according to the rules described above for extension of a helix segment. In this way we can calculate the total enthalpy and entropy of the hairpin formation, ΔH_h and ΔS_h. The melting temperature of a hairpin is calculated as

$$T_m = \Delta H_h / \Delta S_h. \tag{2.27}$$

It was found that corrections for the type of closing base pair and the terminal mismatch slightly improve the prediction of the free energy of the loop formation (SantaLucia & Hicks 2004). Overall, the existing data of the free energy parameters allow calculation of the hairpin T_m with an average accuracy of 4 °C (SantaLucia & Hicks 2004).

Attention to hairpin structures increased greatly with the development of molecular beacons, which are widely used for detecting DNA and RNA molecules containing

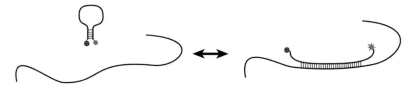

Figure 2.9 A molecular beacon is an oligonucleotide capable of forming a hairpin structure, with a fluorophore and a quencher at the ends. There is no fluorescence emission when the beacon has a hairpin conformation. Binding the oligonucleotide with a complementary single-stranded nucleic acid separates the ends of the oligonucleotide and prevents quenching.

specific sequences (Tyagi & Kramer 1996). A typical molecular beacon consists of an oligonucleotide of about 25 bases that is able to form a hairpin structure with 5–6 bp in the stem and 13–15 nucleotides in the loop. A fluorophore and a quencher are covalently attached to the ends of the oligonucleotide (Fig. 2.9). When the oligonucleotide forms the hairpin and the ends are in close proximity, the fluorescence is quenched. If the oligonucleotide hybridizes with a complementary target, an ssDNA or RNA, the fluorophore and quencher are well separated and the fluorescent emission can be detected. Since even in large concentration the beacons in the hairpin conformation produce low emission, they allow simple detection of very small amounts of specific nucleic acids. A variety of different applications of molecular beacons have been developed over the years (Tan *et al.* 2004).

It should be noted that hybridization of a beacon with a single-stranded nucleic acid depends on the equilibrium between the hairpin and the duplex, while a completely unpaired state of the oligonucleotide has negligible probability under the assay conditions. Our current knowledge of the thermodynamic stability of DNA duplexes and hairpins, outlined above, allows us to design molecular beacons with required properties.

2.2.7   Long DNA molecules

The two-state approximation definitely is not applicable as a description of DNA melting for molecules larger than 30 bp in length. However, melting of DNA molecules a few hundred base pairs in length or larger can be well described as a set of two-state transitions of DNA segments 100–500 bp in length. Localization of these segments in a DNA molecule is determined by its sequence, the cooperativity of the transition and the difference in the stability of the base pairs. The width of a two-state intramolecular transition involving n base pairs, ΔT_{m}, is specified by the equation

$$\Delta T_{\mathrm{m}} = 4RT_{\mathrm{m}}^2/n\,\langle\Delta H\rangle, \qquad (2.28)$$

where $\langle\Delta H\rangle$ is the average enthalpy of melting of one base pair (compare this with Eq. (2.24), which specifies ΔT_{m} for the bimolecular transition). Thus, for DNA segments 100–500 bp in length ΔT_{m} should be in the range of 0.2–1 °C. It is convenient to present the melting curve of a long DNA molecule as the temperature derivative of the fraction of melted bases, since in this case melting of individual segments corresponds to individual peaks of the curve. An example of such a differential melting curve is

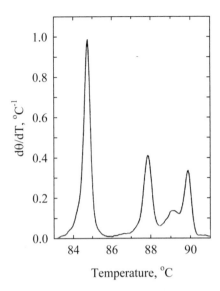

Figure 2.10 The differential melting profile of a 1291 bp DNA, the Sma I restriction fragment of the ColE1 plasmid. The figure was redrawn from Vologodskii *et al.* (1984), with permission.

shown in Fig. 2.10. Clearly, reliable measurement of differential melting curves requires the precise measurement of $\theta(T)$ (Lyubchenko *et al.* 1976).

The entire melting interval of long DNA molecules is usually between 2 and 15 °C, depending on the distribution of AT and GC base pairs along the DNA (Lazurkin *et al.* 1970). Of course, if the DNA consists of hundreds of thousand of base pairs, the individual peaks of the differential melting curve overlap and the melting curve looks very smooth, similar to the one diagramed in Fig. 2.4.

The fine structure of DNA melting curves was discovered in the mid-1970s (Ansevin *et al.* 1976, Gotoh *et al.* 1976, Lyubchenko *et al.* 1976). Even before the experimental observation, the fine structure was predicted theoretically in the computer simulation for randomly chosen DNA sequences (Lyubchenko *et al.* 1976). Later the nature of the fine structure was confirmed experimentally by mapping the individual steps of the transition under the electron microscope (Borovik *et al.* 1980). Thus, the theory of the helix–coil transition in DNA provided a very good qualitative description of the phenomenon. It was very interesting to test if the theory could predict the specific structure of the curve for a DNA molecule with a particular sequence. This became possible when the first phage DNAs were sequenced. Comparison of the fine structure of the measured and computed melting curves brought mixed results, showing good agreement for some molecules and pronounced discrepancies for others (Lyubchenko *et al.* 1978). Clearly, the theoretical model of the transition that only accounts for different stabilities of AT and GC base pairs required a refinement. First of all, accounting for different stacking free energies over different base-pair steps had to be added to the model. It should be noted that only years later was the corresponding set of the thermodynamic parameters obtained from the studies of oligonucleotide melting (Table 2.1).

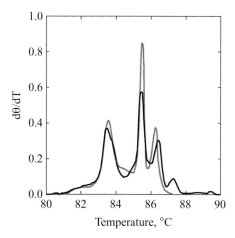

Figure 2.11 Comparison of calculated and measured melting curves of a 2528 bp fragment. The fragment was obtained by cutting the replicative form of phage fd DNA with BspRI endonuclease. Both experimental (black) and theoretical (gray) melting profiles correspond to [Na⁺] of 0.195 M. The theoretical profile was calculated for the parameters shown in Table 2.3. The figure was redrawn from Vologodskii *et al.* (1984), with permission.

Two sets of the needed parameters were obtained by minimizing the difference between calculated and experimental melting curves (Gotoh & Tagashira 1981, Vologodskii *et al.* 1984). Since the early studies of DNA melting showed that the entropy change associated with the helix elongation, ΔS, is nearly the same for AT and GC base pairs (see above), it was natural to assume that only enthalpy should depend on the type of base-pair step (Vologodskii *et al.* 1984, Doktycz *et al.* 1992). Therefore, the values of ΔH_{AT} and ΔH_{GC} were replaced in the refined model by $\Delta H_{AT} + \delta H_k$ and $\Delta H_{GC} + \delta H_k$, where δH_k is the enthalpy correction associated with a base-pair step of type k (the newly added base pair, AT or GC, specifies the first term). Thus, the enthalpy change due to forming a particular helical segment, ΔH, should be calculated as

$$\Delta H = n_{AT}\Delta H_{AT} + n_{GC}\Delta H_{GC} + \sum_{k=1}^{10} n_k \delta H_k, \tag{2.29}$$

where n_k is the number of base-pair steps of type k in the segment.

The obtained sets of 10 new parameters, δH_k, definitely improved the agreement between the calculated and measured melting curves (Gotoh & Tagashira 1981, Vologodskii *et al.* 1984, Delcourt & Blake 1991), although some discrepancy remains. Clearly, this discrepancy reflects limitations of the nearest-neighbor model. An example of a comparison between calculated and measured melting curves is shown in Fig. 2.11.

One important point was clarified during these studies. There are ten different base-pair steps in the double helix, so the researchers initially tried to find ten independent parameters δH_k (Gotoh & Tagashira 1981). Soon it became clear, however, that the theoretical melting curves depend on eight linear combinations of these parameters rather than on the ten independent variables (Vologodskii *et al.* 1984). This is a common

Table 2.3 The values of invariants, δH_I, that specify the free-energy correction for the dinucleotide model relative to the base-pair model (Vologodskii *et al.* 1984).

Invariant no.	δH_I	Corresponding polynucleotide	δH_I, kcal/mol
1	δH_{AA}	poly(A) $\cdot$ poly(T)	-0.08
2	δH_{GG}	poly(G) $\cdot$ poly(C)	0.18
3	$(\delta H_{AT} + \delta H_{TA})/2$	poly(AT) $\cdot$ poly(AT)	0.02
4	$(\delta H_{GC} + \delta H_{CG})/2$	poly(GC) $\cdot$ poly(GC)	-0.15
5	$(\delta H_{AC} + \delta H_{CA})/2$	poly(AC) $\cdot$ poly(GT)	-0.04
6	$(\delta H_{AG} + \delta H_{GA})/2$	poly(AG) $\cdot$ poly(CT)	0.11
7	$(\delta H_{AT} + \delta H_{TC} + \delta H_{CA})/3$	poly(ATC) $\cdot$ poly(GAT)	0.00
8	$(\delta H_{AG} + \delta H_{GC} + \delta H_{CA})/3$	poly(AGC) $\cdot$ poly(GCT)	-0.01

feature of DNA properties that can be represented as a sum of contribution from the base-pair steps (Gray & Tinoco 1970, Goldstein & Benight 1992). This feature is due to the fact that in long dsDNA there are two conditions on the number of base-pair steps that always hold:

$$n_{AT} + n_{AC} + n_{AG} = n_{TA} + n_{CA} + n_{GA} \tag{2.30}$$

$$n_{GT} + n_{GA} + n_{GC} = n_{TG} + n_{AG} + n_{CG}. \tag{2.31}$$

To prove these equations let us mark, for example, all AX and XA steps in a DNA sequence with different colors (here X is any letter except A). It is easy to see that an AX step always follows an XA step and vice versa. This means that in a sufficiently long DNA sequence the number of AX and XA steps is nearly equal, and this is what Eq. (2.30) states. Although we can write similar equations for the other three letters, it turns out that only two of this kind of equation are linearly independent. Since cooperatively melted regions of DNA usually exceed 100 bp, Eqs. (2.30) and (2.31) are well satisfied in the melting process.

Equations (2.30) and (2.31) mean that there are only eight rather than ten independent variables that specify the stability of extended helical regions. This fact also means that there is no way to find the ten variables δH_k unambiguously by minimizing the difference between experimental and computed melting curves. Indeed, it was shown that very a large change of a set of ten parameters does not change the melting curves, if a certain eight linear combinations of these parameters are not changed. We can say that these combinations represent invariants of the sets of ten parameters. The values of these eight invariants rather than ten individual parameters δH_k specify the fine structure of the melting curves. The invariants can be chosen in various ways (Goldstein & Benight 1992). One way is to select eight polynucleotide duplexes whose stability corresponds to eight invariants, δH_I (Vologodskii *et al.* 1984). The values of these invariants are given in Table 2.3.

It is not so convenient to perform calculations by using the values of the invariants rather than δH_k for individual base-pair steps. One way to handle this problem is to set the values δH_k so that they satisfy the invariants (which give eight linear equations for

δH_k) and two additional arbitrarily chosen equations. The latter two equations can be taken as

$$\delta H_{\mathrm{AT}} = \delta H_{\mathrm{TA}}$$
$$\delta H_{\mathrm{GC}} = \delta H_{\mathrm{CG}}. \tag{2.32}$$

It is interesting that the absolute values of ΔH_l in Table 2.3 are much smaller than the difference between ΔH_{AT} and $\Delta H_{\mathrm{GC}}(\approx 1\ \mathrm{kcal/mol})$. Therefore, the dependence of ΔH on DNA sequence is well specified by the values of ΔH_{AT} and ΔH_{GC}, while the values of δH_k should be considered as a small correction (assuming that the values of δH_k and ΔH_l have similar magnitudes).

We should note that the values of ΔH_{AT}, ΔH_{GC} and δH_k do not represent the free-energy contribution from hydrogen bonding and staking, correspondingly, as some authors suggested. Since the two interactions appear simultaneously upon addition of a base pair to a helical region, there is no way to determine the two free-energy contributions separately from the melting experiments. The division of ΔH into two parts only reflects the fact that in the first approximation the values of δH_k can be neglected. The actual contribution of the stacking interaction and hydrogen bonding to the free energy of the helix elongation, ΔG_k, was estimated by Frank-Kamenetskii and co-workers in a different kind of experiment (Protozanova et al. 2004, Yakovchuk et al. 2006). The researchers investigated stacking–unstacking equilibrium at single-stranded nicks of the double helix. They concluded that in a broad range of solution conditions the hydrogen bonding has a destabilizing effect in AT base pairs, contributing 0.5 kcal/ml to ΔG_k, and makes no contribution in GC base pairs (Yakovchuk et al. 2006).

2.2.8   Oligonucleotide model of DNA melting versus polymer model

The theoretical models describing the thermodynamic stability of helical regions in DNA oligonucleotides and in long DNA molecules look rather different, but careful analysis shows that the models and their sets of parameters are actually very similar (Owczarzy et al. 1997, SantaLucia 1998). The main difference between the models is different numbers of the helix elongation parameters, eight and ten. There are different ways to bypass this difference if we want to compare the two descriptions. SantaLucia did it by extracting ten stacking parameters from the sets of eight invariants, by using a mathematical method that minimizes the difference between the obtained ten parameters (SantaLucia 1998). He concluded that there is a good agreement between the sets of parameters used in the oligonucleotide and polymer models, and both models work rather well in describing the oligonucleotide melting temperatures. The latter observation shows that, within a reasonable approximation, Eqs. (2.30) and (2.31) are valid for the melting of oligonucleotide duplexes as well, so even oligonucleotide thermal stability is essentially specified by eight invariants, δG_l (SantaLucia 1998). Here we use another way of comparison. To avoid remaining ambiguity in the extension of the set of eight parameters to the set of ten, we reduced the set of ten parameters to the set of eight. In this way we able to compare three sets of invariants. The first set, for the polymer

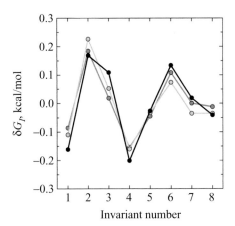

Figure 2.12 Comparison between three sets of melting free-energy corrections associated with different base-pair steps. The plotted values of δG_I correspond to eight polynucleotides in Table 2.2. The polynucleotide-based δG_I were taken from Table 2.3 as δH_I (dark gray), the DNA dumbbell-based δG_I were calculated from the data in Doktycz *et al.* (1992) (light gray) and the oligonucleotide-based δG_I were calculated from the parameter set shown in Table 2.1 (black). The three sets of δG_I correspond to NaCl concentrations of 0.195 M, 0.115 M and 1 M, correspondingly. Lines are drawn for convenience only.

model, was taken directly from Table 2.3, since in this case δG_I coincides with δH_I. The second set of δG_I was obtained from Table 2.1, which presents the parameters of the oligonucleotide model. We first calculated the average value of ΔG_{AT} and ΔG_{GC} and then used these values to calculate δG_I. The third set of δG_I was based on melting data of relatively short DNA dumbbells (Doktycz *et al.* 1992). These three sets of δG_I, presented in Fig. 2.12, are in remarkable agreement. The comparison confirms that we have a very good theoretical description of DNA melting.

There is one more point where the oligonucleotide data look different from the data for large DNA molecules. The entropy of base-pair melting obtained for oligonucleotides (Table 2.1) depends on the type of base-pair step, while in the polymer set of parameters it was found to be sequence independent (and slightly higher in absolute value). This discrepancy is hardly meaningful, however, since a small error in one parameter can be compensated by an error in another parameter. It is important to test, however, how well the values of the entropies allow one to calculate the free energies of melting at very different temperatures. Thus, we can perform a simple test of the two sets of parameters.

Let us take imaginary DNA with 50% GC, with random sequence, in 1 M NaCl. According to Eq. (2.9) the DNA melting temperature is 100 °C, so the free-energy change due to the melting is equal to zero at that temperature. Using the polynucleotide set of parameters we obtain that at 37 °C (310 K) the average ΔG is equal to $\Delta S(T - T^0) = 25 \times (-63) = -1575$ cal/mol. This is in a reasonably good agreement with the oligonucleotide-based data, -1405 cal/mol, which were obtained at temperatures close to 37 °C (SantaLucia 1998). Still, the discrepancy can be reduced if we account for ΔC_p, as was suggested by Rouzina and Bloomfield (1999a, 1999b).

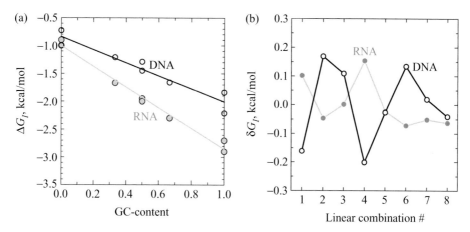

Figure 2.13 Comparison of the free energies of base-pair melting for DNA and RNA duplexes, at 1 M NaCl, 37 °C. (a) The values of ΔG_I, listed in Table 2.4, represent linear combinations of ΔG for individual base-pair steps. The same presentation of the data was used for DNA duplexes. The straight lines correspond to the best fit of each set of data by the base-pair model. (b) The differences in the values of ΔG_I between the dinucleotide model and the base-pair model. The points were obtained by subtracting the values of ΔG_I corresponding to the straight lines in (a) from the exact values of ΔG_I shown by the circles in that panel. Lines are drawn for convenience only. The plots are based on the data for RNA duplexes from Freier *et al.* (1986) and for DNA duplexes from SantaLucia and Hicks (2004).

We can also extrapolate the oligonucleotide data to high temperature. If we take the parameters from Table 2.1 and calculate ΔG for 100 °C, for the same DNA with 50% GC, we obtain ΔG of -17 cal/mol, so in this case agreement is perfect and no correction due to ΔC_p is needed. Thus, maybe the oligonucleotide-based set of parameters is slightly better than the set based on the polymers. The most important conclusion, however, is that we know the values of ΔG with a very good accuracy over a very wide range of temperatures.

2.2.9   Thermal stability of DNA duplexes versus RNA and hybrid DNA·RNA duplexes

Due to an additional hydroxyl group in the sugar ring, the conformation of RNA duplexes corresponds to A rather than B form under physiological conditions. Since the B form of DNA is more stable under these conditions than the A form, one could expect that RNA duplexes will be less stable than DNA duplexes. On the contrary, it was found that RNA duplexes are more stable than DNA duplexes. Studies of the thermal stability of RNA duplexes were pioneered by Tinoco and co-workers in the early 1970s (Tinoco *et al.* 1971, Tinoco *et al.* 1973). The set of thermodynamic parameters that specifies the stability of RNA duplexes was greatly improved in the 1980s by Turner and co-workers, when synthetic RNA oligonucleotides became relatively easily available (Freier *et al.* 1986). A comparison of the free energies of the helix elongation for DNA and RNA duplexes, ΔG, is presented in Fig. 2.13. Assuming that the free energies of ten individual base-pair steps are determined less accurately than the values of their eight linear

Table 2.4 The values of the linear combinations of ΔG, ΔG_I, that specify the free energy of RNA helix formation by the corresponding polynucleotides. The data for individual dinucleotide steps were taken from Freier *et al.* (1986) and correspond to 1 M NaCl, 37 °C.

Linear combination no.	ΔG_I	Corresponding polynucleotide	ΔG_I, kcal/mol
1	ΔG_{AA}	poly(A) · poly(U)	−0.9
2	ΔG_{GG}	poly(G) · poly(C)	−2.9
3	$(\Delta G_{AU} + \Delta G_{UA})/2$	poly(AU) · poly(AU)	−1.0
4	$(\Delta G_{GC} + \Delta G_{CG})/2$	poly(GC) · poly(GC)	−2.7
5	$(\Delta G_{AC} + \Delta G_{CA})/2$	poly(AC) · poly(GU)	−1.95
6	$(\Delta G_{AG} + \Delta G_{GA})/2$	poly(AG) · poly(CU)	−2.0
7	$(\Delta G_{AU} + \Delta G_{UC} + \Delta G_{CA})/3$	poly(AUC) · poly(GAU)	−1.67
8	$(\Delta G_{AG} + \Delta G_{GC} + \Delta G_{CA})/3$	poly(AGC) · poly(GCU)	−2.30

combinations (see above), the linear combinations, ΔG_I, are used for the comparison. These linear combinations are listed in Table 2.4.

One can see from Fig. 2.13(a) that RNA duplexes are more stable than corresponding DNA duplexes at all values of GC-content. To a good approximation the values of ΔG_I for RNA duplexes decrease linearly with GC-content increase, as they do for DNA duplexes. This decrease is much larger for RNA duplexes, however, than for DNA molecules. Thus, though RNA duplexes are more stable than DNA duplexes in general, the difference is especially large for duplexes with high GC-content.

The straight lines in Fig. 2.13(a) correspond to reduction of the dinucleotide model to the base-pair model. Although the base-pair model works rather well for both DNA and RNA duplexes, it is definitely less accurate than the dinucleotide model. The differences between the values of ΔG_I corresponding to the two models are plotted separately in Fig. 2.13(b). As one would expect, there is no correlation between these differences for RNA and DNA molecules. It is interesting to note that the corrections to the base-pair model for RNA are even smaller than those for DNA molecules.

In many cellular processes, such as initiation of replication and reverse transcription, DNA and RNA molecules form hybrid double helices. Melting of these hybrid DNA:RNA helices was studied in detail, following the progress in RNA synthesis. Already the early works showed that the thermal stability of the hybrid helices depends not only on the GC-content of the duplexes but also on the distribution of bases in two strands. It was found that the hybrid helices are substantially more stable when more purines are located in the RNA chain (Martin & Tinoco 1980, Roberts & Crothers 1992). This definitely complicates quantitative analysis of the thermal stability of these duplexes. The most comprehensive study of the issue was reported by Lesnik and Freier (1995). The data obtained in this work and two other studies are presented in Fig. 2.14. We see from Fig. 2.14(a) that, if purine and pyrimidine bases are nearly equally distributed between DNA and RNA strands, the hybrid duplexes are less stable than RNA and DNA ones. Also, the dependence of the duplex stability on GC-content is weaker than that for RNA and DNA duplexes. The data in Fig. 2.14(b) show the difference in free energy of

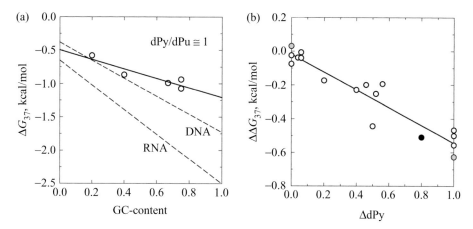

Figure 2.14 Stability of hybrid DNA:RNA duplexes. (a) The free energy of duplex formation at 37 °C, 0.1 M NaCl, per base pair. The shown data (o) correspond to duplexes with nearly equal fractions of pyrimidine and purine bases in the DNA strands. For comparison the corresponding data for DNA and RNA duplexes are shown by the dashed lines. (b) Dependence of the difference in free energy of duplex formation for matched pairs of DR and RD hybrids on the difference in the fractions of pyrimidines in their DNA strands. The data were taken from Lesnik and Freier (1995) (white circles), Gyi et al. (1996) (black circles), and Ratmeyer et al. (1994) (gray circles). The solid lines correspond to the best linear approximations of the experimental data.

duplex formation for matched pairs of DR and RD hybrids, such as d(GCTTCTCTTC) · r(GAAGAGAAGC) and r(GCUUCUCUUC) · d(GAAGAGAAGC), against the difference in the fractions of pyrimidines in their DNA strands. Clearly, purines in the DNA strand destabilize the hybrid to approximately the same extent as AT/U base pairs. We can conclude from the figure that, if we want to make a hybrid duplex more stable than its DNA and RNA analogs, it has to contain only AT base pairs, with all adenines in the RNA strand. It seems that such a duplex, $rA_n:dT_n$, should be more stable than the corresponding DNA duplex but still less stable than the duplex made by the RNA strands.

2.2.10   Reversibility of DNA melting

It was found in the early studies of DNA melting that the reverse process, DNA renaturation, is so slow that the renaturation curves do not follow the denaturation ones for long DNA molecules. Melting of the entire DNA molecule that involves complete separation of the complementary strands is not reversible under real experimental conditions. The estimation shows that the helix nucleation time should be longer than 10^5 s even at high concentrations of Na^+ (Anshelevich et al. 1984). For long DNA molecules, however, the main part of the melting process does not involve a complete separation of the strands, and here one could expect the process to be reversible. It turned that it is not always the case. Hoff and Roos were the first to study the melting reversibility for homogeneous samples of DNA molecules in detail (Hoff & Roos 1972). They found that at $[Na^+]$ close

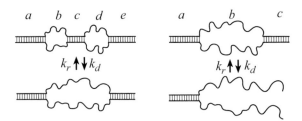

Figure 2.15 Diagrams of the slow denaturation steps in DNA melting. Both of the steps shown involve disappearance of a helical region. Correspondingly, the reverse process involves nucleation of a new helical region, which is slow at low concentrations of Na$^+$ due to strong electrostatic repulsion between the complementary strands.

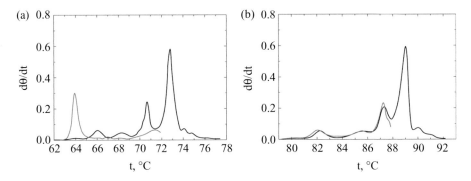

Figure 2.16 Reversibility of DNA melting. Complete differential melting curves of T7 DNA fragment 1459 bp in length at [Na$^+$] of 0.02 M (a) and 0.2 M (b) are shown by black lines. The renaturation curves (gray lines) were obtained after denaturation of this DNA by up to 40%. This figure was reproduced from Figs. 8 and 9 of Perelroyzen *et al.* (1981) by digitizing and replotting, with permission.

to 0.02 M only early stages of the melting are reversible and suggested a model that explains this lack of reversibility. According to this model, irreversibility emerges when a new helical region has to be formed in the process of renaturation. This is diagrammatically shown in Fig. 2.15. Renaturation is in this case analogous to reassociation of the separated DNA strands and has a low rate constant because of the strong electrostatic repulsion of the strands at a low [Na$^+$]. If renaturation does not involve formation of additional helical regions but occurs through extension of the existing ones, the melting process should be reversible.

This model was confirmed experimentally by combining the denaturation curves and mapping of the melting under electron microscopy (EM) (Perelroyzen *et al.* 1981). The experiments showed that at [Na$^+$] of 0.02 M the early steps of DNA melting are reversible, but the steps diagramed in Fig. 2.15 are not. The denaturation and renaturation curves can differ dramatically under these ionic conditions (Fig. 2.16(a)). However, at [Na$^+$] of 0.2 M the melting becomes reversible (Fig. 2.16(b)), until complete separation of DNA strands, which corresponds to the largest peak on the differential melting curve.

Analysis shows, however, that irreversible denaturation curves are rather close to the equilibrium ones. At the equilibrium melting temperature of a helical segment, T_m, the rate constant of denaturation, k_d, has to be equal to the rate constant of renaturation, k_r. At low ion concentration k_r is very small due to strong electrostatic repulsion between ssDNA segments. Under such conditions k_d is also very small (equal to k_r). Therefore, the segment does not have enough time to melt out during the characteristic time of the experiment, Δt. The value of k_d is mainly specified by the value of s^N for a DNA segment N base pairs in length (see Eq. (4.13)). Since the value of s^N decreases extremely fast with increasing temperature, at a temperature slightly higher than T_m the rate constant k_d becomes of the order of $1/\Delta t$. The segment melting is observed at this temperature, T_{exp}. The estimation shows that at $[\text{Na}^+]$ of 0.01 M the difference $T_{exp} - T_m$ should be in the range of 0.2–0.4 °C (Anshelevich *et al.* 1984). The renaturation of the segment proceeds in a different way since k_r is small and depends weakly on temperature. Only when one of the surrounding segments (*b* or *d* in Fig. 2.15) is zipped up at its equilibrium melting temperature can the complementary strands corresponding to segment *c* renaturate as well.

2.3   B–A transition

In a concentrated solution of ethanol a DNA molecule can adopt the A form of the double helix. The A–B transition can be observed when the ethanol fraction in solution increases, under the condition that the concentration of sodium does not exceed 1 mM. At higher concentration of NaCl or in the presence of other additional ions DNA molecules form aggregates and precipitate before the B–A transition starts (this precipitation is widely used in DNA purification or buffer exchange procedures). At low concentration of sodium ions the aggregation is blocked by the electrostatic repulsion between the DNA segments. The ethanol can be replaced by some other alcohols, and it was found that to a good approximation the activity of water mainly controls the transition (Malenkov *et al.* 1975).

Circular dichroism (CD) is usually used to monitor the transition in solution. CD remains a completely empirical method, so there is no way to say anything certain about DNA conformation from its CD spectrum. In many cases, however, there are ways to establish the correspondence between the structure and the CD spectrum. In the case of A-DNA it was found that at high concentration of ethanol the CD spectrum of DNA has a strong similarity with the CD spectrum of RNA, which always exists in the A form (see Section 1.2.2). In addition, it was shown that the same CD spectrum is observed in thin films of DNA at 75% relative humidity, when X-ray analysis shows that DNA is in the A form (Tunis-Schneider & Maestre 1970). The change of CD spectrum upon the B–A transition and the transition curve are shown in Fig. 2.17.

It was found that the B–A transition is not influenced, within the accuracy of the measurements, by temperature (Ivanov *et al.* 1974, Usatyi & Shlyakhtenko 1974). It is hardly possible, due to DNA aggregation, to observe the B–A transition in solutions where sodium ions are replaced by other ions widely used in biophysical studies (Ivanov

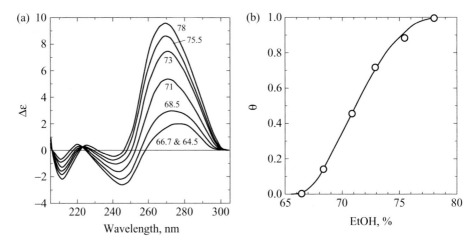

Figure 2.17 The B–A transition in DNA. (a) Circular dichroism spectra of DNA in water–ethanol solutions that correspond to the B–A transition. The volume percent of ethanol is shown near each curve. The solution contains 0.5 mM NaCl. (b) The transition curve calculated from the changes of CD at 270 nm is shown on the right. The plots were reproduced from Fig. 1 of Ivanov *et al.* (1974) by digitizing and replotting, with permission.

et al. 1974). Clearly, sodium ions, as compared with the majority of other ions, facilitate the B–A transition.

It was mentioned above that the B–A transition is well described by the Ising model. There are two parameters in the model, the free energy of A form elongation, ΔG, and the free energy of A form nucleation, ΔG_{nucl}. The value of ΔG depends on the water activity in solution, a. In the vicinity of the transition one can approximate the dependence by a linear function:

$$\Delta G = (a - a_0)RT/Q, \tag{2.33}$$

where Q is a proportionality coefficient and a_0 is the water activity in the middle of the transition. To a good approximation the value of Q does not depend on DNA sequence. This conclusion was made, first of all, from the studies of the transition in oligonucleotide duplexes 10–12 bp in length (Minchenkova *et al.* 1986, Ivanov & Krylov 1992). The B–A transition in such short duplexes occurs as a "two-state" transition and therefore its width depends on the value of Q and the duplex length only. It was found that the transition width remains nearly the same for duplexes with different sequences. The value of Q in units of water activity is equal to 0.14. This and other studies showed, however, that the second parameter of Eq. (2.33), a_0, depends on DNA sequence. It was found that the dinucleotide approximation provides a reasonably good approximation of the sequence dependence of a_0 (Tolstorukov *et al.* 2001). Dinucleotide step GG/CC turned to be the most A-philic, while step AA/TT is the most resistant to the B–A transition. Assuming that Eq. (2.33) is valid over a broad range of water activity, one can estimate the values of ΔG at pure water. It was found in this way that the values of ΔG_{GG} and ΔG_{AA} in water are equal to 0.32 and 1.09 kcal/mol, respectively (Tolstorukov *et al.*

2001). Thus, the difference in ΔG between different dinucleotide steps is very substantial. It certainly should affect the binding affinity of DNA segments with proteins that transform the segments into the A form in DNA–protein complexes.

The sequence dependence of the B–A transition complicates the determination of the nucleation parameter, ΔG_{nucl}. Different approaches have been used to estimate the parameter, which turned out to be close to 4 kcal/mol (correspondingly, the free energy of a single AB junction is equal to 2 kcal/mol) (Ivanov *et al.* 1985, Ivanov *et al.* 1996). Thus all parameters of the Ising model describing the transition are known, and one can use Eqs. (2.1)–(2.6) to estimate desirable statistical features of the B–A transition in DNA with a given sequence. Similarly to the helix–coil transition in DNA, the B–A transition in a long DNA molecule consists of a set of "two-state" transitions in specific DNA segments. The corresponding calculation for a long DNA molecule with a random sequence and GC-content of 50% shows that the average length of these segments is of the order of 10 bp (Ivanov *et al.* 1996).

2.4   B–Z transition

The Z form of DNA was discovered for the d(CGCGCG) · d(CGCGCG) duplex (Wang *et al.* 1979), and it turned out that poly[d(GC)] · poly[d(GC)] is a polynucleotide that very readily undergoes the B–Z transition. The majority of studies of the B–Z transition in linear DNA have used this polynucleotide or its methylated analogues, since methylation of poly[d(GC)] strongly favors Z-DNA formation (see Rich *et al.* (1984) and references therein). The transition of poly[d(GC)] · poly[d(GC)] from the B to Z form occurs at a high concentration of sodium ions (2.4 M) or in the presence of certain bi- or trivalent ions. It is interesting that the B–Z transition in this polynucleotide was observed by Pohl and Jovin seven years before Rich and co-workers discovered the Z form (Pohl & Jovin 1972). Remarkably, based on the inversion of the CD spectrum (see Fig. 2.18), Pohl and Jovin hypothesized (correctly, as we now know) that the transition they observed was due to the formation of a left-handed duplex. However, they could not prove it, since there is no direct way to extract DNA structural features from a CD spectrum. Still, the spectrum provides a very sensitive tool to monitor the appearance of a particular form, if we know the spectrum of this form. CD spectrum remains the simplest way to monitor both the B–A and B–Z transitions (Fig. 2.18), although other spectroscopic methods can be used for Z-DNA detection as well (see Rich *et al.* (1984) for details). The left-handed helix can also be detected by binding the Z-form-specific antibodies with DNA molecules (Di Capua *et al.* 1982, Nordheim *et al.* 1982, Rich *et al.* 1984).

An important feature of the B–Z transition is that it can occur at physiological ionic conditions in segments of negatively supercoiled DNA (see Chapter 6). Studies of the transition in the supercoiled DNA molecules have demonstrated that the transition to the Z form can occur in DNA segments with a broad variety of sequences. However, the free-energy cost of the emergence of this DNA form in a segment with an arbitrary sequence is substantially higher than in d(GC)$_n$ segments. The following statistical-mechanical treatment of the B–Z transition in linear DNA helps to understand the formation of

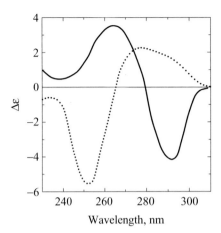

Figure 2.18 CD spectra of the B and Z forms of poly[d(GC)] · poly[d(GC)]. The spectrum of the B form (dotted line) is obtained in a 0.2 M solution of NaCl; the spectrum of the Z form (solid line) was obtained by increasing the NaCl concentration to 3 M and keeping the solution for 3 h at room temperature. The plots were reproduced from Fig. 2 of Pohl & Jovin (1972) by digitizing and replotting, with permission.

the Z form in DNA in a general case. This treatment has a certain similarity to the statistical-mechanical treatment of the B–A transition, although there is an important difference.

In statistical-mechanical terms, the most important feature of Z-DNA is that the repetitive unit of the Z helix consists of two adjacent base pairs rather than a single base pair, and the base pairs have different conformations in this repetitive unit (see Section 1.2.3). Thus, each base pair can be in either of two different states in the Z form, and the statistical-mechanical treatment of the B–Z transition should account for this fact. The first of these states (I) is characterized by the *syn* conformation for purine and the *anti* conformation for complementary pyrimidine. The second state (II) corresponds to the *syn* conformation for pyrimidine and the *anti* conformation for purine. State I is associated with much lower free energy, since the *syn* conformation is significantly more stable for purines than for pyrimidines. Still, although state II is characterized by a higher free energy, it can appear in some sequences because the regular structure of the Z helix requires strict alternation of *syn* and *anti* conformations in each strand of the double helix. Also, nucleoside conformations in each base pair have to be different in the Z helix, *syn–anti* or *anti–syn*. In regular purine–pyrimidine sequences all base pairs always appear in state I, and such sequences are the most probable candidates for formation of Z-DNA. Thus, the free energy of the B–Z transition for a base pair, ΔG, depends not only on the type of the base pair but also on its state, I or II, in the Z helix. Of course, there is also the free energy of the Z-form nucleation, ΔG_{nucl}. However, in addition to the B–Z boundaries another type of boundary can appear in the Z form: the Z–Z boundary, which corresponds to a break of the regular *syn–anti* alternation pattern in each strand. This boundary, which can be regarded as a change of phase in the Z form, is associated with the free energy ΔG_{ZZ}.

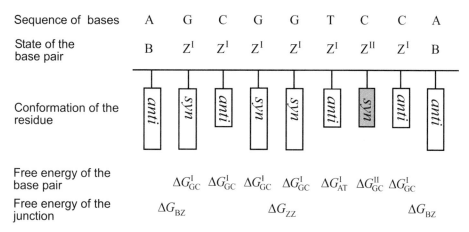

Sequence of bases	A	G	C	G	G	T	C	C	A
State of the base pair	B	Z^I	Z^I	Z^I	Z^I	Z^I	Z^{II}	Z^I	B

Conformation of the residue: *anti* *syn* *anti* *syn* *syn* *anti* *syn* *anti* *anti*

Free energy of the base pair: ΔG^I_{GC} ΔG^I_{GC} ΔG^I_{GC} ΔG^I_{GC} ΔG^I_{AT} ΔG^{II}_{GC} ΔG^I_{GC}

Free energy of the junction: ΔG_{BZ} ΔG_{ZZ} ΔG_{BZ}

Figure 2.19 The free energy of a DNA segment in the Z form. The sequence of bases in the chosen strand of DNA is shown in the top row. The chosen states of the base pairs, B, Z^I or Z^{II}, are shown in the second row; in the case of the Z form they are specified by *syn* or *anti* conformations of the nucleosides in the chosen strand. In this example all the base pairs of the Z helix are in state Z^I except the third base pair from the right, which is in the state Z^{II} (marked in gray). The free energy of the Z segment represents the sum over the segment base pairs and three junctions. There are two B–Z junctions in the segment and one Z–Z junction, which corresponds to the change of phase in the Z form.

Two practically equivalent models were suggested for statistical-mechanical description of the transition (Ellison *et al.* 1985, Mirkin *et al.* 1987). Below we will follow a slightly simpler model that has a smaller number of parameters (Mirkin *et al.* 1987). To illustrate the free-energy calculation for a segment in the Z form, let us consider the example of a Z helix formed by a GCGGTCC segment flanked by B-form segments (Fig. 2.19). Since the sequence and conformations of the residues in one strand completely define the sequence and conformations of the residues in the complementary strand, we specify below only one arbitrarily chosen strand. Below Z^I and Z^{II} denote two possible states of a base pair in the Z form. Instead of the nucleation free energy, ΔG_{nucl}, we use here the free energy of a single B–Z junction, $\Delta G_{BZ} = \Delta G_{nucl}/2$.

The total free energy of the transition to this particular conformation is

$$5\Delta G^I_{GC} + \Delta G^{II}_{GC} + \Delta G^I_{AT} + 2\Delta G_{BZ} + \Delta G_{ZZ}.$$

Of course, the chosen conformation of the Z segment represents only one of $2^7 = 128$ possibilities (two Z conformations for each base pair), and some of them have free energy comparable with the free energy of the shown conformation. In general, the total free-energy change for the B–Z transition in this segment, ΔG_{total}, has to be calculated as

$$\Delta G_{total} = -RT \ln \left(\sum_{k=1}^{127} \exp[-\Delta G(k)/RT] \right), \qquad (2.34)$$

where $\Delta G(k)$ corresponds to the free-energy change for the transition to conformation k.

Table 2.5 The values of the free-energy parameters of the B–Z transition in TBE buffer. The shown values were taken from a few cited studies and corrected for the proper expression for the free energy of supercoiling in this buffer (see Section 6.3.3).

Parameter	Value, kcal/mol	Reference
ΔG_{GC}^I	0.41	(Peck & Wang 1983)
ΔG_{AT}^I	2.1	(Haniford & Pulleyblank 1983, Mirkin *et al.* 1987)
ΔG_{GC}^{II}	4.7	(Ellison *et al.* 1985)
ΔG_{AT}^{II}	6.6	(Ellison *et al.* 1985)
ΔG_i^{BZ}	7.5	(Peck & Wang 1983)
ΔG_i^{ZZ}	7.0	(Mirkin *et al.* 1987)

Thus, the model of the B–Z transition allows for three possible states of each base pair and contains six parameters: ΔG_{GC}^I, ΔG_{GC}^{II}, ΔG_{AT}^I, ΔG_{AT}^{II}, ΔG_{BZ}, ΔG_{ZZ}. Under regular ionic conditions close to physiological ones all these parameters have positive values. Therefore, the Z form is not adopted by linear DNA under these conditions. However, the B–Z transition under physiological conditions can be observed in negatively supercoiled DNA (Nordheim *et al.* 1982, Singleton *et al.* 1982). The studies of the transition in supercoiled DNA were mainly performed in TBE buffer (89 mM Tris, 89 mM boric acid, 2 mM EDTA), and the corresponding set of transition parameters for this buffer is shown in Table 2.5.

It should be noted that to determine the parameters from experiments with supercoiled DNA molecules one needs to know how the supercoiling free energy, G_s, depends on the superhelix density, σ (see Section 6.3.3). In the 1980s the free energy of supercoiling, $G_s(\sigma)$, was determined only for physiological ionic conditions, and the parameter determination was based on the assumption that $G_s(\sigma)$ is the same in TBE buffer. It was shown later, both by computer simulation and experimentally, that the assumption is incorrect (Vologodskii & Cozzarelli 1994, Rybenkov *et al.* 1997). The data shown in Table 2.5 are based on the cited experimental studies and $G_s(\sigma)$ corrected for TBE buffer (see Section 6.3.3 for details).

There is only one study where ΔG_{GC}^I was determined for ionic conditions close to physiological (TBE buffer + 0.1 M NaCl) (Peck & Wang 1983). Following the authors' assumption that ΔG_j^{BZ} remains the same in the latter ionic conditions, we can conclude, using ΔG_j^{BZ} from Table 2.5, that ΔG_{GC}^I is equal to 0.53 kcal/mol at these conditions. This value is notably higher than in TBE buffer. This estimation is based on a few assumptions, however. More data on the issue are needed to make the parameter estimations for physiological ionic conditions more reliable.

We can conclude from Table 2.5 that even in TBE buffer the probability of spontaneous appearance of the Z form in a segment of linear DNA is very low. For example, for the segment $d(GC)_3 \cdot d(GC)_3$ surrounded by base pairs in the B form, this probability is smaller than 10^{-12}. It should be even lower at physiological ionic conditions. Another important conclusion from the table is that insertion of an AT base pair in a $d(GC)_n \cdot d(GC)_n$ segment results in relatively small free-energy increase of the segment's Z form,

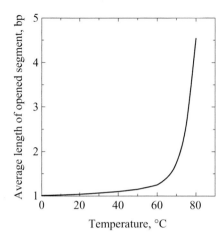

Figure 2.20 The average size of an opened region of the double helix as a function of temperature. The plot is based on the calculation by Lukashin *et al.* (1976). The DNA melting temperature in those calculations was 87 °C.

if the pair does not disturb the regular alternation of purines and pyrimidines in each DNA strand.

2.5   Thermal fluctuations of DNA secondary structure

At physiological ionic conditions and usual physiological temperature the structure of the double helix corresponds to the B form. Still, due to thermal fluctuations there is a certain probability to find any chosen base pair in an alternative conformation. These transient states of the base pairs are very important, since they may be recognized by proteins interacting with specific DNA conformations. The transient opening of the base pairs is necessary for chemical modifications of DNA inside the cell such as methylation. The alternative conformations may contribute to DNA damage inside the cell. In this section we consider how we can address the problems related to the thermal fluctuations of DNA structure and what we know about these fluctuations (reviewed by Frank-Kamenetskii & Prakash (2014)).

    Let us first analyze what we can conclude about thermal fluctuation from the studies of conformational transitions in the double helix. The first conclusion is that the size of segments in the transient states should be small, 1–2 bp in length (Fig. 2.20). This follows from the fact that for all alternative structures the free energy of the elongation of an alternative state is positive under physiological conditions, so the fluctuations are less probable for larger DNA segments. This conclusion is supported by the experimental observation that there is no correlation between the lifetimes of open states in adjacent base pairs in DNA duplexes (Gueron *et al.* 1987).

    We could proceed with the use of the transition models to estimate the probabilities of different fluctuations. These estimations could be not very accurate, however. Indeed, during the transitions, due to their cooperativity all DNA segments in alternative conformations consist of at least a few base pairs. Correspondingly, the parameters of the transitions, first of all the nucleation free energy, were determined from the experimental data that correspond to these extended segments of the alternative conformations. Although the boundaries between different structures are relatively short (see Kang & Wells (1985), for example), the conformations of 1–2 bp at the boundaries are somewhat different from the conformations of the internal base pairs. Thus, there is no reason to believe that the estimations of the nucleation parameters have to be valid for segments 1 bp in length. The corresponding calculations of the fluctuation probabilities require an independent test. Such a test was provided first by the quantitative analysis of DNA unwinding by formaldehyde (Lukashin *et al.* 1976). Successful quantitative description of the process gave a solid support to the estimation of the probability of base-pair opening based on the theory of the helix–coil transition.

Let us consider this theoretical description of the base-pair opening. Since a base pair in the helical state eliminates direct interaction between surrounding base pairs, it is sufficient to consider opening of the central base pair in the middle of a segment of three base pairs. Following Eq. (2.12) we obtain for the probability of the base-pair opening, P,

$$P = s_{i-1}\sigma s_{i+1}/(s_{i-1}s_i s_{i+1} + s_{i-1}\sigma s_{i+1}) \cong \sigma/s_i. \tag{2.35}$$

Equation (2.35) predicts that the probability of opening should be around 10^{-5} for AT pairs and around 10^{-6} for GC pairs at room temperature and physiological ionic conditions (≈ 0.2 M NaCl).

The most direct estimation of the probability of base-pair opening was accomplished by the groups of Gueron and Russu (Gueron *et al.* 1987, Kochoyan *et al.* 1987, Leroy *et al.* 1988, Moe & Russu 1990, Coman & Russu 2005). The estimates are based on comparison of the rates of imino proton exchange for base pairs in the middle of oligonucleotide duplexes with the rate in free nucleotides. Their results are in very good agreement with the above estimation based on the helix–coil theory. The researchers found that the base-pair opening probability is equal to 10^{-5}–10^{-6} and 10^{-6}–10^{-7} for AT and GC pairs, respectively. Of course, the estimations assume that base-pair opening is a necessary condition for the exchange of imino protons. The assumption seems very reasonable, since the imino protons participate in the hydrogen bonds between the complementary bases. It remains unclear, however, what kind of base-pair opening is detected in the experiments. At least two types of conformation, diagramed in Fig. 2.21, are possible.

Kinks represent another possible structural disruption of the double helix. Kinks preserve the base pairing but eliminate the stacking interaction in the base-pair step (Fig. 2.22). Kinks were first suggested by Crick and Klug (Crick & Klug 1975). Later they were found in many DNA complexes with ligands and proteins (Sobell *et al.* 1977, Suzuki & Yagi 1995, Werner *et al.* 1996, Dickerson 1998) and were observed in the MD simulations (Lankas *et al.* 2006). The double helix is sharply bent at a kink, so the formation of kinks is stimulated by strong bending stress of the double helix in DNA–protein complexes. The data suggest that kinks appear in very small DNA circles, smaller than 70 bp in length (Du *et al.* 2008).

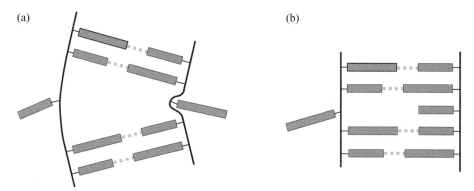

Figure 2.21 Two types of conformation with a disrupted base pair. (a) Both bases of the disrupted pair are exposed to the surrounding solution. Equation (2.35) corresponds to the probability of such a conformation. (b) Only one base is flipped out from the helix while the other retains stacking interaction with adjacent base pairs. Such base-pair opening was observed in the crystal structure of a DNA–protein complex (reviewed by Roberts and Cheng (1998), and obtained in molecular dynamic simulations (Giudice *et al.* 2003, Huang & MacKerell 2004). The disruption does not cause a sharp change of the double-helix direction.

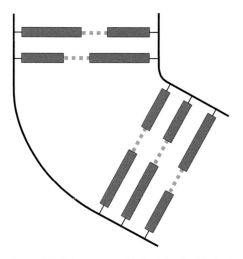

Figure 2.22 Diagram of a kink of the double helix. There is no stacking interaction between two adjacent base pairs at the kink, although the hydrogen bonds in both base pairs are preserved.

Both kinking and base-pair opening are stimulated by the single-stranded nicks of the double helix. The origin of this effect is quite clear: if the strand is nicked, there is no conformational requirement for closing the loop formed by the complementary strands, associated with the kink or base-pair opening. Therefore, single-stranded nicks greatly increase the entropic gain associated with kink formation and base-pair opening. Formation of kinks at the nicks was studied quantitatively for various base-pair steps (Protozanova *et al.* 2004, Yakovchuk *et al.* 2006). These studies showed that the probability of kink formation at near-physiological ionic conditions and temperature of 37 °C is in the range of 0.01–0.3, with the highest value for the TA/TA base-pair step and the lowest value for the GC/GC step.

Single-stranded nicks strongly affect the probability of opening of the surrounding base pairs. To estimate this probability let us assume that the kink is formed. The additional free energy for opening a base pair exposed to the kink is equal to the free energy of the helix propagation. Using the data in Table 2.1 we conclude that the probability of base-pair opening at the nick should be in the range of 0.0004–0.1, depending on the nucleotide sequence of the involved segment. Thus, the effect of the nick on the probability of base-pair opening is huge, up to four orders of magnitude. It is not surprising that proteins that interact with ssDNA preferentially bind the double helix at nicks.

There are no reliable experimental evaluations for the probabilities of transient formation of base pairs in A and Z forms. The only way to get a rough estimation of the probabilities is to use the theoretical models and the parameters of the B–A and B–Z transitions. The corresponding calculation predicts that the probability of finding a base pair in the A form in physiological conditions is in the range of 10^{-4}–10^{-2}. The probability of finding two base pairs of the GC step in the Z form should be around 10^{-12} (since the repeating unit of the Z form is equal to 2 bp, it seems reasonable to calculate the probability for two base pairs rather than one). The probability of the Z form for AT base pairs should be even smaller.

A special case of transient deviation from the B form is the formation of Hoogsteen base pairs. In these pairs purine bases are in *syn* conformation relative to the sugar rather than in *anti*, as they are in the B form (see Fig. 1.6). Although Hoogsteen AT and GC base pairs can form a double helix with nearly the same helical repeat as the one in B-DNA (Abrescia *et al.* 2002), the helix has not been observed in DNA solutions. However, the Hoogsteen base pairing was observed in a crystal of DNA duplex consisting of AT base pairs (Abrescia *et al.* 2002), and in DNA complexes with proteins and ligands (Ughetto *et al.* 1985, Seaman & Hurley 1993, Nair *et al.* 2005). Recently, a report appeared that transient Hoogsteen base pairs were detected in solution in a regular DNA duplex (Nikolova *et al.* 2011). The probability of their formation at physiological conditions was close to 0.005, for both AT and GC base pairs at an internal CA/TG step of an oligonucleotide duplex. The lifetime of this transient state was equal to 0.3 and 1.5 ms for AT and GC base pairs, respectively.

2.6   Correlations between the states of base pairs

The average size of DNA segments that melt as a whole is in the range of 100–500 bp. Therefore, at a temperature near the melting transition, opening (closing) of one base pair can affect the states of other base pairs as far away as hundreds of nucleotides. In other words, there is very long correlation between the states of base pairs within this temperature range. This correlation length reduces dramatically, however, as the temperature decreases below the melting interval. At physiological temperature, around

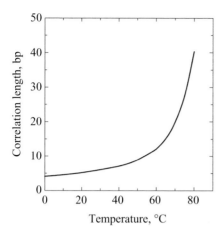

Figure 2.23 Correlation length for base-pair opening at temperatures below the melting interval. The plot is based on the computational data from Lukashin *et al.* (1976), which are in good agreement with available experimental data (see the text). The DNA melting temperature is equal to 87 °C.

37 °C, the correlation is limited to a few base pairs. If we somehow open a base pair inside a sufficiently long DNA segment, it strongly increases the probability of opening for the adjacent base pairs, the base pairs next to the adjacent ones and, to smaller extents, a few more close base pairs. The total length of this influence can be called the correlation length for the transient opening of base pairs (Lukashin *et al.* 1976). The correlation length can be calculated by using the theory of the helix–coil transition in DNA (Fig. 2.23). One can see from the figure that the correlation length does not exceed 10 bp at temperatures below 50 °C for DNA segments with average GC-content. Of course, the correlation length is larger in AT-rich DNA segments. Very short correlation length was also observed in NMR studies of base-pair opening in oligonucleotide duplexes (Kochoyan *et al.* 1987, Leroy *et al.* 1988). It was found in these studies that the probability of opening at 15 °C does not depend on the distance from the duplex end if this distance exceeds 3–4 bp.

From a structural point of view, the short correlation length between the states of DNA base pairs is due to the remarkable conformational flexibility of ssDNA. Because of this flexibility only a few base pairs of the double helix are needed to completely dump local conformational disturbances. This is well illustrated by the perturbations caused by chemical modifications of the regular DNA structure. Formation of cyclobutane thymine dimers due to absorbance of UV radiation is a good example of such a disturbance. The intrastrand dimer contains a four-membered ring arising from the coupling of the C=C double bonds of thymines. The planes of the thymines in the dimer form an angle of about 40°, and the distance between the covalently bound edges of the bases is close to 0.15 nm. Therefore, the dimer introduces substantial local perturbation in the double-helix structure (McAteer *et al.* 1998, Park *et al.* 2002). However, DNA structure 2 bp away from the dimer is hardly changed (Fig. 2.24). Careful comparison of the dimer-containing

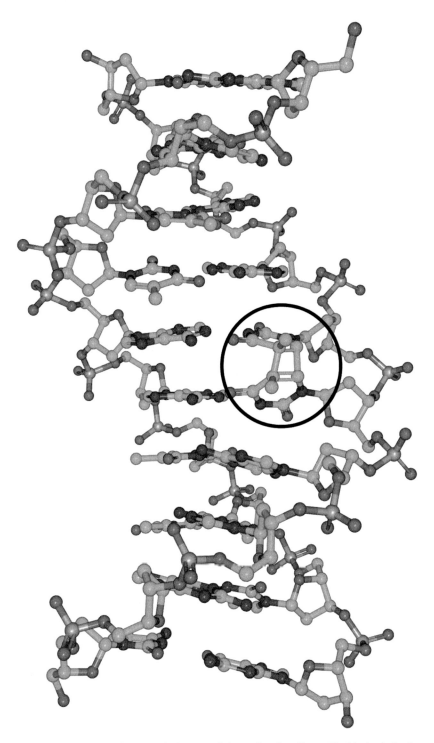

Figure 2.24 Structure of a DNA duplex containing a thymine dimer. The black circle shows the position of the dimer. The structure was obtained by X-ray analysis of the decamer crystals (Park *et al.* 2002). A black-and-white version of this figure will appear in some formats. For the color version, please refer to the plate section.

and unmodified duplexes shows that the structures are nearly indistinguishable when the distance from the dimer exceeds 4 bp (McAteer *et al.* 1998).

Thus, at physiological conditions there is only a short correlation between transient openings of base pairs. Still, from time to time claims appear in the literature that very long correlations between states of base pairs were observed. None of these claims have been sustained, and the above consideration shows that they should not. To the same category belongs the suggestion that a DNA molecule at physiological temperature carries information about the borders of segments that melt cooperatively, so the borders may coincide with the borders of genes or other functional areas. This would mean that DNA somehow knows at physiological temperature about the correlations of base-pair openings in the melting interval. This definitely contradicts what has been reliably established about the physical properties of the double helix.

2.7   Binding small ligands with DNA

Many small molecules can bind with DNA. They can interact with DNA due to their positive electric charge, and due to their ability to establish nonelectrostatic interaction with the DNA backbone or with the bases. Here we consider interaction of DNA with molecules that are larger in size than simple ions, and that bind DNA without strong sequence specificity. Such compounds, capable of reversible noncovalent binding, are called ligands. Many ligands are strong mutagens. Many antibiotics and chemotherapeutics interacting with DNA belong to this category. Some of these ligands absorb UV or visible light and are used for staining, to visualize the bands after electrophoretic separation of DNA molecules. Among these molecules are aminoacridines and aromatic hydrocarbons, oligopeptides and even small proteins. Interaction of ligands with DNA has been a subject of biophysical studies for many years.

There are several ways to detect binding of ligands to DNA. In many cases the binding changes the optical properties of a ligand, so the binding can be monitored via absorption, optical activity or fluorescence of the compounds (reviewed by Biver (2012)). In some cases binding of a ligand can change the DNA bending rigidity, length of the molecules and, correspondingly, their hydrodynamic properties. Studying the latter properties allowed Lerman to suggest that acridines interact with DNA by intercalation between the adjacent base pairs (Lerman 1961). Later, the model of intercalation was confirmed by direct structural studies of the complexes between oligonucleotide duplexes and aromatic compounds, initially for RNA duplexes (Sobell *et al.* 1977) and later for DNA molecules as well (reviewed by Adams (2002)). The intercalation not only elongates the double helix, it also unwinds it, so intercalating compounds cause relaxation of negative supercoiling and are widely used for this goal. In early years ethidium bromide was the most popular compound used for DNA unwinding (Bauer & Vinograd 1968). It is worth noting that an incorrect value of the unwinding angle, ϕ, was used in early studies of DNA supercoiling: the correct value of ϕ for ethidium bromide is equal to $26°$ (Wang 1974). The ligand was also widely used for staining DNA-containing gels. Later, however, it was replaced for DNA unwinding by chloroquine, which equilibrates

Figure 2.25 Nonspecific binding of ligands that occupy more than one unit on a polynucleotide. The polynucleotide units are shown in dark gray and the ligands in light gray. Some unoccupied units are not available for binding: for example, two units on the left-hand side of the diagram.

with DNA faster and therefore provides better resolution of DNA samples in the gels. Ethidium has been also pushed aside as the staining agent by dyes that provide much better sensitivity of DNA detection, such as SYBR Green.

Let us consider a quantitative description of ligand binding with ds or ssDNA. In the simplest model the bound ligands do not interact one with another, except that the polynucleotide site occupied by one ligand becomes inaccessible for other molecules of the ligand. We assume that all binding sites are equivalent and that each bound ligand occupies n consecutive base pairs or nucleotides. The equation for binding ligand L to a polynucleotide site S can be written as

$$L + S \leftrightarrow LS, \tag{2.36}$$

where LS corresponds to the occupied site. The equilibrium constant of the reaction is defined as

$$K = \frac{[LS]}{[L][S]}. \tag{2.37}$$

It is convenient to introduce the fraction of occupied nucleotide units

$$v = [LS]/C_p, \tag{2.38}$$

where C_p is the total concentration of the nucleotide units. If $n = 1$, the concentration of free units available for ligand binding can be expressed as $C_p - [LS]$, and Eq. (2.37) can be presented in form of the Scatchard plot,

$$\frac{v}{[L]} = K(1 - v). \tag{2.39}$$

Equation (2.39) shows that the value of $v/[L]$, which can be determined experimentally, should be a linear function of v, and the intercept of the corresponding plot should be equal to K. However, Eq. (2.39) does not usually work, indicating that the ligand binding process does not obey this simplest model. There are two assumptions of the above model that may be invalid. First, one has to take into account that the ligand can occupy more than a single unit of the polynucleotide; that is, n can be larger than 1. Second, bound molecules of a ligand can interact with one another if they bind at adjacent sites. Even the case of noninteractive binding creates substantial difficulties for the analysis, since the number of accessible binding sites depends not only on the number of bound ligands, but also on their specific positions (Fig. 2.25).

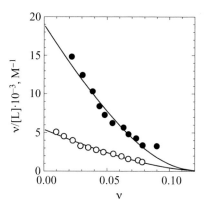

Figure 2.26 Scatchard plots for the binding of ε-Dnp–Lys–(Lys)$_z$ to poly(rA) · poly(rU) for $z = 5$ ($\circ$) and $z = 6$ ($\bullet$). The curves are the best fit to Eq. (2.40) and correspond to $n = 6.6$, $K = 5.4 \times 10^3$ M^{-1} for $z = 5$; $n = 7.1$, $K = 1.9 \times 10^4$ M^{-1} for $z = 6$. The plot was reproduced from Fig. 8a of McGhee and von Hippel (1974) by digitizing and replotting, with permission.

Zasedatelev *et al.* were the first to present a complete analysis of the model that accounts for both the size of the ligand and an interaction between the bound molecules (Zasedatelev *et al.* 1971). The most elegant analysis of the model was suggested by McGhee and von Hippel (1974), who used an approach based on a chain of conditional probabilities, similar to one introduced by Poland for the description of the helix–coil transition in DNA (Poland 1974). For the case of noninteracting ligands occupying n polynucleotide units they obtained the following equation:

$$\frac{v}{[L]} = K(1 - nv) \cdot \left[\frac{1 - nv}{1 - (n-1)v}\right]^{n-1}. \tag{2.40}$$

For $n = 1$ Eq. (2.40) coincides with Eq. (2.39). However, for $n \geq 2$ the equation does not give a straight line in coordinates ($v/[L]$, v). Instead, the plot always falls below the line for $n = 1$ (Fig. 2.26). Still, the intercept on the $v/[L]$ axis is equal to K, so the value of K can be reliably determined from the plot.

The equation that describes the binding becomes more complex if we want to account for the interaction between the bound ligands. Assuming that the binding constant for a ligand molecule that forms a contact with one already bound molecule is $K\omega$, McGhee and von Hippel obtained

$$\frac{v}{[L]} = K(1 - nv)\left[\frac{(2\omega + 1)(1 - nv) + v - R}{2(\omega - 1)(1 - nv)}\right]^{n-1}\left[\frac{1 - (n+1)v + R}{2(1 - nv)}\right]^2, \tag{2.41}$$

where

$$R = \sqrt{[1 - (n+1)v]^2 + 4\omega v(1 - nv)}. \tag{2.42}$$

Equations (2.41) and (2.42) are converted to Eq. (2.40) if $\omega = 1$. For $\omega > 1$ the Scatchard plots go above the corresponding case of noninteractive ligands. Equations (2.41) and (2.42) have three fitting parameters, K, n and ω. Although the value of K can be found from the intercept of the plot on the $v/[L]$ axis, for $\omega \geq 5$ the needed extrapolation of the plot becomes very steep, making accurate determination of K difficult. Determination of other parameters requires their general optimization to fit the experimental data (McGhee & von Hippel 1974).

The above models of the ligand binding do not account for a possible dependence of ligand binding affinity on DNA sequence or GC-content. Although one can consider more general models that account for a sequence dependence of the binding, these are hardly practical since they will contain too many adjustable parameters. On the other hand, it has been found that, in cases where sequence dependence of binding affinity had been documented, the model described above fails to fit the experimental data (McGhee & von Hippel 1974).

References

Aboul-ela, F., Koh, D., Tinoco, I., Jr. & Martin, F. H. (1985). Base–base mismatches. Thermodynamics of double helix formation for dCA₃XA₃G + dCT₃YT₃G (X, Y = A, C, G, T). *Nucleic Acids Res.* 13, 4811–24.

Abrescia, N. G., Thompson, A., Huynh-Dinh, T. & Subirana, J. A. (2002). Crystal structure of an antiparallel DNA fragment with Hoogsteen base pairing. *Proc. Natl. Acad. Sci. U. S. A.* 99, 2806–11.

Adams, A. (2002). Crystal structures of acridines complexed with nucleic acids. *Curr. Med. Chem.* 9, 1667–75.

Allawi, H. T. & SantaLucia, J., Jr. (1997). Thermodynamics and NMR of internal G·T mismatches in DNA. *Biochemistry* 36, 10581–94.

(1998a). Nearest-neighbor thermodynamics of internal A·C mismatches in DNA: sequence dependence and pH effects. *Biochemistry* 37, 9435–44.

(1998b). Nearest neighbor thermodynamic parameters for internal G·A mismatches in DNA. *Biochemistry* 37, 2170–9.

(1998c). NMR solution structure of a DNA dodecamer containing single G·T mismatches. *Nucleic Acids Res.* 26, 4925–34.

(1998d). Thermodynamics of internal C·T mismatches in DNA. *Nucleic Acids Res.* 26, 2694–701.

Amirikyan, B. R., Vologodskii, A. V. & Lyubchenko Yu, L. (1981). Determination of DNA cooperativity factor. *Nucleic Acids Res.* 9, 5469–82.

Ansevin, A. T., Vizard, D. L., Brown, B. W. & McConathy, J. (1976). High-resolution thermal denaturation of DNA. I. Theoretical and practical considerations for the resolution of thermal subtransitions. *Biopolymers* 15, 153–74.

Anshelevich, V. V., Vologodskii, A. V., Lukashin, A. V. & Frank-Kamenetskii, M. D. (1984). Slow relaxational processes in the melting of linear biopolymers. A theory and its application to nucleic acids. *Biopolymers* 23, 39–58.

Antao, V. P., Lai, S. Y. & Tinoco, I., Jr. (1991). A thermodynamic study of unusually stable RNA and DNA hairpins. *Nucleic Acids Res.* 19, 5901–5.

Arnold, F. H., Wolk, S., Cruz, P. & Tinoco, I., Jr. (1987). Structure, dynamics, and thermodynamics of mismatched DNA oligonucleotide duplexes d(CCCAGGG)₂ and d(CCCTGGG)₂. *Biochemistry* 26, 4068–75.

Bauer, W. & Vinograd, J. (1968). The interaction of closed circular DNA with intercalative dyes. I. The superhelix density of SV40 DNA in the presence and absence of dye. *J. Mol. Biol.* 33, 141–71.

Biver, T. (2012). Use of UV–Vis spectrometry to gain information on the mode of binding of small molecules to DNAs and RNAs. *Appl. Spectrosc. Rev.* 47, 272–325.

Blagoi, Y. P., Sorokin, V. A., Valeyev, V. A., Khomenko, S. A. & Gladchenko, G. O. (1978). Magnesium ion effect on the helix–coil transition of DNA. *Biopolymers* 17, 1103–18.

Blommers, M. J., Walters, J. A., Haasnoot, C. A., Aelen, J. M., van der Marel, G. A., van Boom, J. H. & Hilbers, C. W. (1989). Effects of base sequence on the loop folding in DNA hairpins. *Biochemistry* 28, 7491–8.

Borovik, A. S., Kalambet, Y. A., Lyubchenko, Y. L., Shitov, V. T. & Golovanov, E. I. (1980). Equilibrium melting of plasmid ColE1 DNA: electron-microscopic visualization. *Nucleic Acids Res.* 8, 4165–84.

Breslauer, K. J., Frank, R., Blocker, H. & Marky, L. A. (1986). Predicting DNA duplex stability from the base sequence. *Proc. Natl. Acad. Sci. U. S. A.* 83, 3746–50.

Cantor, C. R. & Schimmel, P. R. (1980). *Biophysical Chemistry*. New York: Freeman.

Coman, D. & Russu, I. M. (2005). A nuclear magnetic resonance investigation of the energetics of basepair opening pathways in DNA. *Biophys. J.* 89, 3285–92.

Conceição, A. S., Minetti, C. A., Remeta, D. P., Dickstein, R. & Breslauer, K. J. (2010). Energetic signatures of single base bulges: thermodynamic consequences and biological implications. *Nucleic Acids Res.* 38, 97–116.

Craig, M. E., Crothers, D. M. & Doty, P. (1971). Relaxation kinetics of dimer formation by self complementary oligonucleotides *J. Mol. Biol.* 62, 383–401.

Crick, F. H. & Klug, A. (1975). Kinky helix. *Nature* 255, 530–3.

Delcourt, S. G. & Blake, R. D. (1991). Stacking energies in DNA. *J. Biol. Chem.* 266, 15160–9.

Devoe, H. & Tinoco, I., Jr. (1962). The hypochromism of helical polynucleotides. *J. Mol. Biol.* 4, 518–27.

Di Capua, E., Engel, A., Stasiak, A. & Koller, T. (1982). Characterization of complexes between recA protein and duplex DNA by electron microscopy. *J. Mol. Biol.* 157, 87–103.

Dickerson, R. E. (1998). DNA bending: the prevalence of kinkiness and the virtues of normality. *Nucleic Acids Res.* 26, 1906–26.

Doktycz, M. J., Goldstein, R. F., Paner, T. M., Gallo, F. J. & Benight, A. S. (1992). Studies of DNA dumbbells. I. Melting curves of 17 DNA dumbbells with different duplex stem sequences linked by T4 endloops: evaluation of the nearest-neighbor stacking interactions in DNA. *Biopolymers* 32, 849–64.

Doktycz, M. J., Paner, T. M., Amaratunga, M. & Benight, A. S. (1990). Thermodynamic stability of the 5′ dangling-ended DNA hairpins formed from sequences 5′-(XY)₂GGATAC(T)₄GTATCC-3′, where X, Y = A, T, G, C. *Biopolymers* 30, 829–45.

Du, Q., Kotlyar, A. & Vologodskii, A. (2008). Kinking the double helix by bending deformation. *Nucleic Acids Res.* 36, 1120–8.

Eickbush, T. H. & Moudrianakis, E. N. (1978). The compaction of DNA helices into either continuous supercoils or folded-fiber rods and toroids. *Cell* 13, 295–306.

Ellison, M. J., Kelleher, R. J., 3rd, Wang, A. H., Habener, J. F. & Rich, A. (1985). Sequence-dependent energetics of the B–Z transition in supercoiled DNA containing nonalternating purine–pyrimidine sequences. *Proc. Natl. Acad. Sci. U. S. A.* 82, 8320–4.

Fixman, M. & Freire, J. J. (1977). Theory of DNA melting curves. *Biopolymers* 16, 2693–704.

Frank-Kamenetskii, M. D. (1971). Simplification of the empirical relationship between melting temperature of DNA, its GC-content and concentration of sodium ions in solution. *Biopolymers* 10, 2623–4.

Frank-Kamenetskii, M. D. & Prakash, S. (2014). Fluctuations in the DNA double helix: a critical review. *Phys. Life Rev.* 11, 153–70.

Freier, S. M., Kierzek, R., Jaeger, J. A., Sugimoto, N., Caruthers, M. H., Neilson, T. & Turner, D. H. (1986). Improved free-energy parameters for predictions of RNA duplex stability. *Proc. Natl. Acad. Sci. U. S. A.* 83, 9373–7.

Giudice, E., Varnai, P. & Lavery, R. (2003). Base pair opening within B-DNA: free energy pathways for GC and AT pairs from umbrella sampling simulations. *Nucleic Acids Res.* 31, 1434–43.

Goddard, N. L., Bonnet, G., Krichevsky, O. & Libchaber, A. (2000). Sequence dependent rigidity of single stranded DNA. *Phys. Rev. Lett.* 85, 2400–3.

Goldstein, R. F. & Benight, A. S. (1992). How many numbers are required to specify sequence-dependent properties of polynucleotides? *Biopolymers* 32, 1679–93.

Gotoh, O., Husimi, Y., Yabuki, S. & Wada, A. (1976). Hyperfine structure in melting profile of bacteriophage lambda DNA. *Biopolymers* 15, 655–70.

Gotoh, O. & Tagashira, Y. (1981). Stabilities of nearest-neighbor doublets in double-helical DNA determined by fitting calculated melting profiles to observed profiles. *Biopolymers* 20, 1033–43.

Gray, D. M. & Tinoco, I. (1970). A new approach to the study of sequence-dependent properties of polynucleotides. *Biopolymers* 9, 223–44.

Gruenwedel, D. W. (1974). Salt effects on the denaturation of DNA. III. A calorimetric investigation of the transition enthalpy of calf thymus DNA in Na_2SO_4 solutions of varying ionic strength. *Biochim. Biophys. Acta* 340, 16–30.

Gueron, M., Kochoyan, M. & Leroy, J. L. (1987). A single mode of DNA base-pair opening drives imino proton exchange. *Nature* 328, 89–92.

Gyi, J. I., Conn, G. L., Lane, A. N. & Brown, T. (1996). Comparison of the thermodynamic stabilities and solution conformations of DNA. RNA hybrids containing purine-rich and pyrimidine-rich strands with DNA and RNA duplexes. *Biochemistry* 35, 12538–48.

Haniford, D. B. & Pulleyblank, D. E. (1983). Facile transition of poly[d(TG) × d(CA)] into a left-handed helix in physiological conditions. *Nature* 302, 632–4.

Hoff, A. J. & Roos, A. L. (1972). Hysteresis of denaturation of DNA in the melting range. *Biopolymers* 11, 1289–94.

Huang, N. & MacKerell, A. D. J. (2004). Atomistic view of base flipping in DNA. *Phil. Trans. R. Soc. A* 362, 1439–60.

Integrated DNA Technologies. http://www.idtdna.com/analyzer/Applications/OligoAnalyzer/ [October 17, 2014].

Ivanov, V. I. & Krylov, D. (1992). A-DNA in solution as studied by diverse approaches. *Methods Enzymol.* 211, 111–27.

Ivanov, V. I., Krylov, D. & Minyat, E. E. (1985). Three-state diagram for DNA. *J. Biomol. Struct. Dyn.* 3, 43–55.

Ivanov, V. I., Minchenkova, L. E., Burckhardt, G., Birch-Hirschfeld, E., Fritzsche, H. & Zimmer, C. (1996). The detection of B-form/A-form junction in a deoxyribonucleotide duplex. *Biophys. J.* 71, 3344–9.

Ivanov, V. I., Minchenkova, L. E., Minyat, E. E., Frank-Kamenetskii, M. D. & Schyolkina, A. K. (1974). The B to A transition of DNA in solution. *J. Mol. Biol.* 87, 817–33.

Kalnik, M. W., Norman, D. G., Li, B. F., Swann, P. F. & Patel, D. J. (1990). Conformational transitions in thymidine bulge-containing deoxytridecanucleotide duplexes. Role of flanking sequence and temperature in modulating the equilibrium between looped out and stacked thymidine bulge states. *J. Biol. Chem.* 265, 636–47.

Kalnik, M. W., Norman, D. G., Zagorski, M. G., Swann, P. F. & Patel, D. J. (1989). Conformational transitions in cytidine bulge-containing deoxytridecanucleotide duplexes: extra cytidine equilibrates between looped out (low temperature) and stacked (elevated temperature) conformations in solution. *Biochemistry* 28, 294–303.

Kang, D. S. & Wells, R. D. (1985). B–Z DNA junctions contain few, if any, nonpaired bases at physiological superhelical densities. *J. Biol. Chem.* 260, 7783–90.

Kochoyan, M., Leroy, J. L. & Gueron, M. (1987). Proton exchange and base-pair lifetimes in a deoxy-duplex containing a purine–pyrimidine step and in the duplex of inverse sequence. *J. Mol. Biol.* 196, 599–609.

Kozyavkin, S. A., Mirkin, S. M. & Amirikyan, B. R. (1987). The ionic strength dependence of the cooperativity factor for DNA melting. *J. Biomol. Struct. Dyn.* 5, 119–26.

Krueger, A., Protozanova, E. & Frank-Kamenetskii, M. D. (2006). Sequence-dependent base pair opening in DNA double helix. *Biophys. J.* 90, 3091–9.

Lankas, F., Lavery, R. & Maddocks, J. H. (2006). Kinking occurs during molecular dynamics simulations of small DNA minicircles. *Structure* 14, 1527–34.

Lazurkin, Y. S., Frank-Kamenetskii, M. D. & Trifonov, E. N. (1970). Melting of DNA: its study and application as a research method. *Biopolymers* 9, 1253–306.

Lerman, L. S. (1961). Structural considerations in the interaction of DNA and acridines. *J. Mol. Biol.* 3, 18–30.

Leroy, J. L., Kochoyan, M., Huynh-Dinh, T. & Gueron, M. (1988). Characterization of base-pair opening in deoxynucleotide duplexes using catalyzed exchange of the imino proton. *J. Mol. Biol.* 200, 223–38.

Lesnik, E. A. & Freier, S. M. (1995). Relative thermodynamic stability of DNA, RNA, and DNA:RNA hybrid duplexes: relationship with base composition and structure. *Biochemistry* 34, 10807–15.

Lukashin, A. V., Vologodskii, A. V., Frank-Kamenetskii, M. D. & Lyubchenko, Y. L. (1976). Fluctuational opening of the double helix as revealed by theoretical and experimental study of DNA interaction with formaldehyde. *J. Mol. Biol.* 108, 665–82.

Lyubchenko, Y. L., Frank-Kamenetskii, M. D., Vologodskii, A. V., Lazurkin, Y. S. & Gause, G. G., Jr. (1976). Fine structure of DNA melting curves. *Biopolymers* 15, 1019–36.

Lyubchenko, Y. L., Vologodskii, A. V. & Frank-Kamenetskii, M. D. (1978). Direct comparison of theoretical and experimental melting profiles for RF II phiX174 DNA. *Nature* 271, 28–31.

Malenkov, G., Minchenkova, L., Minyat, E., Schyolkina, A. & Ivanov, V. (1975). The nature of the B–A transition of DNA in solution. *FEBS Lett.* 51, 38–42.

Marmur, J. & Doty, P. (1959). Heterogeneity in deoxyribonucleic acids: I. Dependence on composition of the configurational stability of deoxyribonucleic acids. *Nature* 183, 1427–9.

Martin, F. H. & Tinoco, I., Jr. (1980). DNA–RNA hybrid duplexes containing oligo(dA:rU) sequences are exceptionally unstable and may facilitate termination of transcription. *Nucleic Acids Res.* 8, 2295–9.

McAteer, K., Jing, Y., Kao, J., Taylor, J. S. & Kennedy, M. A. (1998). Solution-state structure of a DNA dodecamer duplex containing a Cis-syn thymine cyclobutane dimer, the major UV photoproduct of DNA. *J. Mol. Biol.* 282, 1013–32.

McGhee, J. D. & von Hippel, P. H. (1974). Theoretical aspects of DNA–protein interactions: co-operative and non-co-operative binding of large ligands to a one-dimensional homogeneous lattice. *J. Mol. Biol.* 86, 469–89.

Melchior, W. B., Jr. & Von Hippel, P. H. (1973). Alteration of the relative stability of dA–dT and dG–dC base pairs in DNA. *Proc. Natl. Acad. Sci. U. S. A.* 70, 298–302.

Minchenkova, L. E., Schyolkina, A. K., Chernov, B. K. & Ivanov, V. I. (1986). CC/GG contacts facilitate the B to A transition of DNA in solution. *J. Biomol. Struct. Dyn.* 4, 463–76.

Mirkin, S. M., Lyamichev, V. I., Kumarev, V. P., Kobzev, V. F., Nosikov, V. V. & Vologodskii, A. V. (1987). The energetics of the B–Z transition in DNA. *J. Biomol. Struct. Dyn.* 5, 79–88.

Moe, J. G. & Russu, I. M. (1990). Proton exchange and base-pair opening kinetics in 5′-d(CGCGAATTCGCG)-3′ and related dodecamers. *Nucleic Acids Res.* 18, 821–7.

Morden, K. M. & Maskos, K. (1993). NMR studies of an extrahelical cytosine in an A.T rich region of a deoxyribodecanucleotide. *Biopolymers* 33, 27–36.

Nair, D. T., Johnson, R. E., Prakash, L., Prakash, S. & Aggarwal, A. K. (2005). Human DNA polymerase iota incorporates dCTP opposite template G via a G.C + Hoogsteen base pair. *Structure* 13, 1569–77.

Nakano, M., Moody, E. M., Liang, J. & Bevilacqua, P. C. (2002). Selection for thermodynamically stable DNA tetraloops using temperature gradient gel electrophoresis reveals four motifs: d(cGNNAg), d(cGNABg),d(cCNNGg), and d(gCNNGc). *Biochemistry* 41, 14281–92.

Nikolova, E. N., Kim, E., Wise, A. A., O'Brien, P. J., Andricioaei, I. & Al-Hashimi, H. M. (2011). Transient Hoogsteen base pairs in canonical duplex DNA. *Nature* 470, 498–502.

Nordheim, A., Lafer, E. M., Peck, L. J., Wang, J. C., Stollar, B. D. & Rich, A. (1982). Negatively supercoiled plasmids contain left-handed Z-DNA segments as detected by specific antibody binding. *Cell* 31, 309–18.

Owczarzy, R., Moreira, B. G., You, Y., Behlke, M. A. & Walder, J. A. (2008). Predicting stability of DNA duplexes in solutions containing magnesium and monovalent cations. *Biochemistry* 47, 5336–53.

Owczarzy, R., Vallone, P. M., Gallo, F. J., Paner, T. M., Lane, M. J. & Benight, A. S. (1997). Predicting sequence-dependent melting stability of short duplex DNA oligomers. *Biopolymers* 44, 217–39.

Owczarzy, R., You, Y., Moreira, B. G., Manthey, J. A., Huang, L., Behlke, M. A. & Walder, J. A. (2004). Effects of sodium ions on DNA duplex oligomers: improved predictions of melting temperatures. *Biochemistry* 43, 3537–54.

Owen, R. J., Hill, L. R. & Lapage, S. P. (1969). Determination of DNA base compositions from melting profiles in dilute buffers. *Biopolymers* 7, 503–16.

Park, H., Zhang, K., Ren, Y., Nadji, S., Sinha, N., Taylor, J. S. & Kang, C. (2002). Crystal structure of a DNA decamer containing a *cis-syn* thymine dimer. *Proc. Natl. Acad. Sci. U. S. A.* 99, 15965–70.

Patel, D. J., Kozlowski, S. A., Ikuta, S. & Itakura, K. (1984). Deoxyadenosine–deoxycytidine pairing in the d(C-G-C-G-A-A-T-T-C-A-C-G) duplex – conformation and dynamics at and adjacent to the dA·dC mismatch site. *Biochemistry* 23, 3218–26.

Patel, D. J., Kozlowski, S. A., Marky, L. A., Rice, J. A., Broka, C., Dallas, J., Itakura, K. & Breslauer, K. J. (1982). Structure, dynamics, and energetics of deoxyguanosine.thymidine wobble base pair formation in the self-complementary d(CGTGAATTCGCG) duplex in solution. *Biochemistry* 21, 437–44.

Peck, L. J. & Wang, J. C. (1983). Energetics of B-to-Z transition in DNA. *Proc. Natl. Acad. Sci. U. S. A.* 80, 6206–10.

Perelroyzen, M. P., Lyamichev, V. I., Kalambet, Y. A., Lyubchenko, Y. L. & Vologodskii, A. V. (1981). A study of the reversibility of helix–coil transition in DNA. *Nucleic Acids Res.* 9, 4043–59.

Peyrard, M. & Bishop, A. R. (1989). Statistical mechanics of a nonlinear model for DNA denaturation. *Phys. Rev. Lett.* 62, 2755–8.

Peyret, N., Seneviratne, P. A., Allawi, H. T. & SantaLucia, J., Jr. (1999). Nearest-neighbor thermodynamics and NMR of DNA sequences with internal A.A, C.C, G.G, and T.T mismatches. *Biochemistry* 38, 3468–77.

Pohl, F. M. & Jovin, T. M. (1972). Salt-induced cooperative conformational change of a synthetic DNA – equilibrium and kinetic studies with poly(dG-dC). *J. Mol. Biol.* 67, 375–96.

Poland, D. (1974). Recursion relation generation of probability profiles for specific-sequence macromolecules with long-range correlations. *Biopolymers* 13, 1859–71.

Protozanova, E., Yakovchuk, P. & Frank-Kamenetskii, M. D. (2004). Stacked–unstacked equilibrium at the nick site of DNA. *J. Mol. Biol.* 342, 775–85.

Ratmeyer, L., Vinayak, R., Zhong, Y. Y., Zon, G. & Wilson, W. D. (1994). Sequence specific thermodynamic and structural properties for DNA.RNA duplexes. *Biochemistry* 33, 5298–304.

Rice, S. A. & Doty, P. (1957). The thermal denaturation of desoxyribose nucleic acid. *J. Am. Chem. Soc.* 79, 3937–47.

Rich, A., Nordheim, A. & Wang, A. H.-J. (1984). The chemistry and biology of left-handed Z-DNA. *Annu. Rev. Biochem.* 53, 791–846.

Roberts, R. J. & Cheng, X. (1998). Base flipping. *Annu. Rev. Biochem.* 67, 181–98.

Roberts, R. W. & Crothers, D. M. (1992). Stability and properties of double and triple helices: dramatic effects of RNA or DNA backbone composition. *Science* 258, 1463–6.

Rouzina, I. & Bloomfield, V. A. (1999a). Heat capacity effects on the melting of DNA. 1. General aspects. *Biophys. J.* 77, 3242–51.

(1999b). Heat capacity effects on the melting of DNA. 2. Analysis of nearest-neighbor base pair effects. *Biophys. J.* 77, 3252–5.

Rybenkov, V. V., Vologodskii, A. V. & Cozzarelli, N. R. (1997). The effect of ionic conditions on DNA helical repeat, effective diameter, and free energy of supercoiling. *Nucleic Acids Res.* 25, 1412–18.

SantaLucia, J., Jr. (1998). A unified view of polymer, dumbbell, and oligonucleotide DNA nearest-neighbor thermodynamics. *Proc. Natl. Acad. Sci. U. S. A.* 95, 1460–5.

SantaLucia, J., Jr., Allawi, H. T. & Seneviratne, P. A. (1996). Improved nearest-neighbor parameters for predicting DNA duplex stability. *Biochemistry* 35, 3555–62.

SantaLucia, J., Jr. & Hicks, D. (2004). The thermodynamics of DNA structural motifs. *Annu. Rev. Biophys. Biomol. Struct.* 33, 415–40.

Seaman, F. C. & Hurley, L. (1993). Interstrand cross-linking by bizelesin produces a Watson–Crick to Hoogsteen base-pairing transition region in d(CGTAATTACG)$_2$. *Biochemistry* 32, 12577–85.

Senior, M. M., Jones, R. A. & Breslauer, K. J. (1988). Influence of loop residues on the relative stabilities of DNA hairpin structures. *Proc. Natl. Acad. Sci. U. S. A.* 85, 6242–6.

Shiao, D. D. & Sturtevant, J. M. (1973). Heats of thermally induced helix–coil transitions of DNA in aqueous solution. *Biopolymers* 12, 1829–36.

Singleton, C. K., Klysik, J., Stirdivant, S. M. & Wells, R. D. (1982). Left-handed Z-DNA is induced by supercoiling in physiological ionic conditions. *Nature* 299, 312–16.

Sobell, H. M., Tsai, C. C., Jain, S. C. & Gilbert, S. G. (1977). Visualization of drug–nucleic acid interactions at atomic resolution. III. Unifying structural concepts in understanding drug–DNA interactions and their broader implications in understanding protein–DNA interactions. *J. Mol. Biol.* 114, 333–65.

Sugimoto, N., Nakano, S., Yoneyama, M. & Honda, K. (1996). Improved thermodynamic parameters and helix initiation factor to predict stability of DNA duplexes. *Nucleic Acids Res.* 24, 4501–5.

Suzuki, M. & Yagi, N. (1995). Stereochemical basis of DNA bending by transcription factors. *Nucleic Acids Res.* 23, 2083–2091.

Tan, W., Wang, K. & Drake, T. J. (2004). Molecular beacons. *Curr. Opin. Chem. Biol.* 8, 547–53.

Tanaka, F., Kameda, A., Yamamoto, M. & Ohuchi, A. (2004). Thermodynamic parameters based on a nearest-neighbor model for DNA sequences with a single-bulge loop. *Biochemistry* 43, 7143–50.

Tinoco, I., Jr., Borer, P. N., Dengler, B., Levin, M. D., Uhlenbeck, O. C., Crothers, D. M. & Bralla, J. (1973). Improved estimation of secondary structure in ribonucleic acids. *Nat. New Biol.* 246, 40–1.

Tinoco, I., Jr., Uhlenbeck, O. C. & Levine, M. D. (1971). Estimation of secondary structure in ribonucleic acids. *Nature* 230, 362–7.

Tolstorukov, M. Y., Ivanov, V. I., Malenkov, G. G., Jernigan, R. L. & Zhurkin, V. B. (2001). Sequence-dependent B–A transition in DNA evaluated with dimeric and trimeric scales. *Biophys. J.* 81, 3409–21.

Tunis-Schneider, M. J. & Maestre, M. F. (1970). Circular dichroism spectra of oriented and unoriented deoxyribonucleic acid films – a preliminary study. *J. Mol. Biol.* 52, 521–41.

Tyagi, S. & Kramer, F. R. (1996). Molecular beacons: probes that fluoresce upon hybridization. *Nature Biotechnol.* 14, 303–8.

Ughetto, G., Wang, A. H., Quigley, G. J., van der Marel, G. A., van Boom, J. H. & Rich, A. (1985). A comparison of the structure of echinomycin and triostin A complexed to a DNA fragment. *Nucleic Acids Res.* 13, 2305–23.

Usatyi, A. F. & Shlyakhtenko, L. S. (1974). Melting of DNA in ethanol–water solutions. *Biopolymers* 13, 2435–46.

Varani, G. (1995). Exceptionally stable nucleic acid hairpins. *Annu. Rev. Biophys. Biomol. Struct.* 24, 379–404.

Vologodskii, A. V., Amirikyan, B. R., Lyubchenko, Y. L. & Frank-Kamenetskii, M. D. (1984). Allowance for heterogeneous stacking in the DNA helix–coil transition theory. *J. Biomol. Struct. Dyn.* 2, 131–48.

Vologodskii, A. V. & Cozzarelli, N. R. (1994). Conformational and thermodynamic properties of supercoiled DNA. *Annu. Rev. Biophys. Biomol. Struct.* 23, 609–43.

Voskoboinik, A. D., Monaselidze, D. R., Mgeladze, G. N., Chanchalashvili, Z. I., Lazurkin, Y. S. & Frank-Kamenetskii, M. D. (1975). Study of DNA melting in the region of the inversion of relative stability of AT and GC pairs. *Mol. Biol.* 9, 783–90.

Wang, A. H., Quigley, G. J., Kolpak, F. J., Crawford, J. L., van Boom, J. H., van der Marel, G. & Rich, A. (1979). Molecular structure of a left-handed double helical DNA fragment at atomic resolution. *Nature* 282, 680–6.

Wang, J. C. (1974). The degree of unwinding of the DNA helix by ethidium. I. Titration of twisted PM2 DNA molecules in alkaline cesium chloride density gradients. *J. Mol. Biol.* 89, 783–801.

Werner, M. H., Gronenborn, A. M. & Clore, G. M. (1996). Intercalation, DNA kinking, and the control of transcription. *Science* 271, 778–84.

Woodson, S. A. & Crothers, D. M. (1988a). Preferential location of bulged guanosine internal to a G.C tract by 1H NMR. *Biochemistry* 27, 436–45.

(1988b). Structural model for an oligonucleotide containing a bulged guanosine by NMR and energy minimization *Biochemistry* 27, 3130–41.

(1989). Conformation of a bulge-containing oligomer from a hot-spot sequence by NMR and energy minimization. *Biopolymers* 28, 1149–77.

Yakovchuk, P., Protozanova, E. & Frank-Kamenetskii, M. D. (2006). Base-stacking and base-pairing contributions into thermal stability of the DNA double helix. *Nucleic Acids Res.* 34, 564–74.

Zasedatelev, A. S., Gurskii, G. V. & Vol'kenshtein, M. V. (1971). Theory of one-dimensional adsorption. I. Adsorption of small molecules on a homopolymer. *Mol. Biol.* 5, 194–8.

Zhu, J. & Wartell, R. M. (1999). The effect of base sequence on the stability of RNA and DNA single base bulges. *Biochemistry* 38, 15986–93.

Zimm, B. H. & Bragg, J. K. (1959). Theory of the phase transition between helix and random coil in polypeptide chains. *J. Chem. Phys.* 31, 526–31.

3 Equilibrium large-scale conformational properties of DNA

3.1 DNA and basic concepts of polymer statistical physics

In solution, under physiological ionic conditions and temperature, the axis of dsDNA adopts many different conformations, a common property of a typical polymer chain. The number of these conformations is so large, and the probability of each individual conformation is so small, that it does not make much sense to talk about these individual probabilities. Instead, we should describe the conformational properties in statistical terms, evaluating probability distributions and various average characteristics. This approach is a subject of polymer statistical physics that will be introduced in the chapter. The chapter considers the theoretical basis and some experimental methods of evaluating DNA statistical properties. This analysis is important for understanding DNA functioning in biochemical reactions and various biological processes. Remarkably, in the case of DNA many of these large-scale conformational properties can be now computed with remarkable accuracy. Still, the approach does not provide a description of DNA conformations at an atomic level of resolution, which would be very valuable for better understanding of many issues related to DNA-protein interactions.

Polymer statistical physics was born in 1934, when Kuhn, Guth and Mark concluded that a sufficiently long polymer chain forms a random coil in solution with statistical properties of random walks (Guth & Mark 1934, Kuhn 1934). To elaborate this key idea we should review briefly the properties of random walks.

3.1.1 Random walks

Let us consider statistical properties of trajectories produced by a walker who makes each step in a random direction, without any correlation with previous steps. We assume that all steps have equal length, l, and any direction of a step has equal probability. If $\mathbf{l}_i$ is a vector that specifies step i, the total displacement of the walker after n steps, $\mathbf{S}$, can be written as

$$\mathbf{S} = \sum_{i=1}^{n} \mathbf{l}_i .$$ (3.1)

A typical trajectory of such a random walk with the corresponding notations is shown in Fig. 3.1.

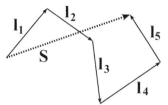

Figure 3.1 Vector representation of the random walk. Five steps of the trajectory are shown by solid lines; the total displacement, **S**, is shown by a dashed line.

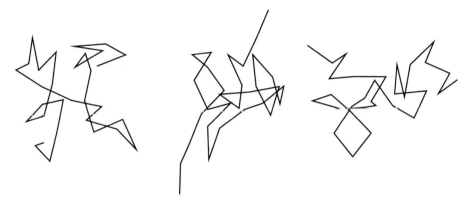

Figure 3.2 The projections of typical 3D trajectories of a random walk 30 steps in length.

Of course, the average value of **S**, $\langle \mathbf{S} \rangle$, through all possible walks of n steps is equal to zero. However, $\langle \mathbf{S}^2 \rangle \equiv \langle S^2 \rangle$ is not equal to zero:

$$\langle S^2 \rangle = \left\langle \left(\sum_{i=1}^{n} \mathbf{l}_i \right)^2 \right\rangle = \sum_{i=1}^{n} \langle \mathbf{l}_i^2 \rangle = nl^2 \qquad (3.2)$$

where we used that $\langle \mathbf{l}_i \mathbf{l}_j \rangle = \langle \mathbf{l}_i \rangle \langle \mathbf{l}_j \rangle = l^2 \delta_{ij}$, and δ_{ij} is the Kronecker delta. We can also express $\langle S^2 \rangle$ through the total length of the trajectory, $L = nl$:

$$\langle S^2 \rangle = Ll. \qquad (3.3)$$

It follows from Eq. (3.2) that the average distance that the walker makes after n steps, $\sqrt{\langle \mathbf{S}^2 \rangle}$, is equal to $l\sqrt{n}$. Examples of random-walk trajectories are shown in Fig. 3.2.

Another important characteristic of random walks is the distribution of the walker's displacements along a particular direction. Since all directions are equivalent, we will calculate the distribution for the displacement along the x axis, $P(S_x, n)$, of the coordinate system specified by unit vectors **i**, **j** and **k**. Clearly,

$$S_x = \sum_{i=1}^{n} l_x^i, \qquad (3.4)$$

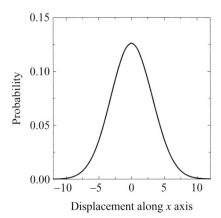

Figure 3.3 The distribution of displacements along the x axis for the random walk of 30 steps, with $l = 1$.

where l_x is the x component of step i. Equation (3.4) allows one to use the central limit theorem to calculate $P(x)$ (see Box 3.1). To apply the theorem one needs to know $\langle l_x \rangle$ and the variance of l_x, $\text{Var}(l_x)$. We will use symbol σ^2 for the variance. It is easy to show that

$$\sigma^2 = \langle (l_x - \langle l_x \rangle)^2 \rangle = \langle (l_x)^2 \rangle = l^2/3. \tag{3.5}$$

Thus, for distribution $P(S_x, n) \equiv P(x, n)$ we obtain

$$P(x, n) = \frac{\sqrt{3}}{\sqrt{2n\pi l}} \exp[-3x^2/2nl^2]. \tag{3.6}$$

Figure 3.3 shows the distribution for a random walk of 30 steps. One can see that $P(x, n)$ decreases sharply for $x > \sqrt{n}$, although the probability of each trajectory is exactly the same. The decrease of $P(x, n)$ is due to the fact that there are many more compact trajectories than extended ones. The latter property of the random walks is definitely the most important one for our subject.

Using Eq. (3.6) we can obtain the distribution of the end-to-end distance for the random walks, $P(R, n)$, where $R = \sqrt{x^2 + y^2 + z^2}$. This distribution function, multiplied by dR, specifies the probability that R will be between R and $R + dR$, irrespective of the direction. To obtain $P(R, n)$ we can assume that $P(x, n)$, $P(y, n)$ and $P(z, n)$ are independent one from another, so

$$P(x, y, z, n) = P(x, n)P(y, n)P(z, n). \tag{3.7}$$

In general, this assumption is incorrect, since a trajectory strongly extended in one direction has limited options for extensions in the perpendicular directions. However, for not so extreme extensions approximation (3.7) is sufficiently accurate. Probability $P(R, n)dR$ is equal to the sum of $P(x, y, z, n)dx\, dy\, dz$ over all volume elements $dx\, dy\, dz$ that correspond to a spherical shell of radius R and thickness dR. Since the volume of

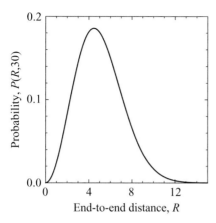

Figure 3.4 The probability distribution of the end-to-end distance for a random walk of 30 steps.

this shell is $4\pi R^2 \, dR$, we conclude that

$$P(R, n) = \left(\frac{\sqrt{3}}{l\sqrt{2\pi n}}\right)^3 \exp\left[-\frac{3(x^2 + y^2 + z^2)}{2nl^2}\right] 4\pi R^2$$

$$= \frac{1}{l^3}\sqrt{\frac{54}{\pi n^3}} \exp\left[-\frac{3R^2}{2nl^2}\right] R^2 \tag{3.8}$$

The distribution $P(R, n)$ for random walks of 30 steps is shown in Fig. 3.4. The maximum of this distribution is close to $\langle S^2 \rangle$ rather than at $R = 0$, as it is for $P(x, n)$. The origin of this difference between $P(x, n)$ and $P(R, n)$ is the volume of the spherical shell, $4\pi R^2 \, dR$, that grows when R increases.

Box 3.1

The normal distribution

The normal or Gaussian probability distribution of a random variable x, $P(x)$, is specified as

$$P(x) \rightarrow \frac{1}{\sqrt{2\pi}\sigma} \exp\left[-\frac{(x - a)^2}{2\sigma^2}\right], \tag{B3.1.1}$$

where x accepts any value between $-\infty$ and $+\infty$, and a and σ are the distribution parameters. The distribution has a bell-like shape with a maximum at $x = a$ and the distribution variance of σ^2 (Fig. B3.1.1). The distribution is symmetrical relative to a, that is $P(x - a) = P(a - x)$. The probability that $-\sigma \leq x - a \leq \sigma$, $\int_{a-\sigma}^{a+\sigma} P(x)\,dx$, is approximately equal to 0.68.

The normal distribution is very important because, according to the *central limit theorem*, the distribution of the sum of a few identically distributed random variables always approaches normal distribution.

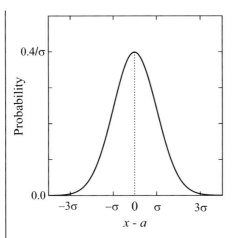

Figure B3.1.1 The normal distribution.

Central limit theorem

Let us assume that S_n is a sum of n independent random variables X_k with identical distributions, so that $\langle X_k \rangle = \mu$ and $\mathrm{Var}(X_k) = \sigma^2$, so

$$S_n = X_1 + X_2 + \cdots + X_n.$$

Then

$$P(S_n) \to \frac{1}{\sqrt{2\pi n}\sigma} \exp\left[-\frac{(S_n - n\mu)^2}{2n\sigma^2} \right] \text{ when } n \to \infty. \qquad (\text{B3.1.2})$$

Thus, the limiting distribution for S_n is the normal distribution.

3.1.2 Discrete wormlike chain as coarse-grained model of DNA

We will introduce now a realistic although very simple model of dsDNA and show that the major statistical properties of the model are similar to the properties of the random walk.

DNA is a very regular and rigid polymer at the scale of a dozen base pairs. However, at the scale of thousands of base pairs the molecule adopts many different conformations, and they are very irregular. This irregularity originates from small thermal fluctuations of angles between adjacent base pairs. The fluctuations accumulate when we move along the molecule contour, resulting in irregular conformations of the DNA axis on a larger scale. To analyze the large-scale statistical properties of DNA we can use a simple model of the double helix. Since we are interested now only in conformations of the DNA axis, we can represent the molecule by a chain of straight segments, one segment for each base pair (Fig. 3.5). We assume that the molecule has minimum energy if the angles between the directions of adjacent segments, θ_i, are equal to zero, that is the minimum energy corresponds to a straight conformation of the model chain. The chain bending

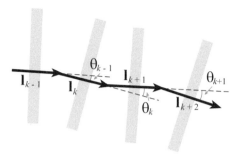

Figure 3.5 The coarse-grained model of DNA. Each straight segment of the model chain (shown by black arrows) corresponds to one base pair (shown by the gray bars); the direction of each segment is perpendicular to the base-pair plane.

energy, E_b, is specified by the equation

$$E_b = \sum_{i=1}^{N-1} \frac{g}{2l}\theta_i^2,$$

(3.9)

where g is the bending rigidity constant, which is defined by DNA internal properties, and N is the number of base pairs in the molecule. In the case of dsDNA the value of g is large, so typical values of angles θ_i are small. In this case the model can be considered as a discrete version of the *wormlike chain*, a model widely used in polymer statistical physics (Bresler & Frenkel 1939, Kratky & Porod 1949, Landau & Lifshitz 1951). We will use the abbreviation WLC for this model. Let us consider the major properties of WLC.

We start from the distribution of angles between adjacent segments, $P(\theta)$, in an unrestricted linear chain whose energy is specified by Eq. (3.9). The number of conformations of two adjacent segments with angles between their directions in the interval $[\theta, \theta + d\theta]$ is proportional to $\sin\theta d\theta$. Indeed, the angles correspond to the ring on the sphere whose radius is $\sin\theta$ (Fig. 3.6). Distribution $P(\theta)$ is specified by the Boltzmann distribution

$$p(\theta) = A\sin\theta \, \exp\left(-\frac{g\theta^2}{2lk_BT}\right),$$

(3.10)

where A is the normalization coefficient. The distribution (3.10) is plotted in Fig. 3.7.

Due to thermal motion, directions of the chain segments are random. However, since the energy of the chain depends on the angles between adjacent segments, the directions of the segments are correlated. The strongest correlation is between the directions of adjacent segments that are separated by one random angle. The correlation is weaker for segments $(i-1)$ and $(i+1)$ because there are two random angles between them. In general, the correlation between the directions of two segments reduces with increasing distance between them along the chain contour. This correlation can be specified by the average dot product of the vectors, $\langle l_i l_j \rangle$. For two vectors which have independent random orientations the product is equal to zero. This means that there is no correlation

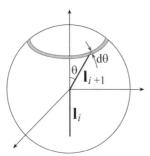

Figure 3.6 On the number of different states of two adjacent segments of the discrete WLC. The number of conformations of two adjacent segments with angles between their directions in the interval $[\theta, \theta + d\theta]$ is proportional to the surface of the ring (shown by gray) of a sphere whose center is at the segment junction. The surface of the ring is equal to $2\pi \sin\theta d\theta$ if the sphere radius is equal to unity.

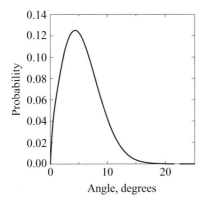

Figure 3.7 The distribution of angles between adjacent segments of WLC, $P(\theta)$. The distribution has only one parameter, the bending rigidity divided by the segment length, g/l. The shown distribution corresponds to the value of g/l for dsDNA at physiological ionic conditions (see below).

in their directions. Let us calculate $\langle \mathbf{l}_i \mathbf{l}_j \rangle$ for our model. According to the definition of θ_i,

$$\langle \mathbf{l}_i \mathbf{l}_{i+1} \rangle = l^2 \langle \cos \theta_i \rangle. \tag{3.10}$$

Because $\langle \cos \theta_i \rangle$ does not depend on i in this model, we can skip the index. Let us consider now the value of $\langle \mathbf{l}_i \mathbf{l}_{i+2} \rangle$ (Fig. 3.8). Both $\mathbf{l}_i$ and $\mathbf{l}_{i+2}$ can be written as sums of two components, one of which is perpendicular to $\mathbf{l}_{i+1}$ and the other is parallel to $\mathbf{l}_{i+1}$. It is clear that the average values of the perpendicular components are equal to zero just from the symmetry. The average values of the parallel components are equal to $l \langle \cos \theta \rangle$. Thus,

$$\langle \mathbf{l}_i \mathbf{l}_{i+2} \rangle = l^2 \langle \cos \theta \rangle^2. \tag{3.11}$$

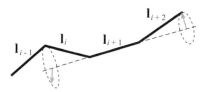

Figure 3.8 Calculation of $\langle \mathbf{l}_i \mathbf{l}_{i+2} \rangle$. The averaged components of $\mathbf{l}_i$ and $\mathbf{l}_{i+2}$ that are perpendicular to $\mathbf{l}_{i+1}$ (shown by the gray vectors) are equal to zero.

Using mathematical induction one can prove that

$$\langle \mathbf{l}_i \mathbf{l}_{i+k} \rangle = l^2 \langle \cos \theta \rangle^k. \tag{3.12}$$

Because $\langle \cos \theta \rangle$ is less than 1,

$$\langle \cos \theta \rangle^k \to 0 \quad \text{when } k \to \infty. \tag{3.13}$$

This shows that, if two segments are well separated one from another along the chain contour, their directions are nearly independent.

Equation (3.12) allows calculation of the average square of the end-to-end distance for the model chain (details of the calculations can be found in Cantor & Schimmel (1980)):

$$\langle \mathbf{S}^2 \rangle = \left\langle \left(\sum_{i=1}^{N} \mathbf{l}_i \right)^2 \right\rangle$$

$$= Nl^2 + 2 \left\langle \left(\sum_{i=1}^{N-1} \mathbf{l}_i \mathbf{l}_{i+1} \right) \right\rangle + 2 \left\langle \left(\sum_{i=1}^{N-2} \mathbf{l}_i \mathbf{l}_{i+2} \right) \right\rangle + 2 \left\langle \left(\sum_{i=1}^{N-3} \mathbf{l}_i \mathbf{l}_{i+3} \right) \right\rangle + \cdots$$

$$= Nl^2 + 2l^2 \sum_{k=1}^{N-1} (N-k) \langle \cos \theta \rangle^k = Nl^2 \left(\frac{1 + \langle \cos \theta \rangle}{1 - \langle \cos \theta \rangle} - \frac{2 \langle \cos \theta \rangle (1 - \langle \cos \theta \rangle^N)}{n(1 - \langle \cos \theta \rangle)^2} \right). \tag{3.14}$$

If $N \gg 1/(1 - \langle \cos \theta \rangle)$, Eq. (3.14) can be reduced to

$$\langle \mathbf{S}^2 \rangle = Nl^2 \frac{1 + \langle \cos \theta \rangle}{1 - \langle \cos \theta \rangle} = Ll \frac{1 + \langle \cos \theta \rangle}{1 - \langle \cos \theta \rangle}, \tag{3.15}$$

where $L = Nl$ is the contour length of the chain. Comparing Eq. (3.15) with Eq. (3.3) for the end-to-end distance of the random walk, we see that they look very similar. The only difference is that the length of the segment is multiplied by $\frac{1 + \langle \cos \theta \rangle}{1 - \langle \cos \theta \rangle}$ in Eq. (3.15). Thus, Eq. (3.15) can be written as

$$\langle \mathbf{S}^2 \rangle = Lb, \tag{3.16}$$

where b is a constant called *the Kuhn statistical segment*, or just *the statistical segment* (sometime it is also called *the statistical length*). The value of b is calculated as

$$b = l \frac{1 + \langle \cos \theta \rangle}{1 - \langle \cos \theta \rangle}. \tag{3.17}$$

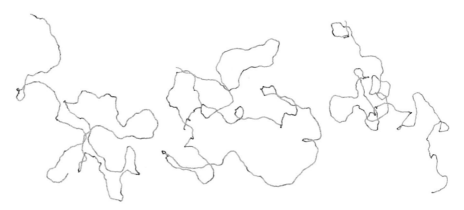

Figure 3.9 The projections of typical simulated conformations of dsDNA molecules 8460 bp (30 Kuhn segments) in length. There is clear similarity between these conformations and the random walk trajectories of 30 steps shown in Fig. 3.2.

Thus, we see that for sufficiently large N statistical properties of our model chain are equivalent to the statistical properties of the random walk, if we substitute the chain segments by the statistical segments. Comparison of typical conformations of the random walk (Fig. 3.2) and WLC (Fig. 3.9) of the same length, measured in number of Kuhn statistical segments, qualitatively illustrate the equivalence. Although we obtained Eq. (3.16) for a particular model chain, it expresses the main property of polymers. The only condition for its validity is that correlation in directions of the chain segments reduces when separation between the segments increases. In general, conformations of a long polymer chain have very strong similarity with random walk trajectories. Thus, the model of the random walk represents the simplest model of a polymer chain; it is called the *freely jointed chain*. If we are interested in a global property of a polymer, such as the average size of the polymer coil in solution, we can analyze this property theoretically for the freely jointed chain, and then transfer the result to our actual polymer. All we need is to know the value of b for the polymer. The value of b depends on the nature of the polymer, and specifies the number of its repeating units that correspond to one statistical segment. The value of b can be determined experimentally by comparing measured and calculated properties of a polymer chain. It is important for further consideration that for dsDNA b is much larger than l ($b/l \approx 300$).

The fact that long polymer chains form random coils in solution explains the remarkable elasticity of polymer materials. Indeed, to increase the end-to-end distance of a polymer we do not need to extend the length of the chain segments. Instead, we only need to change their orientation, increasing the alignment along the extending force. In this way the end-to-end distance of a long chain can be changed by many times! The polymer resists the extension since the number of chain conformations is smaller in an extended state, so its entropy is lower and the free energy is higher. Thus, the elasticity of polymers has an entropic nature.

Another concept widely used in polymer statistical physics is the chain *persistence length, a*. There are a few equivalent definitions of the persistence length. One of them

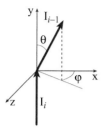

Figure 3.10 The angles specifying mutual orientation of adjacent segments of a model chain. Although in the simple discrete WLC considered here the energy depends only on angle θ, angle φ specifies the second degree of freedom. Analysis of many issues related to the statistical-mechanical treatment of the model requires proper accounting for the existence of this second variable.

says that the value of a is equal to the average value of projection of the end-to-end vector of a long chain on the direction of the first segment:

$$a = \frac{\mathbf{l}_1}{l} \left\langle \sum_{i=1}^{n} \mathbf{l}_i \right\rangle \quad \text{as } n \to \infty. \tag{3.18}$$

Using Eq. (3.12) we obtain from Eq. (3.18) that

$$a = l \sum_{i=1}^{\infty} \langle \cos \theta \rangle^{i-1} = l/(1 - \langle \cos \theta \rangle). \tag{3.19}$$

As we saw earlier, the correlation between the directions of the chain segments reduces when the separation between them along the chain contour increases. The correlation is specified by $\langle \cos \Theta \rangle$, where Θ is the angle between chain segments separated by the distance s along the chain contour. To calculate $\langle \cos \Theta \rangle$ we can use the fact that $l \ll a$, so $1 - l/a \cong \exp(-l/a)$. Then, it follows from Eqs. (3.12) and (3.19) that for segments i and $i + k$ $\langle \cos \Theta \rangle$ can be expressed as

$$\langle \cos \Theta \rangle = \exp(-s/a). \tag{3.20}$$

Equation (3.20) says that for two segments of the chain, separated by distance a along the chain contour $\langle \cos \Theta \rangle = 1/e$.

One more important equation establishes a relationship between the bending rigidity, g, and a. To obtain it let us consider the bending energy at a single joint of the chain as a function of angle θ associated with the joint (see Eq. (3.9)), and take the conformational average from both sides:

$$\langle E_b \rangle = \frac{g}{2l} \langle \theta^2 \rangle. \tag{3.21}$$

According to the basic law of statistical mechanics the average energy of thermal motion per degree of freedom is $k_B T/2$. Two angles, θ and ϕ, specify the mutual orientation of adjacent segments (Fig. 3.10), although the chain energy in this model does not depend on ϕ. Therefore, we conclude that $\langle E_b \rangle = k_B T$. Since we consider a very rigid chain,

$1 - \langle \cos \theta \rangle \cong \langle \theta^2 \rangle / 2$. Thus, Eq. (3.21) can be transformed to

$$a = g/k_B T. \tag{3.22}$$

The value of a for dsDNA depends on ionic conditions and temperature. The former dependence is due to the electrostatic contribution to DNA bending rigidity. At ionic conditions close to physiological ones and a temperature of 20 °C the value of a is close to 48 nm (Hagerman 1988, Taylor & Hagerman 1990, Vologodskaia & Vologodskii 2002, Geggier & Vologodskii 2010). This value does not change notably if $[Na^+] > 0.01$ M or $[Mg^{2+}] > 0.001$ M.

It is worth noting again that Eqs. (3.19)–(3.22) are approximations for a discrete WLC, whose accuracy is better for more rigid chains (smaller values of l/a). These are exact equations for the continuous WLC, which is obtained from the discrete WLC if we take the limit $l \to 0, \theta \to 0$, under the condition that $l/\theta^2 \to a$. Of course, the segments of a real polymer chain have a finite size, so the continuous WLC is not a better model for a polymer than the discrete one. Its only advantage is that the continuous WLC allows us to obtain some very useful relations in compact analytical form. In the last decades, however, computer simulation has become more and more important as an instrument for studying properties of polymers, and dsDNA in particular. This happens because for the majority of problems additional features of polymer chains have to be taken into account to describe properties of the chains properly. These features can be accounted for in computer simulations but not in an analytical treatment. In addition, computer simulations allow consideration of problems that are too difficult for analytical treatment.

In general, the simulation of equilibrium properties is based on a Monte Carlo (MC) sampling of the equilibrium ensemble of a chain conformation. This sampling generates a set of conformations where the probability of appearance of conformation i, $P(i)$, is proportional to its Boltzmann statistical weight:

$$P(i) \propto \exp(-E_i/k_B T), \tag{3.23}$$

where E_i is the total energy of conformation i. E_i can be independent of the chain conformation, as it is for the freely jointed chain, or can be specified by Eq. (3.9), or by any other equation. It can also include volume interactions considered below. There are efficient methods of sampling the equilibrium ensemble (Vologodskii 2006). To obtain the average value or the distribution of a property of interest, the value of this property is calculated for each conformation of the generated set and then the data are treated statistically.

3.1.3   The volume interaction

There is one important difference between trajectories of a random walk and conformations of a polymer chain. Two segments of a polymer cannot occupy the same volume of space. This restriction is called the effect of excluded volume. In general, however, interaction between two segments of a polymer that are close in space is not always reduced to the mutual exclusion from the same volume: the segments can also repel or attract one another at larger distances, so we should talk about volume interaction. The term assumes all interactions between the chain segments that are not transmitted along a polymer contour. The importance of the volume interaction depends on their magnitude and the length of the polymer. If the polymer solution is not sufficiently diluted, an intermolecular interaction can also contribute to the volume interaction. In the case of dsDNA the thickness of the molecule is small, compared with its persistence length. However, the double helix is a strongly charged polyelectrolyte, so there is an electrostatic repulsion between the chain segments when they are close one to another. This repulsion is screened by counterions, so the repulsion strongly depends on ionic conditions. Also, in the presence of multivalent ions DNA segments can attract one another. We consider this in detail below. With few exceptions, volume interactions between segments of the double helix have to be taken into account when we analyze DNA conformational properties.

3.2   Conformational parameters of DNA

3.2.1   Assumptions of the DNA model

The WLC, complemented by the volume interaction, can serve as a very good model to analyze large-scale statistical properties of DNA. To justify this statement we have to consider in detail the approximations made when using WLC as a coarse-grained model of DNA.

1. One assumption of WLC is that the minimum-energy conformation corresponds to zero bend angles between adjacent segments. DNA can have some intrinsic curvature, so the minimum-energy conformation of the DNA axis deviates from a straight line. Of course, the intrinsic curvature has to be specified by DNA sequence. For two reasons, however, the intrinsic curvature does not affect much the statistical properties of typical DNA molecules.

    The first reason is that small intrinsic bends are very difficult to distinguish from the fluctuational bends in DNA molecule with a typical sequence. It was suggested by Trifonov *et al.* (1988) and confirmed by theoretical and computational analysis that the experimentally measured value of the persistence length, a, has contributions from both thermal fluctuations and intrinsic bends (Schellman & Harvey 1995, Katritch & Vologodskii 1997, Bensimon *et al.* 1998, Nelson 1998). The two contributions are specified by so-called dynamic and static persistent lengths, a_{dyn} and a_{stat}, respectively, which define the value of a (Trifonov *et al.* 1988):

$$\frac{1}{a} = \frac{1}{a_{dyn}} + \frac{1}{a_{stat}}. \tag{3.24}$$

    The value of a_{dyn} should be calculated according Eq. (3.18) with the averaging only over thermal fluctuations assuming that the DNA is intrinsically straight. The value of a_{stat} does not account for thermal fluctuations and its calculation has to be performed by averaging in Eq. (3.18) only over all possible sequences, assuming that the temperature is equal to 0 K. Thus, if a DNA molecule of interest is sufficiently large, we can use WLC to obtain a good description of its conformational properties, performing the modeling for the effective value of a measured experimentally.

The second reason that justifies application of WLC for DNA modeling is that the contribution of intrinsic bends to a is small, with the exception of special sequences. This was shown by measuring a for "intrinsically straight" DNA and comparing it with the value of a for a typical DNA sequence. The "intrinsically straight" DNA consists of specially designed segments 10 bp in length. In each segment the sequence of the first five bases is repeated by the sequence of the second five bases (Bednar et al. 1995). Since the DNA helical repeat is close to 10 bp/turn, in such segments the intrinsic bends in the first five base-pair steps are nearly compensated by the bends in the opposite directions in the second five steps. Thus, the measured persistence length of such DNA depends, to a good approximation, on thermal fluctuations only, and is close to the value of a_{dyn}. It turns out that the value of a_{dyn} estimated in this way is very close to the value of a for a typical DNA sequence, that is, the contribution of a_{stat} to a is very small (Vologodskaia & Vologodskii 2002). The contribution of a_{stat} to a becomes notable only for DNA molecules with a large fraction of tracts of four or more adenines in row. The A-tracts are associated with a large intrinsic curvature (reviewed by Diekmann (1987), Crothers et al. (1990) and Hagerman (1990)).

2. It is assumed in WLC that the chain flexibility is isotropic, that is, the bending energy is independent of the bend direction. This is definitely not the case for dsDNA, where the bending rigidity in the directions of the helix grooves, g_1, is a few times smaller than the bending rigidity in the perpendicular directions, g_2 (Zhurkin et al. 1979). In addition to this, there is no reason to believe that the bending rigidities in the directions of the major and minor grooves of the double helix are the same. A model accounting for DNA bending anisotropy was suggested by Schellman (1974). However, if we are interested in conformational properties of DNA segments larger than 20–40 bp in length, there is no need to account for this anisotropy. The helical structure of DNA causes fast self-averaging of the bending anisotropy, so it is not essential for large-scale properties of DNA molecules (Levene & Crothers 1986, Schurr et al. 1995). The value of g which is used as a parameter of WLC is actually the average value over the values of g_1 and g_2 (Landau & Lifshitz 1986):

$$\frac{1}{g} = \frac{1}{2g_1} + \frac{1}{2g_2}. \tag{3.25}$$

The anisotropy has to be taken into account if we want to analyze statistical properties of DNA fragments shorter than 20 bp, or specific DNA conformations in DNA–protein complexes. There are no reliable data on the values of g_1 and g_2, however.

3. It is assumed in WLC that the bending rigidity is the same at each junction of two segments. In terms of DNA this means that the bending rigidity of the double helix does not depend on DNA sequence. This is definitely an approximation: DNA bending rigidity does depend on the sequence, although the effect is smaller than much research has suggested. Recently, the sequence variations of g were estimated

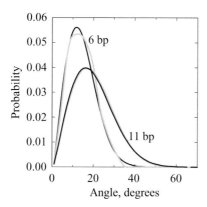

Figure 3.11 The distributions of angles between the chain ends, $P(\Theta)$, for two very different models of DNA bending. The models are specified by the distributions of angles between adjacent segments, $P(\theta)$. In the first model, WLC, the bending energy is proportional to θ^2 and $P(\theta)$ is specified by Eq. (3.10) ($P(\Theta)$ is shown by the black lines). In the second model $P(\theta)$ is the δ-function, $\delta(\theta - \theta_0)$ ($P(\Theta)$ is shown by the gray lines). The values of $\langle \cos \theta \rangle$ for the two models are the same. One can see from the figure that the difference in the corresponding $P(\Theta)$ is small even for fragments of 6 bp and practically disappears for fragments of 11 bp.

in the dinucleotide approximation (Geggier & Vologodskii 2010). It was found that the rigidity for individual base-pair steps, g_i, deviates from the average value by no more than 20%. In a quasi-random sequence good self-averaging of g occurs on the scale of 100 bp, so

$$\frac{1}{g} = \frac{1}{16} \sum_{i=1}^{16} \frac{1}{g_i}. \tag{3.26}$$

Note that, although there are only 10 distinguishable base-pair steps in the double helix, the averaging in Eq. (3.26) has to be taken over 16 possible dinucleotides.

4. The quadratic approximation for the bending potential as a function of θ is used in the WLC model (Eq. (3.9)). The main justification for this choice of the potential is that it corresponds to the first nonzero term of the Maclaurin expansion of the exact potential, if we assume that the bending energy has a minimum at $\theta = 0$. Also, the choice of the potential is not important for the majority of the large-scale conformational properties of DNA. Indeed, due to the central limit theorem of the probability theory, the distribution of angles between the ends of a DNA fragment larger than 10 bp in length depends mainly on the variance of θ, $\langle \theta^2 \rangle$, rather than on the distribution of angles between adjacent segments, $P(\theta)$ (in its simplest form the theorem is given in Box 3.1). This point is illustrated in Fig. 3.11, which shows the distribution of angles between the first and last segments, $P(\Theta)$, for chains with two very different distributions $P(\theta)$. We see from the figure that the distributions $P(\Theta)$ for two cases are slightly different for a chain of 6 bp, but this difference nearly disappears for the chain of 11 bp.

Thus, despite its simplicity, the WLC model, complemented by the volume inter-actions, can be used as a very good model for DNA large-scale statistical properties.

It should be noted, however, that deviations of the bending potential from the quadratic one can become notable for strongly bent DNA conformations (Vologodskii & Frank-Kamenetskii 2013). Because ssDNA is about two orders of magnitude more flexible than the double helix, under sufficiently large bending stress, short breaks of the regular structure should appear in dsDNA. These breaks may have form of kinks where only stacking interaction between two adjacent base pairs is broken but all base pairing is preserved (Crick & Klug 1975), or they may include isolated opened base pairs. The disruptions serve as flexible hinges and partially reduce bending stress in the intact strongly bent DNA segments. There are reliable data that the disruptions appear in torsionally unstressed DNA circles smaller than 75 bp in length (Du *et al.* 2008), but they may appear even in slightly larger circles. The critical DNA curvature that results in disruption formation has been an issue of intensive investigation over the last decade (reviewed by Vologodskii & Frank-Kamenetskii (2013)).

3.2.2   DNA persistence length

To model DNA conformational properties by WLC we need to know only one parameter, DNA persistence length. Here we summarize what we know today about the persistence length of dsDNA and its dependence on the solution conditions.

Peterlin was the first to apply WLC to the analysis of DNA conformational properties and estimate its persistence length (Peterlin 1953) from the light-scattering data obtained by Doty and Bunce (1952). From that moment, numerous studies were dedicated to determination of DNA persistence length. The early works used mainly DNA samples with large variation of molecular weights that compromised the accuracy. The studies were mainly based on the measurements of DNA sedimentation constants and light scattering of the samples. These properties do not allow for a simple theoretical analysis, so the obtained results vary substantially. Therefore, we do not discuss these studies below. Discovery of restriction enzymes and accumulation of sequencing data allowed researchers to use precisely specified DNA samples, and this made the determination of *a* more accurate (Kovacic & van Holde 1977). Still, to avoid accounting for the excluded-volume effect on the conformations of the molecules, researchers had to use relatively short DNA fragments, up to a few hundred base pairs in length. Due to the high bending rigidity of DNA the properties of such molecules are not very sensitive to the value of *a*, so the accuracy obtained in these studies was limited.

Important progress in the accuracy of *a* determination was achieved by using elec-trooptical methods that allow measurement of rotational relaxation of short DNA molecules (reviewed by Hagerman (1988)). Due to the high sensitivity of the rota-tional diffusion coefficient to the shape of the molecules, the method, complemented by careful theoretical analysis, allowed for much more reliable and accurate determination of DNA persistence length (Hagerman 1981, Hagerman & Zimm 1981). These studies showed that DNA persistence length is very close to 50 nm if the concentration of Na^+ ions exceeds 0.01 M or the concentration of Mg^{2+} ions exceeds 0.5 mM (Hagerman

1981). The value of a increases substantially at lower concentration of $[Na^+]$ and $[Mg^{2+}]$ (Hagerman 1981, Hagerman 1988). A limitation of the approach is that it does not allow the measurements to be made at high concentrations of ions.

Many studies used EM to evaluate DNA persistence length. Sometimes the results of this approach were in a reasonable agreement with the data obtained by other methods; sometimes they were far apart. In general, EM is not suited for this kind of problem since the value of a depends on ionic conditions. The ionic conditions are not defined after depositing the DNA sample on a grid and drying it for EM investigation. Also, the approach assumes that DNA conformations are equilibrated on the two-dimensional (2D) surface during the deposition. The validity of this assumption can depend on the deposition procedure and it is almost impossible to test it.

Single-molecule force–extension experiments (see below) offered, in particular, a new way to determine DNA persistence length. The method has two definite advantages. First, not a single point but the force–extension dependence has to be fitted by the theoretical dependence to determine the value of a. The theoretical force–extension dependence has only one parameter, the value of a (if the force is not so large that nonentropic elasticity of the double helix has to be taken into account). This makes the determination of a much more reliable. Second, since DNA molecules are strongly extended in these experiments, the conformations are not disturbed by the excluded-volume effect. Still, the available data on the determination of a based on the force–extension experiments (Bustamante et al. 1994, Baumann et al. 1997, Wang et al. 1997) seem to differ somewhat from the data obtained by other methods.

One of the difficulties of accurate determination of DNA persistence length by the majority of methods is the relatively low sensitivity of the measured properties to the value of a. This is especially true for short DNA fragments, whose conformational ensemble is not influenced by the excluded-volume effect. This is illustrated in Fig. 3.12, where two computed distributions of the end-to-end distance for DNA fragments 200 bp in length are shown. The distributions were calculated for two values of a, 45 and 50 nm. As we see from the figure, they are hardly distinguishable.

There is a property of the fragments, however, which is extremely sensitive to the value of a: the probability that the fragment ends will be in close proximity to one another. This probability increases 2.5-fold if a is reduced from 50 to 45 nm (see the inset in Fig. 3.12). Although the probability of such conformations is very low, it can be measured experimentally. Baldwin and co-workers were the first to develop a procedure to measure the effective concentration of one end of a short DNA fragment in the vicinity of the other end (Shore et al. 1981, Shore & Baldwin 1983a). The measurements of this effective concentration, the j-factor, are based on the cyclization of DNA fragments with short sticky ends by DNA ligase. Due to the extreme sensitivity of the j-factor of short DNA fragments to their persistence length, the measurements of the j-factors of short DNA fragments became the basis of the most accurate method for determination of a and other conformational parameters of DNA. We will describe this method in detail later in a special section of this chapter. Of course, to obtain the value of a from the measured value of the j-factor we need to know how it depends on a. For the WLC model this problem has been very accurately solved analytically (Shimada & Yamakawa 1984).

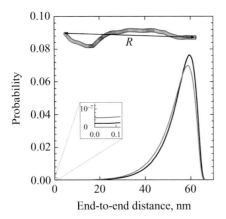

Figure 3.12 The distribution of the end-to-end distance, R, for short DNA fragments. The shown distributions were computed for DNA fragments 200 bp in length assuming that the value of a is equal to 50 nm (black line) and 45 nm (gray line). The fragments of this length have, mainly, extended conformations (a typical conformation is shown at the top of the figure), so the position of the distribution maximum is close to the fragment contour length, 68 nm. For typical values of R the distributions are hardly affected by the change of a. However, the probabilities of closing the fragment ends ($R \cong 0$), shown in the inset, differ by a factor of 2.5 for the two values of a.

It was found by the cyclization method that the DNA persistence length in a solution containing 10 mM $MgCl_2$ is equal to 47 ± 1 nm (Shore & Baldwin 1983a, Shimada & Yamakawa 1984, Taylor & Hagerman 1990, Vologodskaia & Vologodskii 2002). Application of the method is restricted, however, by solution conditions where DNA ligase is active, that is, presence of magnesium ions and not very high temperature.

Many earlier theoretical and experimental studies were dedicated to the dependence of DNA persistence length on ionic conditions (see Hagerman (1988) for a review). Although it seems well established that a is practically independent of ion concentration if $[Na^+] > 10$ mM or $[Mg^{2+}] > 0.5$ mM, at lower ion concentration the value of a increases substantially, approaching 80–100 nm. The data based on DNA force–extension experiments seem to be the most reliable here since extended DNA conformations in the experiment were not affected by the excluded-volume effect, which is large at low salt concentration (Smith *et al.* 1992, Vologodskii 1994). The cited studies also showed that at concentration of sodium ions notably lower than 1 mM DNA conformational properties are not well described by the WLC model due to very large electrostatic repulsion between negative charges on the double helix (see below), so the concept of persistent length is hardly applied in this case. Semi-quantitatively, the dependence of a on ionic conditions was predicted theoretically (Odijk 1977, Skolnick & Fixman 1977).

Precision of the persistence length determination by the cyclization-based method allowed researchers to address questions that were not accessible by other techniques, first of all the sequence dependence of DNA persistence length. The approach to this problem was based on cyclization of DNA fragments with specially designed sequences (Geggier & Vologodskii 2010). The researchers found that DNA persistence length

Table 3.1 The values of DNA persistence lengths that correspond to ten different dinucleotide steps a_{XY} (Geggier & Vologodskii 2010). It is important that only eight linear combinations of ten values of $1/a_{XY}$ were uniquely defined in the study. However, the data in the table allow unambiguous calculation of a for any particular fragment larger than 20–40 bp.

Dinucleotide step	Persistence length, a, in nm
AA/TT	50.4
AC/GT	55.4
AG/CT	51.0
AT	40.9
CA/TG	46.7
CC/GG	41.7
CG	56.0
GA/TC	54.4
GC	44.6
TA	44.7

can be well described in the dinucleotide approximation, when a specific value of a is assigned to each of ten dinucleotide steps. In this case the persistence length of a DNA fragment of N base pairs with an arbitrary sequence can be calculated as

$$\frac{1}{a} = \sum_{i=1}^{10} \frac{1}{a_{XY}} v_{XY}, \tag{3.27}$$

where a_{XY} is the persistence length corresponding to step XY and v_{XY} is the fraction of step XY in the fragment. Equation (3.27) can be used to determine the values of a_{XY} for individual base-pair steps, if the values of a are measured for a set of DNA fragments with different sequences. It is known, however, that the rank of the matrix produced by Eqs. (3.27) cannot exceed 8 for circular or very long linear DNA (see Section 2.2.7 and Gray & Tinoco (1970), Goldstein & Benight (1992)). Therefore, only eight linear combinations of $1/a_{XY}$ can be determined by this approach. On the other hand, these eight linear combinations are sufficient for unambiguous calculation of the average a for any given circular or sufficiently long linear DNA (longer than 20–40 bp). Because these linear combinations of $1/a_{XY}$ are not convenient for calculations of a for a particular sequence, it is better to have a set of ten a_{XY}, even if their values are not specified uniquely. Various ways to address this problem have been suggested (Gray & Tinoco 1970, Goldstein & Benight 1992, Vologodskii et al. 1984, Licinio & Guerra 2007). The method based on symmetry properties of DNA bases (Licinio & Guerra 2007) was used by Geggier and Vologodskii (2010), and the corresponding values of a_{XY} are shown in Table 3.1. The values of uniquely defined linear combinations of $1/a_{XY}$ that are valuable for comparison with other approaches can be obtained from the table (Geggier & Vologodskii 2010).

As one could expect, the variations of a in Table 3.1 are modest, although they could be meaningful for some problems related to DNA functioning. Interestingly, the

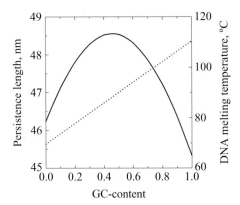

Figure 3.13 Dependence of DNA bending rigidity on the GC-content. The calculated dependence (solid line) is based on the data of Table 3.1 under the assumption that for each value of GC-content the DNA sequence is random. For comparison, the dependence of DNA melting temperature on the GC-content is shown by the dotted line (see Eq. (2.9)).

data in the table contradict the common belief that AT-rich DNA molecules are more flexible than GC-rich molecules. The dependence of DNA persistence length on the GC-content, based on Table 3.1, is shown in Fig. 3.13. We see from the figure that there is no correlation between DNA bending rigidity and its thermal stability, which increases linearly with the GC-content. GC-rich DNA molecules with random sequences are even slightly more flexible than AT-rich molecules.

The information presented in Table 3.1 is sufficient to calculate the value of a for any DNA fragment. It is not sufficient, however, for calculating the bending energy of a particular DNA conformation. For the latter goal we need the anisotropic bending rigidities of individual base-pair steps. It seems, however, that more detailed information on the sequence dependence of DNA bending rigidity will not be forthcoming soon. There was a hope that the needed set of rigidity constants might be extracted from the statistical analysis of numerous crystal structures of DNA–protein complexes (Olson et al. 1998, Balasubramanian et al. 2009). This approach to the parameter evaluation is based on the assumption that the variations of the helix parameters over different crystal structures correspond to thermal fluctuations of these parameters in free DNA in solution. It is hardly possible to justify the assumption. Thus, it is not surprising that the results obtained by this approach have very limited correlation with the solution data presented in Table 3.1 (Geggier & Vologodskii 2010). It seems at the moment that the only way to obtain more detailed information on DNA conformational flexibility is MD simulation of DNA duplexes. Still, the current accuracy of the force field used in the simulation is insufficient for this goal (see Section 1.7.3).

The cyclization-based method also allowed information to be obtained on the temperature dependence of DNA persistence length. The reduction of a with temperature rise is not negligable (Fig. 3.14). This should be taken into account when the relevant experimental data obtained at different temperatures are compared.

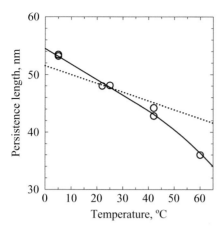

Figure 3.14 The temperature dependence of DNA persistence length (Geggier *et al.* 2011). The solid line is hand-drawn through the experimental points. The dotted line corresponds to Eq. (3.22) and the assumption that the DNA bending rigidity, *g*, is temperature independent.

3.2.3 Electrostatic interaction between DNA segments and effective diameter of the double helix

Volume interaction between DNA segments, the subject of this section, can be defined as an interaction between segments that are well separated along the chain contour but close in space. In the case of dsDNA this interaction is mainly caused by the electrostatic repulsion between the negatively charged segments. This electrostatic repulsion is screened in solution by small counterions distributed around DNA segments. This screening and, correspondingly, the electrostatic repulsion strongly depends on ionic conditions in the solution. Accurate theoretical treatment of the counterion distribution around the double helix is a very challenging problem that has not been solved yet (see the reviews Frank-Kamenetskii *et al.* (1987), Anderson & Record (1990) for earlier studies). However, as we will see below, we know enough about this distribution to properly account for the electrostatic repulsion between DNA segments. Still, before considering the electrostatic interaction explicitly, we will use a simpler empirical approach to the problem, based on the concept of DNA effective diameter.

The concept of the effective diameter was introduced in DNA studies by Stigter (1977). He considered two DNA models, a chain of cylinders with the diameter of the double helix and the electrostatic repulsion between them, and a chain of uncharged impenetrable cylinders of larger diameter. He wanted to mimic the statistical properties of the first model by the properties of the second model with a proper diameter of the uncharged cylinders. The diameter of these cylinders he called the DNA effective diameter. In doing so the smooth electrostatic potential is replaced by the hard-core potential of the impenetrable bodies, a usual approximation in the statistical mechanics. The actual values of the effective diameter, d_{eff}, are defined through the second virial coefficient of the cylinders, B_2,

$$B_2 = \frac{1}{2} \int [1 - \exp(-U(\mathbf{r})/kT)] \, d\mathbf{r}, \tag{3.28}$$

where $U(\mathbf{r})$ is the potential defining the interaction between the cylinders that depends on their mutual separation and orientation, specified by the set of variables $\mathbf{r}$. The value of d_{eff} is chosen so that V_2 for the charged DNA segments and for the uncharged hard-core cylinders of the same length are the same. Although it is not evident that the model chain with the hard-core potential provides a sufficiently good approximation for DNA statistical properties, it does for the great majority of DNA conformational properties. We will illustrate this below.

Lerman and co-workers were the first who managed to determine the values of d_{eff} from experimental measurements (Brian *et al.* 1981). In their study the researchers determined the equilibrium distribution of relatively short DNA fragments along the centrifuge tube. Clearly, this distribution should depend on the interaction between the fragments, if the DNA concentration is sufficiently high. The researchers obtained the theoretical expression for the distribution of uncharged cylindrical segments of diameter d, and by comparing it with the measured data determined the values of d_{eff}. It was found that the value of d_{eff} increases dramatically when [Na$^+$] reduces. The advantage of this approach is that d_{eff} is the only parameter of the distribution due to the small size of DNA fragments used (assuming that the total concentration of DNA fragments is known).

Another experimental approach to determination of d_{eff} is based on the measurements of the equilibrium probability of knots, P. It was found in computer simulations that P is extremely sensitive to the diameter of a polymer chain (Le Bret 1980, Klenin *et al.* 1988). Although P depends on the chain bending rigidity as well, the latter dependence is relatively weak in the range of its changes for the double helix. Therefore, the measurement of P and comparison of it with the corresponding computations represents an excellent opportunity for determination of d_{eff}. The equilibrium probabilities of the simplest knots, trefoils, were determined after closing linear DNA molecules with long sticky ends and electrophoretic separation of circular molecules with different topologies (see Section 6.7.3). The experiments were performed at different ionic conditions by Shaw and Wang (1993) and by Rybenkov *et al.* (1993). Remarkably, the obtained values of d_{eff} were in very good agreement with the results obtained by Lerman and co-workers (see Fig. 3.15(a)). Figure 3.15(b) shows the values of d_{eff} for solutions containing both Na$^+$ and Mg^{2+} (Rybenkov *et al.* 1997).

We see from Fig. 3.15 that at low concentration of salt the value of d_{eff} is a few times greater than the DNA geometric diameter. This is why the volume interactions between the segments have to be taken into account in the analysis of DNA conformational properties.

Now we can consider how the values of DNA effective diameter, obtained from the analysis of DNA conformational properties, agree with our knowledge of the electrostatic interaction between DNA segments. To this end we have to start from the basic properties of electrolyte solutions.

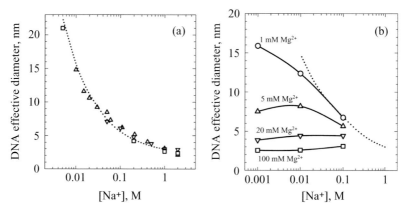

Figure 3.15 DNA effective diameter, d_{eff}, as a function of ionic conditions. (a) The values of d_{eff} are based on the equilibrium sedimentation of short DNA fragments ($\square$, Brian *et al.* 1981), and on the equilibrium fraction of knots in the solution of NaCl ($\triangle$, Rybenkov *et al.* 1993, and ∇, Shaw & Wang 1993). The dotted line corresponds to the theoretical calculations of d_{eff} based on the Poisson–Boltzmann equation (Stigter 1977). (b) The values of d_{eff} in solution containing both Na$^+$ and Mg^{2+} ions (Rybenkov *et al.* 1997). The dotted line is reproduced from (a).

Suppose that we have a solution of small ions, such as Na$^+$ and Cl$^-$. Due to the electrostatic interaction the distributions of these ions in space are correlated. There is a higher probability to find a negative ion than a positive one in the vicinity of a positive ion, and vice versa. The distribution of other ions around a chosen one averaged over time generates the screened electrostatic potential, $U(r)$, usually called the Debye–Hückel potential:

$$U(r) = ze\frac{\exp(-\kappa r)}{r},\tag{3.29}$$

where r is the distance from the chosen ion, e is the charge of proton, ze is the charge of the chosen ion, and κ is the Debye screening parameter,

$$\kappa = \left(\frac{8\pi e^2}{Dk_B T}\right)^{1/2} I^{1/2},\tag{3.30}$$

where D is the dielectric constant. The numerical coefficient in Eq. (3.30) assumes that e and k_B are in cgs units. The value of κ depends on the ionic strength, I, defined as

$$I = \sum_i C_i z_i^2/2,\tag{3.31}$$

where C_i is the concentration of type i ions in the solution, measured as the number of ions per cm^3. If I is converted into moles/liter, at room temperature in aqueous buffer κ is equal to $3.316 I^{1/2}$/nm. Equation (3.29) is valid under the condition that the C_i are sufficiently small, so that the average separation between the adjacent ions, $C_i^{-1/3}$, is larger than the Bjerrum length. The Bjerrum length corresponds to a distance between two elementary charges at which the electrostatic energy of their interaction is equal to k_BT,

$$\lambda_B = \frac{e^2}{Dk_BT}. \tag{3.32}$$

In water at room temperature λ_B is equal to 0.71 nm, so the condition is satisfied for most of the solutions used in biological studies. However, the average separation between the charges of a polyion can be much smaller than λ_B. For the B form of DNA this separation projected on the helix axis is only 0.17 nm, so the above analysis cannot be applied directly to dsDNA.

   Two major approaches to DNA electrostatic properties dominate in the literature. One is based on the counterion condensation theory, suggested by Manning (1969). Manning's idea originates from consideration of the energy of a small counterion located in the close vicinity of the polyion, where the polyion potential is not screened by other small ions. The polyion is modeled by a charged line with the linear charge density v. A simple calculation shows that the energy of the counterion diverges if v is larger than a limiting value v^*:

$$v^* = e/\lambda_B. \tag{3.33}$$

To overcome this problem, Onsager and later Manning suggested that if $v > v^*$, small counterions condense on the polyion to reduce its v to v^* (Manning 1969). With this reduced value of v, the electrostatic potential of the polyion, $U(r)$, can be expressed in the Debye–Hückel approximation specified by Eq. (3.29),

$$U(r) = \int_{-L}^{L} \exp(-\kappa\sqrt{x^2+r^2})\frac{v^*dx}{\sqrt{x^2+r^2}}, \tag{3.34}$$

where r is the distance from the polyion modeled by the charged straight line and the variable x corresponds to the coordinate on the polyion. The integral converges fast to the limiting value when we increase L. The attractive feature of the condensation theory is that it allows relatively simple analytical treatment of many important problems associated with electrostatic properties of polyions. One should keep in mind, however, that a polyion, and especially DNA, has very notable thickness on the length scale of small ions. Therefore, modeling the double helix as an infinitely thin line may be a rather rough approximation. Still, over the last decades the condensation theory has become very popular, despite the fact that many studies have shown that it does not provide a good approximation of DNA electrostatic properties (Stigter 1978, Gueron & Weisbuch 1980, Stigter 1995, Vologodskii & Cozzarelli 1995, Korolev et al. 1998).

   The second approach is based on the Poisson–Boltzmann (PB) equation for electrolytes (see Landau & Lifshitz (1951), for example). The treatment based on the PB equation does not assume that $v < v^*$. Also, DNA is modeled in this approach by a charge cylinder whose diameter corresponds to the diameter of the double helix. Still, there are approximations beyond this treatment; in particular, it does not account for the physical volume of small ions. MC simulations of the counterion distribution show that the volume of small ions becomes essential at high concentrations of them in solution

Table 3.2 The values of DNA effective diameter, d_{eff}, and effective charge density, v_{eff}, for various concentrations of NaCl (Stigter 1977).

$[Na^+]$, M	κ, nm^{-1}	d_{eff}, nm	v_{eff}, e/nm
0.01	0.329	15.7	2.43
0.02	0.465	11.2	2.96
0.05	0.735	7.4	4.18
0.1	1.04	5.6	6.01
0.2	1.47	4.4	9.90
0.5	2.32	3.4	25.7
1.0	3.29	2.95	74.2

(Le Bret & Zimm 1984, Korolev *et al.* 1999). The equation does provide, however, a good description of at least some electrostatic properties of DNA.

Stigter applied the PB equation to calculate the electrostatic potential of a DNA segment modeled by a charged cylinder. Then he used the potential to calculate the second virial coefficient, B_2, of DNA segments (Stigter 1977). The calculation used the fact that the main contribution to B_2 comes from the large separation between the segments (since large separation corresponds to larger volume of integration in Eq. (3.28)). Stigter found that at sufficiently large distances from the cylinder its electrostatic potential is well described by the Debye–Hückel potential for a charged line with some effective charge density, v_{eff} (see Eq. (3.34)). He calculated the values of v_{eff} for different concentrations of ions by matching the potential based on the PB equation for the cylinder and the one given by Eq. (3.34) in the overlap region far from the DNA surface. The values of v_{eff} found in this way are shown in Table 3.2. It should be emphasized that they do not have a direct physical meaning and only give the correct electrostatic potential beyond the vicinity of the double-helix surface. One can note that these effective charge densities are strikingly different, especially at large concentration of the monovalent salt, from Manning's charge density v^*, which is equal to 1.4 e/nm. Correspondingly, the DNA electrostatic potentials predicted by the condensation theory and by the PB equation are very different, except in the case of very low concentration of monovalent ions.

The potential specified by Eq. (3.34) was used to compute the equilibrium fraction of trefoil knots (Vologodskii & Cozzarelli 1995) determined experimentally for different concentrations of NaCl in solution (Rybenkov *et al.* 1993). Both Manning's value v^* and the values of v_{eff} were used in the computation, the results of which are shown in Fig. 3.16. Comparison between the computed and measured probabilities unambiguously shows that the potential based on the PB equation provides good agreement with the experimental results. On the other hand, the potential based on the condensation theory strongly underestimates the electrostatic repulsion between DNA segments at large distances from the DNA surface, at all studied ionic conditions. Another experimental observation that strongly contradicts the condensation theory is discussed in the next section.

It should be emphasized that the calculated fractions of trefoils are not sensitive to the shape of the potential in the vicinity of the DNA surface, since the close contacts between the segments are excluded by either potential. Therefore, the agreement between the calculation based on the PB equation and the experimental data only shows that the

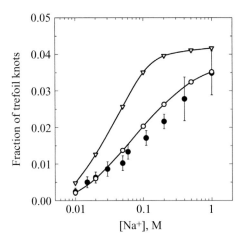

Figure 3.16 Comparison between the measured and computed fractions of trefoil knots. The computations used the electrostatic potential based on the PB equation (○) and on the condensation theory (▽). The experimental data (●) were obtained by Rybenkov *et al.* (1993).

PB equation provides a good approximation of the potential at distances from the DNA surface that exceed $1/\kappa$. There are data showing that the potential in the vicinity of the DNA surface, at interaxial distances smaller than 1 nm, is not described by the PB equation (Rau *et al.* 1984). It was noted before that the PB equation does not account for correlations between small ions, which is necessary in the vicinity of the DNA surface. Also, the water is treated implicitly in this approach as a medium with a dielectric constant. This is an oversimplification, owing to strong interaction between water molecules and ions and water and DNA. It seems that the assumption that Mg^{2+} and Ca^{2+} form stable hydration clusters gives a better description of their interaction with DNA (Gavryushov 2009). However, it is clear now that explicit treatment of water molecules is needed for accurate description of ion distribution around DNA. With the growing power of computers such treatment becomes possible, first of all by the MD simulations (Li *et al.* 2011, Yoo & Aksimentiev 2012, Giambasu *et al.* 2014).

It is interesting to compare conformational properties of DNA molecules calculated with the hard-core potential and the one specified by Eq. (3.34). The second virial coefficient for straight charged segments with the potential specified by Eq. (3.34) was first calculated by Onsager (1949). His expression for B_2 looks identical to the corresponding expression for uncharged hard cylinders with some effective diameter, d_{eff}. Using this result, Stigter found the values of d_{eff} for different concentrations of sodium ions (Table 3.2). The fractions of trefoil knots calculated for the model of hard cylinders with diameter d_{eff} and for the model with the potential specified by Eq. (3.34) are shown in Fig. 3.17.

Clearly, the two sets of data are in perfect agreement. Therefore, the hard-core approximation for the electrostatic potential is able to provide a very accurate description of DNA conformational properties under conditions where DNA segments repel one another. The only known exception is the conformational properties of highly supercoiled DNA under low concentration of sodium ions. In this case the electrostatic repulsion between the

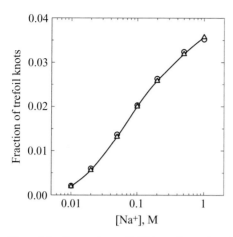

Figure 3.17 Comparison of DNA conformational properties computed for two potentials of the intersegment interaction. The fractions of trefoils for 10 kb DNA molecules were computed for the model of hard cylinders with diameter d_{eff} ($\triangle$) and for the model with the potential specified by Eq. (3.34) with the effective linear charge densities ν_{eff} shown in Table 3.2 ($\circ$). The two potentials give the same second virial coefficients of DNA segments at identical [Na$^+$].

segments is much more important, and details of the potential matter (Vologodskii & Cozzarelli 1995).

3.2.4 DNA helical repeat and torsional rigidity of double helix

The DNA helical repeat is important for many DNA properties related to its super-coiling, cyclization and formation of loops. Therefore, its value and dependence on DNA sequence and solution conditions have been studied in detail. Analyzing X-ray data on DNA structure, Watson and Crick concluded that the DNA helical repeat, the number of base pairs per complete turn of the double helix, γ, is equal to 10.0 bp/turn (Watson & Crick 1953). It took 25 years before Wang found that the value of γ in solution is 10.4 ± 0.1 bp/turn (Wang 1979). This difference between the values of γ in fibers and crystals (see Balasubramanian *et al.* (2009) for a review) and in water solutions was confirmed in the subsequent studies. Probably the difference is due to the intermolecular interactions between the helices in the condensed phase. The difference manifests the fact that structural parameters of the double helix are usually affected by the crystal packing and, therefore, the corresponding data have to be considered with caution.

A long time before Wang managed to measure the absolute value of γ, experiments with circular DNA were used to determine how γ depends on ionic conditions and on the solution temperature. In these experiments the equilibrium distributions of topoisomers of a circular DNA were created at different solution conditions (see Section 6.1). Due to a change of γ, the values of Lk that corresponded to the distribution maxima, Lk_0^1 and Lk_0^2, were also different. This shift between the distribution maxima was measured by running the two sets of topoisomers in a single gel (see Fig. 6.12, which shows how to determine the distribution maximum). Both distributions are centered around

$Lk_0^i = N/\gamma_i$, where N is the number of base pairs in the DNA molecule. Therefore, the value of $\delta Lk_0 = Lk_0^2 - Lk_0^1$ is directly related to the change in DNA helical repeat, $\delta\gamma$. It is easy to see that the value $\delta\gamma$ can be calculated as $\delta\gamma = -\gamma^2/N\,\delta Lk_0$, where $\gamma \cong 10.5$ (see Depew & Wang (1975), for example).

Since in these experiments the total change of Lk_0 was measured for DNA molecules a few kilobase pairs in length, small changes in γ were determined with very high accuracy. These studies showed that the value of γ decreases when the concentration of monovalent ions, $[X^+]$, increases as

$$\gamma = \gamma_X - k_X \lg[X^+], \tag{3.35}$$

where γ_X and k_X are constants whose values depend on the type of monovalent ion (Wang 1969, Shure & Vinograd 1976, Anderson & Bauer 1978, Rybenkov et al. 1997). For Na^+, K^+, Li^+ and NH_4^+, over the 0.01–0.2 M concentration range, $k_X = 0.025 \pm 0.01$ if $[X^+]$ is measured in M (Anderson & Bauer 1978, Rybenkov et al. 1997).

It should be noted that this dependence of γ on the ion concentration contradicts the counterion condensation theory discussed in the previous section. Indeed, it seems reasonable to assume that only ions that are tightly bound with to the double helix can affect the value of γ. Thus, if the number of tightly bound (condensed) ions does not depend on their concentration in solution, the value of γ should be independent of this concentration. This is not what is observed for solutions of the majority of monovalent ions. Therefore, there is no reason to think that the number of monovalent ions tightly bound to DNA remains the same regardless of their concentration in solution.

The value of γ does not change, however, in a solution of 0.5–50 mM $MgCl_2$ (Rybenkov et al. 1997). Binding of Mg^{2+} to DNA definitely saturates at very low concentration of these ions due to their very high DNA binding affinity (Taylor & Hagerman 1990).

The DNA helical repeat increases linearly with increase of solution temperature: $d\gamma/dT = 3.1 \times 10^{-3}/°C$ over a temperature interval of 5–80 °C (Wang 1969, Upholt et al. 1971, Depew & Wang 1975, Duguet 1993).

The DNA helical repeat depends on its sequence as well. Studies of this dependence, pioneered by Wang and co-workers, showed that the values of γ for different regular sequences vary from 10.25 to 10.75 (Peck & Wang 1981, Strauss et al. 1981, Goulet et al. 1987, Geggier & Vologodskii 2010). The available data are summarized in Table 3.3. It should be noted that, although the experiments showed that the γ of poly(dA) · poly(dT) is close to 10.0 (Peck & Wang 1981, Strauss et al. 1981, Goulet et al. 1987), this very low value of γ refers to the special B' conformation of the polynucleotide rather than to isolated AA/TT steps. The value of $\gamma_{AA/TT}$ shown in Table 3.3 refers to the isolated AA/TT steps in the regular B form of DNA.

So far, we have discussed equilibrium values of DNA helical repeat. Of course, under torsional stress the value of γ changes, and this change depends on the DNA torsional rigidity. The torsional rigidity of the double helix, C, influences many DNA conformational properties where the torsional orientation of DNA segments matters, such as the properties of supercoiled DNA. The value of C was first estimated from the data on the fluorescence depolarization of dyes bound to DNA molecules (Wahl et al. 1970, Barkley & Zimm 1979, Millar et al. 1980). The depolarization results, mainly,

Table 3.3 Sequence dependence of DNA helical repeat (Geggier & Vologodskii 2010). The experimentally determined values refer to the dinucleotide approximation, where the helical repeat of a DNA segment with an arbitrary sequence is specified by its dinucleotide steps. The values of γ_{XY} shown in the table correspond to ten different dinucleotide steps. It is important that only eight linear combinations of ten values of $1/\gamma_{XY}$ were uniquely defined by Geggier and Vologodskii (2010). However, the data in the table allow unambiguous calculation of γ for any fragment larger than 20–40 bp in length.

Dinucleotide step	Helical repeat, γ, in bp/turn
AA/TT	10.27
AC/GT	10.58
AG/CT	10.42
AT	10.59
CA/TG	10.60
CC/GG	10.76
CG	10.35
GA/TC	10.36
GC	10.43
TA	10.65

from the torsional diffusion of the double helix. The value of C is obtained from fitting the depolarization curve with the theoretical equation for the DNA dynamics, so it is based on rather complicated analysis (Barkley & Zimm 1979). Over the years the approach was carefully elaborated by Schurr and co-workers (see Fujimoto & Schurr (1990) and references therein). The method gives a value of C close to 1.7×10^{-19} erg·cm (it is a common practice to use these units for C).

Another approach is based on the width of the equilibrium distribution of topoisomers of circular DNA molecules (Benham 1978, Vologodskii et al. 1979, Shore & Baldwin 1983b, Horowitz & Wang 1984, Frank-Kamenetskii et al. 1985). In general, both bending and torsional fluctuations in the DNA molecules contribute to the width of the experimentally measured distribution. The contribution from bending can be computed by using the DNA model described above. However, for DNA molecules of 250 bp or shorter the contribution from bending becomes negligible, and the distribution width is completely specified by the value of C. It turns out that the data for both very small DNA circles and molecules a few thousand base pairs in length are consistent with each other (see Fig. 6.13). The value of C obtained by this approach is equal to 3.1×10^{-19} erg·cm (Frank-Kamenetskii et al. 1985, Shimada & Yamakawa 1988, Klenin et al. 1989, Geggier et al. 2011). Overall, this approach to determination of DNA torsional rigidity has practically no assumptions, and the above value of C should be considered as the most reliable.

During the last decade, the value of C was also determined from the single-molecule force–extension experiments (see Section 3.4). Very sophisticated and beautiful designs were used there to address properties related to DNA torsional deformation (Cluzel et al. 1996, Bryant et al. 2003). While the analysis of the experimental data by Cluzel et al. suggested a value of C close to 3.0×10^{-19} erg·cm (Vologodskii & Marko 1997), Bryant et al. obtained essentially higher values of C, around 4.2×10^{-19} erg·cm (Bryant *et al.* 2003). It is possible that the very high stretching force applied to the DNA in the experiments contributed to the high value of C. On the other hand, at low stretching force fluctuations of writhe, Wr, contribute to the measured value, complicating the analysis (Moroz & Nelson 1998, Lipfert *et al.* 2010).

Similarly to the correlation in the directions of the DNA axis, we can consider the correlation in the torsional orientation of the base pairs. Clearly, due to thermal fluctuations the correlation reduces when the separation between the pairs along the DNA contour, s, increases. One can show that the correlation reduces as $\exp(-s/a_t)$, similar to the correlation of the local direction of the DNA axis (see Eq. (3.20)). The value of a_t is called the torsional persistence length. Its value is directly related to the torsional rigidity:

$$a_t = 2C/k_B T. \tag{3.36}$$

For a C of 3.0×10^{-19} erg·cm, a_t is close to 150 nm.

Owing to thermal fluctuations the angle of the helix rotation, $\phi = 360/\gamma$, permanently changes. It is useful to compare the amplitude of the angle fluctuations with the sequence variation of its average values. It follows from Table 3.3 that the average value of ϕ changes from 33.5° to 35°. The average amplitude of the angle fluctuations, assuming a C of 3.0×10^{-19} erg·cm, is equal to 4° (see Eq. (B6.2.3)). Thus, the amplitude of the angle thermal fluctuations exceeds its variations with DNA sequence. This definitely limits the possibility that the values of ϕ can serve as a basis for recognition of specific sequences by DNA-binding proteins.

3.2.4   Single-stranded DNA

ssDNA is very flexible. This flexibility is key for the remarkable conformational plasticity of dsDNA, which can adopt different helical forms. Due to this flexibility all mismatches and chemical modifications of the bases cause no structural changes a few base pairs away from them. It is not so easy to specify this flexibility quantitatively, however.

Interaction between the segments of dsDNA is divided into two categories. The first category is due to interaction between DNA segments close to one another along the chain contour, and it specifies DNA bending rigidity. The second category is the interaction of DNA segments close in space but separated along the chain contour, the volume interaction. In the case of dsDNA the volume interactions can be well approximated by the notion of the effective diameter of the double helix, which accounts both for the physical thickness of the chain and for electrostatic repulsion between its segments. Such separation of the intersegment interactions into two categories works very well for dsDNA because its persistence length is much larger than its effective diameter, if the ion concentration in solution is not too low. This condition does not hold for ssDNA. Correspondingly, the concepts of persistence length and effective diameter become more ambiguous and can only approximately specify the conformational properties of the single-stranded molecules. Therefore, it is not surprising that different experimental approaches aiming to determine the persistence length of ssDNA brought different results. The values of persistence length obtained in various studies differed one from another because different sets of ssDNA conformations were essential in different methods.

The obtained values of ssDNA persistence length vary from less than 1 to 3 nm at near-physiological concentration of NaCl (Smith *et al.* 1996, Murphy *et al.* 2004, Chi *et al.* 2013). The contour length of ssDNA per nucleotide was found to be close to 0.6 nm. Of course, the obtained values of ssDNA persistence length increased at lower concentration of ions in solution (Murphy *et al.* 2004).

It should be noted that the conformational properties of ssDNA in solution can be affected by the stacking interaction between the bases. This interaction depends, among other factors, on DNA sequence. While it is hardly detectable in $d(T)_n$, the interaction in $d(A)_n$ greatly increases the persistence length at physiological temperature. For this reason $d(T)_n$ was often used in the studies of ssDNA flexibility.

3.3   DNA condensation

Owing to the high bending rigidity of the double helix, DNA coils occupy a lot of space in solution, a few orders of magnitude more than inside cells or virus capsids. For example, under regular ionic conditions, when the DNA persistence length is close to 50 nm, a coil of λ DNA occupies a volume with diameter of about 1000 nm while the diameter of the phage capsid is only 50 nm. In capsids and inside cells DNA is tightly packed, filling about 50% of available space. It seems that such tight packing is hardly possible for free DNA in solution due to the strong electrostatic repulsion between segments of the double helix. It turns out, however, that there are a few ways to achieve very tight compaction of DNA in solution (Bloomfield 1996). The phenomenon has been named DNA condensation.

Lerman was the first to observe DNA condensation in water solution containing simple neutral polymers and sufficiently high concentration of monovalent salt (Lerman 1971). The condensation occurs when the polymer has a sufficiently high degree of polymerization and a short statistical segment. Poly(ethylene oxide), $(EO)_n$, an uncharged flexible polymer, turned out to be very suitable for this goal. The condensation occurs under rather high concentration of $(EO)_n$, when the polymer coils contact one another. It had been known that under such conditions polymers with different flexibilities have a tendency to be separated into two distinct phases due to the excluded volume effect (Flory 1953). In the case of DNA and $(EO)_n$, tight side-by-side packing of DNA segments remarkably reduces the solution volume that is otherwise excluded for the dense coils of $(EO)_n$, and this promotes the condensation. Later theoretical studies suggested a quantitative analysis of the problem (Post & Zimm 1979, Grosberg *et al.* 1982). Of course, the condensation is possible only under the condition that electrostatic repulsion between DNA segments is efficiently screened by counterions. Therefore, high concentration of NaCl is a necessary condition for this type of DNA condensation. In the solution of $(EO)_n$ the condensation takes place at [NaCl] of 0.3 M (Lerman 1971). Condensation changes many physical properties of DNA, such as sedimentation and light scattering, so the measurement of these properties can be used to detect the phenomenon. In his

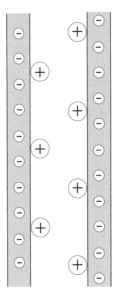

Figure 3.18 Diagram of the ion arrangement in condensed DNA. The positively charged multivalent ions are bound with two negatively charged DNA segments (shown in gray) forming periodic patterns. These patterns on the two DNA segments are shifted relative to one another, increasing the distance between the positively charged ions on the two helices. Such a correlated arrangement of the multivalent ions creates attraction between the DNA segments. Picture adopted from Rouzina and Bloomfield (1996).

first study Lerman measured DNA sedimentation rate, which dramatically increases as a result of the condensation (Lerman 1971).

DNA condensation also occurs in the presence of multivalent ions that carry at least three positive charges: polyamines spermidine^{3+} and spermine^{4+}, when their concentration in solution is in the range of 1 mM (Gosule & Schellman 1976), Co(NH$_3$)$_6^{3+}$ (Widom & Baldwin 1980) and cationic polypeptides such as polylysine (Laemmli 1975). Most basic proteins such as H1 and H5 can also cause condensation (Hsiang & Cole 1977, Garcia-Ramírez & Subirana 1994). It seems natural to assume that these ions form cross-links between negatively charged DNA segments, and there are experimental data supporting this idea (Allison *et al.* 1981). In an alternative and probably more realistic explanation the multivalent ions bound to the double helix form a correlated pattern in the condensate (Rouzina & Bloomfield 1996). As a result of this correlated arrangement of the multivalent ions, two double helices can attract one another, as diagramed in Fig. 3.18. The condensation produced by the multivalent ions in dilute DNA solutions gives very nice toroidal particles (Fig. 3.19). Sometimes, the particles have a globular shape without a visible hole in the middle or form rods (Arscott *et al.* 1990). It is interesting that to a first approximation the size of the particles does not depend on DNA length (Chattoraj *et al.* 1978), although it depends on some other factors (Hud & Vilfan 2005).

DNA precipitation following aggregation at high concentration of ethanol is widely used in the purification of DNA samples. Clearly, the aggregation and condensation are very similar processes, and under certain conditions well-defined rod-like particles are formed in ethanol solutions (Lang 1973).

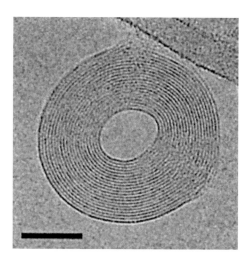

Figure 3.19 Cryoelectron micrographs of DNA toroid formed by λ DNA. The toroids were prepared by adding $Co(NH_3)_6Cl_3$ to DNA solution at room temperature. The scale bar corresponds to 50 nm. Picture reproduced from Fig. 1A of Hud and Downing (2001), with permission.

It seems that the secondary structure of condensed DNA molecules is not significantly different from that of B-DNA, as one can judge from the CD spectrum of condensed DNA (Bloomfield 1996).

An elegant new way of studying DNA condensation in isolated molecules has been introduced by force-extension spectroscopy (Baumann *et al.* 2000). It turned out that the extending force, if it is sufficiently small, does not create a substantial obstacle for DNA condensation, so formation of the condensate in the extended DNA molecule can be detected by monitoring the extension. An example of such an experiment is shown in Fig. 3.20. As one could expect, the force-extension curve is not reversible in this experiment since the extending force represents an obstacle for the DNA contraction needed for the condensate nucleation.

The chiral structure of toroids means that their formation should depend on DNA supercoiling. Indeed, toroids have a substantial *Wr*, so DNA supercoiling should promote their formation. It was predicted that the size of toroids formed by supercoiled molecules would be smaller than that formed by linear or open circular molecules (Grosberg & Zhestkov 1985). Later the effect was confirmed experimentally (Arscott *et al.* 1990).

3.4   Computational and experimental methods specific for DNA studies

There are many general experimental and theoretical methods used in the studies of DNA physical properties, as well as in many other fields of biophysics. We do not describe them in this book. Still, there are specific approaches that have turned out to be especially fruitful in DNA studies. The most important of these DNA specific approaches are considered below.

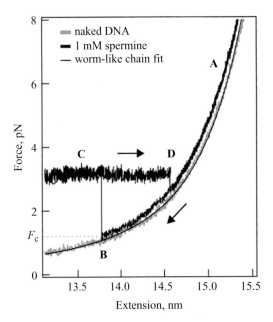

Figure 3.20 Condensation in a single DNA molecule. Force–extension curves of λ DNA with and without 1 mM spermine in the solution. The smooth black line is the force–extension curve for WLC (Eq. (3.49)). The extension of fully stretched DNA is slowly decreased until it reaches ∼85% of the contour length (here 13.8 mm). At this point, (B), the WLC behavior (A) suddenly disappears and the force rises abruptly (within 50–100 ms) to a value of ∼3.1 pN. This marked force jump indicates the collapse of part of the DNA into a condensate. With any further reduction in the extension, the force on the DNA stays roughly constant (C). When the DNA is stretched again, this force balance remains stable up to significantly higher extensions (∼90% of the contour length, here 14.6 mm), until a sharp force reduction down onto the WLC curve signifies the disruption of the last piece of condensed DNA (D). Picture reproduced from Fig. 2 of van den Broek *et al.* (2010), with permission.

3.4.1   Monte Carlo simulation of DNA equilibrium properties

Large DNA molecules adopt great variety of conformations in solutions. Therefore, conformational properties of such molecules have to be described in probabilistic terms. We can talk, for example, about the probability that two selected sites of a molecule will be juxtaposed, or about the probability that the distance between the ends of the molecule has a particular value. We can ask about average values of a particular property over the equilibrium ensemble of the molecule conformations, such as the average extension of the molecule under the action of the stretching force. Of course, the time required for the molecule to reach a conformation with a particular property is also a random variable with a specific distribution. Very often, there is no way to address all these questions experimentally. We can, however, address many of them computationally, using well-tested models and simulation procedures. Therefore, the simulations serve as an important instrument in the studies of different biologically important problems involving DNA molecules.

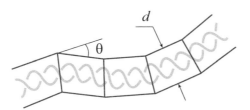

Figure 3.21 A model for simulation of large-scale conformational properties of DNA. The model chain consists of straight cylinders. Each cylinder corresponds to a certain number of base pairs. The diameter of the cylinders depends on ionic conditions, but is always larger than the geometrical diameter of the double helix. The double helix is shown in the figure for scale clarification only.

Throughout this section, DNA conformations will be specified by conformations of the molecular axis. Of course, for many issues information about conformational properties of the DNA axis is insufficient; we need to know conformational properties at an atomic level of resolution. Simulation of this type, MD simulation, is a powerful general method of molecular biophysics, and its detailed analysis is beyond the scope of this book. A brief discussion of the current state of MD simulations of DNA molecules can be found in Section 1.7.3.

For the coarse-grained level of description, DNA can be modeled as a chain of straight segments with the physical properties described earlier in this chapter. Such phenomenological models of the double helix have only a few parameters that have been unambiguously determined from the experimental data. Let us now summarize in detail the model for simulation of DNA equilibrium properties.

A DNA molecule N base pairs in length can be modeled as a chain consisting of m rigid segments that are cylinders of equal length l (Fig. 3.21).

The value of l does not affect the simulation results if it is sufficiently small. Since decreasing l strongly increases the simulation time, it is always attractive to use the largest l suitable for the property of interest. For the majority of DNA conformational properties it is enough to have one straight cylinder for 30 bp. Of course, if we are interested in the properties of a smaller scale, such as the distribution of bend angles between the ends of 20 bp DNA fragments, we have to use shorter segments, up to one segment for each base pair. The bending energy of the chain, E_b, is specified by the angles between the directions of adjacent segments i and $i + 1$, θ_i:

$$E_b = \frac{g}{2l} \sum_{i=1}^{n-1} \theta_i^2, \tag{3.37}$$

where g is the bending rigidity constant and n is the number of segments in the chain. The constant g is the most important parameter of the model. If $l \ll a$, the value of g is proportional to DNA persistence length, a:

$$g = ak_B T. \tag{3.38}$$

The cylinders have a certain diameter, d, and are impenetrable to one another. The value of d corresponds to the effective diameter of the double helix (see Section 3.2). Thus, g and d are the only two parameters of the model.

Using this model and MC methods, one can construct a large set of chain conformations that represents the equilibrium conformational distribution. The procedure for this conformational sampling must satisfy one condition: the probability that conformation i will appear in the set has to be proportional to $\exp(-E_i/k_B T)$. Once the conformational set is constructed, we can perform its statistical analysis to estimate conformational properties of interest.

The model described above does not account for torsional orientations of DNA base pairs. These orientations do not affect the majority of conformational properties of linear or nicked circular DNA molecules, if the DNA is intrinsically straight. In some cases, however, one has to account for the torsional orientations of base pairs explicitly. The corresponding more complex model is described by Katritch and Vologodskii (1997). It also allows introduction of a sequence-dependent DNA bending rigidity and even local anisotropy of the bending rigidity. Of course, the conformational sampling requires much more extensive computations in this case.

The Metropolis MC procedure is usually used for the statistical sampling of chain conformations (Metropolis *et al.* 1953, Frank-Kamenetskii *et al.* 1985). The procedure consists of consecutive deformations of the chain parts. At each step of the procedure a new trial conformation can be accepted or rejected. If the trial conformation is rejected, the current conformation must be added again to the constructed conformational set. The starting conformation is chosen arbitrarily.

At least two types of move should be used in the simulations of linear molecules. In the first type, a subchain is rotated by a randomly chosen angle, φ_1, around the straight line connecting two randomly chosen vertices of the chain (Fig. 3.22(a)). In the second type of displacement a subchain includes one end of the chain, so the end-to-end distance of the chain can be changed. This randomly chosen subchain is rotated by a random angle, φ_2, around a randomly oriented axis passing through the internal end of the subchain (Fig. 3.22(b)). The values of φ_i are usually uniformly distributed over intervals $(-\varphi_{i0}, \varphi_{i0})$, chosen so that about half of the trial conformations of each type are accepted. Other types of move can be used to increase the rate of sampling as long as they do not violate the principle of detailed balance (Binder & Heermann 1997). Whether the new conformation is accepted or rejected is determined by applying the rules of the procedure.

1. If the energy of the new conformation, E_{new}, is lower than the energy of the previous conformation, E_{old}, the new conformation is accepted.
2. If the energy of the new conformation is greater than the energy of the previous conformation, then the new conformation is accepted with the probability $p = \exp[(E_{old} - E_{new})/k_B T]$.

Forbidden conformations, which may appear during the displacements, such as conformations with overlapping segments, should be considered as having infinite energy. According to the acceptance rules such conformations are rejected.

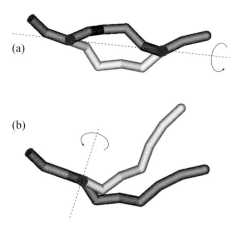

(a)

(b)

Figure 3.22 Trial moves of the model chain during the Metropolis MC simulations. The trial conformations of the chain are shown by lighter gray tone than the current conformations. (a) In the crankshaft move a subchain is rotated by a random angle around the axis connecting two randomly chosen vertices. (b) In the second type of move a randomly chosen subchain includes one end of the chain. The subchain is rotated by a random angle around a randomly oriented axis passing through the internal end of the subchain.

Of course, there is a strong correlation between successive conformations constructed by the Metropolis procedure. Clearly, the constructed conformational set should be sufficiently large to overcome this correlation. Fortunately, modern computers allow us to construct sets of billions of conformations and obtain very accurate estimations of statistical properties.

Circular DNA with a single-stranded nick has no torsional stress (if the molecules are larger than 1 kb). Simulation of such molecules requires only the crankshaft move (see Fig. 3.22(a)). Nicked circular DNA maintains the topology of its axis, however: it can be unknotted or form knots of different types (see Section 6.7). To keep a chosen topology during the simulation procedure one needs to check that it was not changed after each move (a moving segment can cross another segment of the chain). The check is performed by calculating the Alexander polynomial (Vologodskii *et al.* 1974, Frank-Kamenetskii *et al.* 1975). A similar approach can be used for the simulation of two circular chains (Vologodskii *et al.* 1975, Vologodskii & Rybenkov 2009). If both strands of a DNA molecule are closed, the complementary strands form a link with a certain value of the *linking number*, *Lk* (see Chapter 8). *Lk* is a topological property whose value must be retained through all conformational changes. Simulation of such molecules requires explicit or implicit specification of the chain twist (see Vologodskii (2006) for a review).

Although the modern computers allow very efficient conformational sampling, the number of rare conformations in the constructed set may be insufficient to estimate the probability of these conformations accurately. In such cases different kinds of biased sampling can be used. The method of "umbrella sampling" addresses calculation of conformational distributions under specific conformational conditions (McCammon & Harvey 1987, Klenin *et al.* 1991). For example, we want to calculate the distribution of orientations of the two juxtaposed sites of the chain. The sampling can be based in this case by adding to the chain energy an artificial potential, $U(\mathbf{x})$, where $\mathbf{x}$ specifies the distance between the sites. We can choose $U(\mathbf{x})$ so that it greatly increases the probability of the site juxtaposition. However, it does not disturb the distribution of orientations since $U(\mathbf{x})$ has the same value for all conformations where the sites are juxtaposed ($x \approx x_0$). Indeed, the statistical weights of all conformations with the juxtaposed sites will be multiplied by the same factor, $\exp(-U(\mathbf{x}_0)/kT)$. The efficiency of this approach depends on proper choice of $U(\mathbf{x})$ (see Klenin *et al.* (1991), Du *et al.* (2007) as examples). Another approach was developed specifically for calculation of the cyclization probability of very short (or very long) DNA fragments, when this probability of cyclization is very small and direct MC sampling is inefficient (Podtelezhnikov *et al.* 2000).

3.4.2   DNA cyclization

3.4.2.1   *j*-factor

Conversion of linear DNA molecules into circular form is often a necessary step during laboratory manipulation of DNA, it occurs inside the cells and it provides a remarkable tool to study DNA conformational properties. Therefore, we consider this process in detail.

Very often, linear DNA molecules have so-called sticky or cohesive ends, hanging single-stranded segments on both ends of the molecule that are complementary one to another. Sticky ends of two to four nucleotides in length are created by many restriction endonucleases. Some phage DNAs have sticky ends of 12–20 nucleotides in length. Discovery and creation of sequence-specific nicking enzymes (Chan *et al.* 2011) allows us now to make sticky ends of any desired length and sequence (Vologodskii 2012). We will assume here that the molecules have sticky ends that can join one another to form a double-stranded segment, converting linear DNA molecules into circular ones.

The lifetime of the circular state depends on the sticky-end length, sequence, solution temperature and ionic conditions. Wang and Davidson were the first to study DNA cyclization by joining long sticky ends of phage DNAs (Wang & Davidson 1966, Wang & Davidson 1968). They found, in particular, that the rate of cyclization, k_c, is not limited by the rate of diffusion of one end of the molecule to another. This means that, on average, many collisions of the sticky ends precede their joining, and therefore k_c is proportional to the equilibrium concentration of one end of the molecule in the vicinity of the other end, called the *j*-factor, *j*:

$$k_c = j k_a, \tag{3.39}$$

where k_a is the rate constant of bimolecular association of two distinguishable half molecules (with a single sticky end). The concept of the *j*-factor was introduced by Jacobson and Stockmayer as the ratio of the equilibrium constants of the cyclization and dimerization, K_c/K_a (Jacobson & Stockmayer 1950). Equation (3.39) means that the *j*-factor can also be determined as the ratio of the corresponding kinetic constants. It turned out that it is easier to measure the kinetic constants (or their ratio directly) to determine the value of *j*. On the other hand, the value of *j* is completely specified by DNA conformational parameters, as we consider below. Thus, comparison between

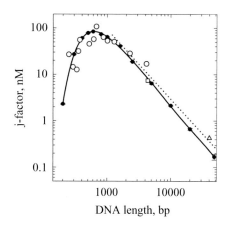

Figure 3.23 *j*-factor of DNA molecules as a function of their length. The solid line and filled circles correspond to the MC calculations, described by Podtelezhnikov and Vologodskii (2000), for a discrete WLC with a persistence length of 48 nm and diameter of 5 nm (corresponds to 0.2 M NaCl and room temperature) The dotted line corresponds to the values for the freely jointed chain with zero diameter of the segments (Eq. (3.40)). The open circles are the experimental data from Shore and Baldwin (1983a). The square is the experimental estimation of *j* for pBR322 DNA (Legerski & Robberson 1985), and the triangle is the estimation for λ DNA (Wang & Davidson 1966). The latter two points were corrected for 20 °C.

measured and theoretical values of *j* can be used to evaluate parameters of the double helix.

It was shown above that global properties of a long polymer chain can be well approximated by the properties of the freely jointed chain, if the effect of the excluded volume is small. The latter effect is definitely small for dsDNA molecules at ionic conditions close to the physiological ones (200 mM NaCl and 10 mM MgCl$_2$). Thus, for DNA molecules larger than 2–3 kb in length the *j*-factor can be sufficiently well estimated from Eq. (3.8) for the probability distribution of the distance between the chain ends, *R*. Considering the limit of $R \to 0$ in Eq. (3.8) we conclude that

$$j = [3/(2\pi Lb)]^{3/2}/N_A, \tag{3.40}$$

where *j* is expressed in M, *b* is the length of the DNA statistical segment ($\approx 10^{-6}$ dm), *L* is the molecule length in dm and N_A is Avogadro's number.

Shimada and Yamakawa obtained a very good approximation for *j*-factors of very short WLCs (Shimada & Yamakawa 1984):

$$j = \frac{4\pi^3 b^3}{N_A L^6} \exp\left(-\frac{\pi^2 b}{L} + 0.514 L/b\right), \tag{3.41}$$

where the *j*-factor is expressed in M, and *L* and *b* have to be expressed in dm. Equation (3.41) can be used for DNA fragments up to 300 bp in length. Accurate estimation of *j* for DNA molecules of any length can be obtained by computation for the WLC with a certain value of the segment diameter (Podtelezhnikov & Vologodskii 2000). The result of such computations that correspond to physiological ionic conditions is shown in Fig. 3.23 along with some experimental measurements of *j*. One can see from this figure that the maximum value of j corresponds to DNA molecules about 600 bp in length.

3.4.2.2   Ligase-assisted cyclization of short DNA fragments

At room temperature the duplexes formed by long sticky ends of phage DNAs are so stable that the molecules remain in circular form for many days. In particular, the samples of DNA with long sticky ends can be electrophoresed to separate different DNA forms. This is not the case, however, for sticky ends created by restriction enzymes. For sticky ends four nucleotides in length the lifetime of the circular state is in the scale of minutes (Revet & Fourcade 1998, Tagashira *et al.* 2002, Geggier *et al.* 2011). Thus, if we want to preserve the circular form, the nicks in the circular molecules created by the sticky ends joining have to be covalently joined (ligated by DNA ligase). If the conditions formulated below are satisfied, we still can measure j-factors of these DNA molecules by monitoring the ligation kinetics. This ligase-assisted cyclization was elaborated in detail by Baldwin and co-workers in the early 1980s (Shore *et al.* 1981, Shore & Baldwin 1983a).

Let us suppose that we have a DNA fragment with sticky ends and want to measure its j-factor. When we add DNA ligase in solution, fragments with joined sticky ends are converted by the enzyme into covalently closed circles. Of course, other ligation products are accumulated as well in the reaction, first of all the linear dimers, then the circular dimers and so on. By analyzing the kinetics of this process one can show (Shore *et al.* 1981, Shore & Baldwin 1983a, Taylor & Hagerman 1990) that the ratio of the monomeric circles, $C(t)$, to dimers, $D(t)$, extrapolated to zero ligation time, t, specifies j:

$$j = 2M_0 \lim_{t \to 0} \frac{C(t)}{D(t)} \qquad (3.42)$$

where M_0 is the initial concentration of the fragment and the coefficient '2' reflects the fact that the sticky ends made by restriction enzymes are usually self-complementary.

It is worth making a couple of technical notes here. First, since the cyclization of linear dimers can proceed much faster than the cyclization of the monomers, it is important to account for both linear and circular dimers of the fragment when evaluating $D(t)$. The second remark is related to the fragments cut from a plasmid. Measuring the j-factors of such fragments is greatly simplified by the fact that there is no need to separate the fragments from the plasmid DNA. Indeed, Eq. (3.42) is valid even when the vector DNA is left in the solution and can be ligated with the fragments.

Equation (3.42) is valid, however, only under the following conditions.

1. The equilibrium fraction of joined sticky ends is small, so it remains proportional to the probability of their juxtaposition, f. At room temperature and ionic conditions required for DNA ligase the sticky ends of four A/T or two A/T and two G/C nucleotides satisfy the condition, although f becomes too large for the sticky ends with four G/C nucleotides (see below).
2. The ligase concentration is sufficiently low, so many events of the association and dissociation of the sticky ends occur, on average, before the ligation event. Under this condition the rate of ligation is proportional to the equilibrium fraction of molecules

with joined but not ligated sticky ends (Shore *et al.* 1981, Shore & Baldwin 1983a, Peters & Maher 2010). Violation of these conditions can cause large mistakes in the *j*-factor estimation (Du *et al.* 2005, Peters & Maher 2010).

3. Conformational distributions of joined sticky ends and adjacent base pairs are similar in the circles and dimers. Only under this condition is Eq. (3.42) valid. The estimation shows that this condition should not hold for cyclization of very short DNA fragments, about 100 bp or shorter (Vologodskii & Frank-Kamenetskii 2013). In such cases, due to large bending stress in the DNA circles, kinks of the double helix have to be formed at the nicks surrounding joined sticky ends. In the linear dimers of the fragments, however, the regular conformation of this region prevails.

When Baldwin and co-workers applied the ligase-assisted determination of *j* to DNA fragments of $\approx$250 bp in length, they found huge scatter of *j* for fragments of nearly the same length (Shore *et al.* 1981). Eventually Shore and Baldwin realized that the scatter is actually the *j*-factor oscillations that reflect the periodicity of the double helix (Shore & Baldwin 1983a). The ligation of the joined sticky ends requires the torsional alignment of the two ends at the nick, and this alignment may cause substantial torsional stress in short DNA fragments. Thus, in general, the *j*-factor of dsDNA has two components, the regular one that accounts for closing of DNA ends and the proper alignment of the directions of their axes, j_0, and the torsional component, j_t, that accounts for the torsional alignment:

$$j = j_0 j_t. \tag{3.43}$$

In this notation it is the j_0 value that is plotted in Fig. 3.23 and expressed by Eqs. (3.40) and (3.41). The value of j_t can be calculated as

$$j_t = \frac{\sum_{Lk=-\infty}^{\infty} P(Lk - Lk_0)}{\int_{-\infty}^{\infty} P(x)dx}, \tag{3.44}$$

where $P(x)$ is the probability that the torsional orientation of the joined ends corresponds to the linking number difference of $Lk - Lk_0$. The value of Lk_0 is equal to the average number of helix turns in the corresponding linear DNA, $Lk_0 = N/\gamma$ (see Section 6.1). The denominator on the right-hand side of Eq. (3.44) is needed for correct normalization of j_t. It provides the condition that for long DNA molecules j_t is equal to unity. Although in general the torsional alignment can be provided by both bending and torsional deformations of the DNA molecule, for molecules up 300 bp in length the contribution of bending deformations is negligible (see Section 6.2). Therefore, for such short DNA molecules $P(x)$ depends only on their torsional rigidity, C, and j_t can be calculated as

$$j_t = \sqrt{\frac{2\pi C}{k_B T L}} \sum_{LK=-\infty}^{+\infty} \exp\left(\frac{2\pi^2 C(Lk - Lk_0)^2}{k_B T L}\right), \tag{3.45}$$

where L is the DNA length. The dependence of *j*-factor on DNA length that accounts for j_t is shown in Fig. 3.24. One can see from the figure that the *j*-factor oscillates with a periodicity of about 10 bp, and the amplitude of the oscillations exceeds one order of magnitude for DNA fragments $\approx$200 bp in length. Clearly, the maximum and minimum

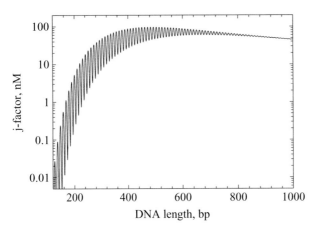

Figure 3.24 Oscillation of j-factor with DNA length for short DNA fragments. The calculated values of j were obtained by multiplying j_0, shown in Fig. 3.23, by j_t, expressed by Eq. (3.45).

values of j correspond to fragment lengths with integer and semi-integer equilibrium numbers of helix turns.

The experimental measurements by Shore and Baldwin for fragments of different lengths showed j-factor oscillations with periodicity close to 10 bp, in full agreement with the theoretical expectation. This was the most direct proof that dsDNA forms a helix in solution with the helical repeat close to 10 bp/turn, as was concluded by Watson and Crick from analysis of the X-ray data.

The WLC model describes DNA cyclization up to a certain limit of very large curvature. If a DNA segment is bent above this limiting curvature, the regular structure of the double helix breaks by forming a kink or open base pair with a sharp localized bend (see Fig. 2.19(a) and Fig. 2.20). This phenomenon was first analyzed by Crick and Klug (1975). Cloutier and Widom were the first to suggest that such breaks can result in much higher values of j for very short DNA fragments (Cloutier & Widom 2004), and later this was confirmed in theoretical and computer calculations (Yan & Marko 2004, Du *et al.* 2005, Wiggins *et al.* 2005). However, the too high ligase concentration used by Cloutier and Widom resulted in great overestimation of j for the fragments of 95 130 bp in length (Cloutier & Widom 2004). Later study showed that circles of these lengths keep a smooth structure and their j-factors are well described by the WLC model (if the circles are not negatively supercoiled) (Du *et al.* 2005, Du *et al.* 2008). The current data suggest that the disruptions of the regular DNA structure only appear in circles of about 75 bp or less, if circles without negative torsional stress can be formed (Du *et al.* 2008). The appearance of the disruptions is greatly enhanced by the negative torsional stress.

It took some time to realize that cyclization of short DNA fragments can serve as a method to determine DNA persistence length with unmatched accuracy (Taylor & Hagerman 1990, Vologodskaia & Vologodskii 2002, Geggier & Vologodskii 2010). For this purpose, the measured j-factors for a set of DNA fragments of different lengths N, covering one period of the j-factor oscillations, have to be fitted by the theoretical values of j. The theoretical values of j depend on three parameters of DNA fragments, a, C and

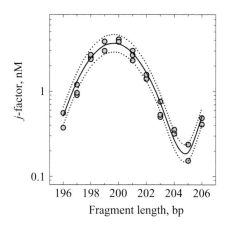

Figure 3.25 Dependence of the j-factor on the fragment length. All fragments have the same sequence except the incremental part. The experimental data (circles) were fitted by the theoretical equations (3.41), (3.43) and (3.45) by adjusting the values of DNA helical repeat, γ, torsional rigidity, C, and fragment persistence length, a. The best fit (solid line) corresponds to $a = 49.5$ nm, $\gamma = 10.50$ bp/turn and $C = 2.4 \times 10^{-19}$ erg·cm. The dotted lines show the calculated j-factors for a of 48.5 and 50.5 nm and the same values of γ and C. The plot was reproduced from Fig. 7 of Vologodskaia and Vologodskii (2002), with permission.

γ, that depend, in general, on the fragment sequence, temperature and ionic conditions. Since each of these parameters specifies a different feature of $j(N)$, the fitting procedure is simple and unambiguous. This is illustrated in Fig. 3.25. One can see from the figure that the value of a was estimated from the experimental data with accuracy close to 2%. Indeed, the theoretical j-factors calculated for a 1 nm smaller or larger than the optimum value do not fit the experimental data. The DNA helical repeat, γ, can be estimated with accuracy close to 0.5%. Although the third parameter, C, can also be estimated with a good accuracy, its value may reflect the torsional rigidity of DNA with an additional single-stranded nick (as there are two nicks in the circle with joined sticky ends). Therefore, the value of C obtained in such an experiment may not be quite relevant.

3.4.2.3   Thermodynamic and kinetics of sticky-end joining

Joining and dissociation of the sticky ends represents a special case of DNA melting/renaturation. Its first thermodynamic analysis was suggested by Wang and Davidson (1966). We use below their model of the process and only modify the treatment to eliminate an adjustable parameter from their analysis. The free energy of DNA cyclization by joining the sticky ends of n nucleotides in length, ΔG_{LC}, can be expressed as

$$\Delta G_{LC} = \sum_i \Delta G_i(T) - RT \ln(j) + \Delta G_{nucl} + \Delta G_1^{st} + \Delta G_2^{st}, \qquad (3.46)$$

where $\Delta G_i(T)$ is the free energy of the helix elongation on adding base pair i to base pair $(i-1)$, ΔG_{nucl} is the free energy of helix nucleation and ΔG_1^{st} and ΔG_2^{st} are the free energy of stacking across the nicks in the circular form of the molecule (compare with Eq. (2.25) for formation of DNA duplexes). The sum in Eq. (3.46) has to be taken over all $(n - 1)$ base-pair steps in the joined sticky ends. The values $\Delta G_i(T)$ and ΔG_{nucl} are known for different ionic conditions and temperatures, as well as the values of ΔG_1^{st} and ΔG_2^{st} (SantaLucia 1998, Protozanova et al. 2004, Yakovchuk et al. 2006, Owczarzy et al. 2008). Equation (3.46) allows calculation of the melting curve of the sticky ends. The fraction of circular form, θ, at temperature T should be calculated as

$$\theta = \exp(-\Delta G_{\mathrm{LC}}/RT)/(1 + \exp(-\Delta G_{\mathrm{LC}}/RT)). \tag{3.47}$$

The melting temperature of the sticky ends, T_{m}, is defined as the temperature at which equilibrium fractions of linear and circular forms are equal ($\Delta G_{\mathrm{LC}} = 0$).

The above treatment corresponds to the two-state approximation of sticky-end melting. It is reasonable to use this approximation since the melting temperature of even long sticky ends is 10–20 °C lower than the melting temperature of DNA. Equations (3.46) and (3.47) predict the experimentally determined T_{m} of λ DNA sticky ends with accuracy of 5 °C (Wang & Davidson 1966). Equation (3.46) predicts strong dependence of T_{m} on ionic conditions, in line with the corresponding dependence of DNA melting (Wang & Davidson 1968). Only qualitative data are available for the stability of short sticky ends (Revet & Fourcade 1998, Dahlgren & Lyubchenko 2002, Tagashira et al. 2002), so it remains to be tested how well the equations predict T_{m} of sticky ends produced by restriction endonucleases.

The value of k_{c} weakly depends on the sticky-end length, but it depends on DNA length, since it is proportional to the j-factor. What is less evident is that the cyclization rate decreases by orders of magnitude with decreasing salt concentration (Wang & Davidson 1966, Rybenkov et al. 1993, Vologodskii 2012). At [NaCl] of 0.01 M the cyclization time of 10 kb DNA at room temperature constitutes hundreds of hours, while it is in the range of minutes at [NaCl] of 0.2 M (Rybenkov et al. 1993). It would be interesting to analyze if this enormous slowing of the cyclization rate is in agreement with the kinetics of duplex formation for short oligonucleotides. It seems that there is a semi-quantitative agreement, although a lack of systematic data does not allow a quantitative comparison to be performed.

3.4.3   Extension of single DNA molecules by force

3.4.3.1   Basic experimental setting and theoretical analysis

The possibility to apply a measurable extending force to a single polymer molecule is amazing. The beauty of the experiments based on this technique has made them extremely popular over the last 20 years. The approach brought many interesting results in the field of DNA conformational properties and DNA–protein interaction. Here we outline the major ideas of the approach and some of the most interesting results obtained with these experiments.

The first force–extension experiment on a single DNA molecule was reported by Smith, Finzi and Bustamante (1992). Although the methods of DNA manipulations and force measurement were very accurate in that study, they were later replaced by more efficient ones. In the simplest typical setting, used today, biotin molecules are covalently attached to one end of each strand of DNA, either to the 3'-end or

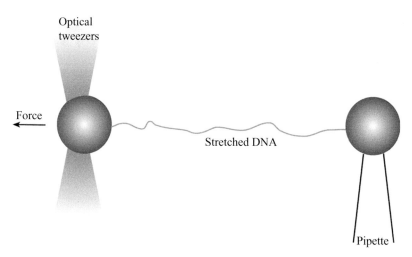

Optical
tweezers

Force

Stretched DNA

Pipette

Figure 3.26 A typical setting of a DNA force–extension experiment. The right-hand bead is suctioned by the pipette; the left-hand one is in the optical trap that attracts it to the focus of the laser beam.

the 5′-end. Then these DNA molecules are captured by the streptavidin-coated beads 1–3 μm in diameter, added in the solution. The streptavidin–biotin interaction is sufficiently strong to sustain large stretching forces applied to the beads. The beads can be optically trapped by the optical tweezers. Sometimes a second bead is suctioned onto the end of a micropipette (Fig. 3.26). In optical tweezers the focused laser beam attracts the bead to its center, and the force acting on the bead is proportional to the bead displacement from the focus. Sometimes a magnetic bead and a magnetic tweezers are used instead of an optical trap. In the latter case the magnetic field exerts a constant force on the bead regardless of its position in the tweezers, but the force can be regulated by the intensity of the magnetic field. The beads are sufficiently large that their positions can be monitored by an optical microscope. Special methods have been developed to monitor the force exerted on the beads by optical and magnetic tweezers (Svoboda & Block 1994, Gosse & Croquette 2002). Usually the DNA molecules are held in a microfluidic channel that simplifies exchange of buffers and addition of various ligands and proteins to the system. Of course, the sample preparation starts with many DNA molecules in the chamber, but at the end those that are not attached to the pipette or optical trap are washed out.

In their pioneering study Smith *et al.* measured relative DNA extension, $\langle x \rangle / L_0$, as function of force over a wide range of $\langle x \rangle / L_0$. Of course, it was very interesting to compare the obtained result with the theoretical predictions. Although researchers had worked for many years on the elastic properties of polymers, they had to deal with complex systems of melts and cross-linked networks. Therefore, only a semi-quantitative comparison between experimental and theoretical data had been possible. For the first time the theory of polymer extension could be compared with the experimental data for a single polymer chain. The theoretical dependence of the extension as a function of the

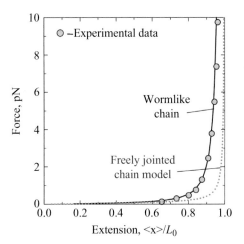

Figure 3.27 Comparison of the experimental and theoretical dependences of DNA extension on the stretching force. The experimental data were taken from Smith *et al.* (1992); the theoretical dependences shown by lines correspond to the freely jointed and WLC models (Bueche 1962, Bustamante *et al.* 1994, Vologodskii 1994).

stretching force was known at the time only for the freely jointed model of a polymer chain:

$$\frac{\langle x \rangle}{L_0} = \coth\left(\frac{Fb}{k_B T}\right) - \frac{k_B T}{Fb}, \tag{3.48}$$

where b is the Kuhn statistical length of the polymer ($b = 2a$). Parameter b was well known, so the researchers were puzzled when they found that for $\langle x \rangle / L_0 > 0.8$ the theoretical predictions were very different from the experimental results (Fig. 3.27). It took two years to understand the reason for this discrepancy. It turned out that the freely jointed chain model is too rough in describing conformational properties of highly extended DNA molecules (Smith *et al.* 1992). When the force–extension dependence was obtained for the WLC model and compared with the experimental data, a remarkable agreement was obtained (see Fig. 3.27). Marko and Siggia obtained a very good approximation for the average extension $\langle x \rangle$ of a DNA molecule of contour length L_0 under force f that is widely used in DNA force–extension studies (Marko & Siggia 1995):

$$Fa/k_B T = \frac{1}{4(1 - \langle x \rangle / L_0)^2} - \frac{1}{4} + \frac{\langle x \rangle}{L_0} \tag{3.49}$$

Equation (3.49) is very accurate if $0.8 \le \langle x \rangle / L_0 \le 0.99$, although for larger extensions one has to account for stretching of DNA contour length to obtain a better approximation of the experimental data (see below) (Odijk 1995).

3.4.3.2   Overstretching dsDNA

One of the most intriguing stories related to the force–extension single-molecule experiments is the story of the overstretching transition in DNA. The transition was discovered simultaneously by two groups, and their papers were published side by side (Cluzel *et al.* 1996, Smith *et al.* 1996). The researchers found that at a large force, around

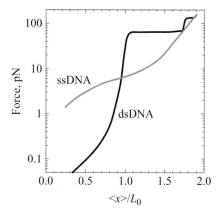

Figure 3.28 The overstretching transition in dsDNA. The transition is exhibited as a horizontal segment of the force–extension curve at 65 pN (black line). An additional transition is observed at 135 pN. The force–extension curve for ssDNA is shown for comparison (gray line). The experimental data were reproduced from Fig. 2 of Bustamante *et al.* (2000), with permission.

65 pN, when the entropic elasticity of the double helix has been exhausted, the molecule undergoes a very cooperative transition that increases its contour length by approximately 70% (Fig. 3.28). It was suggested that at this extending force DNA forms a new double-stranded conformation (Cluzel *et al.* 1996, Smith *et al.* 1996). A few structural models of the new form of DNA were suggested (Cluzel *et al.* 1996, Lebrun & Lavery 1996, Kosikov *et al.* 1999). It took five years before an alternative interpretation of the transition appeared: force-induced DNA melting (Rouzina & Bloomfield 2001). Qualitatively, it was clear that a sufficient extending force has to cause DNA melting. Indeed, the contour length of ssDNA is nearly twice as large as the contour length of B-DNA. Therefore, at sufficiently large force the free energy of two extended single strands has to be lower than the free energy of extended B-DNA. The known energetics of DNA melting and the force–extension properties of ds and ssDNA allowed quite accurate prediction of the force that has to cause DNA melting. The calculations showed that the force-induced melting should take place at the stretching force of 60–80 pN (Rouzina & Bloomfield 2001). In addition, DNA extension at the end of the overstretching transition turned out to be equal to the extension of ssDNA of the same length (see Fig. 3.28).

Over the years, numerous theoretical arguments and experimental observations supporting either model were presented (Williams *et al.* 2002, Cocco *et al.* 2004, Shokri *et al.* 2008), but the issue was not resolved. Although the melting model looked very attractive, it had some problems. The main problem was that DNA complementary strands remained bound to each other under much higher force than the force causing the transition (Cocco *et al.* 2004). This problem, which we call the transition-end problem, did not have any reasonable explanation.

A breakthrough in the studies of the transition was made when the single-molecule force–extension experiments were complemented by the observation of fluorescence of the extended molecules (van Mameren *et al.* 2009). It allowed the researchers to obtain absolutely convincing evidence that they observed DNA melting under a stretching

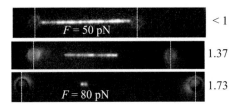

Figure 3.29 Extending λ DNA by force. Specific staining of dsDNA segments with the intercalating dye YOYO shows that the DNA overstretching transition at 65 pN occurs by reducing the single double-stranded region from the molecule ends. The vertical dashed lines highlight the locations of the optically trapped beads. Three fluorescent images correspond to different degrees of the transition. The relative DNA extension is shown on the right. The image was reproduced from van Mameren *et al.* (2009) with permission.

force of 65 pN. In one of the experiments they stained dsDNA segments of the stretched DNA with the intercalating dye YOYO, which does not bind to ssDNA. Thus, in the stretching experiments they were able to observe dsDNA segments but not ss. The obtained fluorescence images of the stretched DNA at different stages of the transition are shown in Fig. 3.29. The images clearly show that there is only one double-stranded region in the molecule over the entire range of the transition. The transition starts and develops from the free single-stranded ends of the molecule (only one strand at each end of the DNA molecule was attached to the beads). This kind of behavior is impossible for a transition between double-stranded forms of DNA, since such transitions have to divide DNA into alternating segments of two forms. The length of these segments is limited to a few hundred base pairs. There is no way that the transition between two double-stranded forms can maintain a single segment of B-DNA over the entire length of 50 kb λ DNA. The only possibility for such extreme behavior is melting DNA from the ends. Force-induced melting from the free single-stranded ends has a large advantage over the melting of internal regions, since in the former case the stretching force extends only one strand rather than two and therefore produces a larger extension and work, correspondingly. In another experiment the researchers were able to directly visualize, using a different kind of staining, free single-stranded coils formed by the melting from the molecule ends (van Mameren *et al.* 2009).

It looked as if this was the end of the story. Still, the transition-end problem was not solved. Soon, however, an absolutely unexpected result was obtained for a DNA sample that did not have free single-stranded ends. In such a system the stretching has to be applied to both DNA strands, and the melting transition has to be shifted to higher forces. According to the theoretical estimates the shift has to be 20–35 pN (Rouzina & Bloomfield 2001, Cocco *et al.* 2004). In the experiment DNA strands were bound to one another at each end of the double helix but DNA was attached to the beads by a single-stranded linker (Paik & Perkins 2011). Therefore, there was no accumulation of the torsional stress during the transition, since fast rotation around single covalent bonds eliminates such stress. The experiment with such a sample showed that the transition was not shifted relative to its position in DNA with free ends (Paik & Perkins 2011). It was the result that one would expect if the transition occurred between two different double-stranded forms. Therefore, a rather improbable picture, suggested seven years earlier by Marko and co-workers, turned out to be correct: two different transitions can occur at 65 pN (Cocco *et al.* 2004). One of them is the force-induced melting; the other is formation of S-DNA, an extended double-stranded conformation. Not only do the transitions occur at nearly the same stretching force, but also they are associated with the same extension of the molecule! It was difficult to imagine that such a scenario could be correct. Successive studies unambiguously showed that factors such as ionic conditions, temperature and DNA GC-content determine which of two transitions take place in a particular experiment (Fu *et al.* 2010, Fu *et al.* 2011, Zhang *et al.* 2012, Zhang *et al.* 2013). In some experiments the researchers were able to observe the transition between S-form and melted DNA molecules (Bosaeus *et al.* 2012). Moreover, finally, the existence of two transitions explained the transition-end problem: under experimental conditions where the great majority of the double helix melts, the most GC-rich segment (the most thermostable) undergoes the B–S transition, keeping DNA strands together. Since the transition between S-form and melted DNA gives very little gain in the extension (Rief *et al.* 1999, Bosaeus *et al.* 2012), it occurs at forces much higher than 65 pN.

3.4.4   Twisting and braiding of dsDNA

In the first DNA force–extension experiments the molecule was attached to the beads (surface/pipette) by one of two strands of the double helix. Soon, however, the researchers learned how to attach both strands to the beads. Then, rotating one bead they were able to introduce torsional stress into the double helix and investigate the force-induced extension of the twisted molecules (Strick *et al.* 1996). The researchers found that torsional stress reduces DNA extension (Fig. 3.30). Previous studies of DNA supercoiling suggested an explanation for this contraction: the linking number difference, ΔLk, introduced by the bead rotation, created Wr in DNA molecules, either as a simple helix formed by the DNA axis or as plectonemic branches. The branches are associated with a larger DNA contraction and therefore formed under lower extending forces. The appearance of the branches was demonstrated both by computer simulation (Vologodskii & Marko 1997) and experimentally, when stretched DNA was stained by a fluorescent dye (van Loenhout *et al.* 2012). In the latter study the researchers managed to measure the lifetime of branches, which turned out to be on the millisecond timescale.

An even more sophisticated object was made and studied by the group of Bensimon and Croquette (Strick *et al.* 1998). The researchers connected two beads by two DNA molecules and then rotated one bead, keeping the other immobilized. Since at each DNA end only one strand was attached to a bead, there was no torsional stress in the individual DNA molecules. However, rotation of the bead created a DNA braid, a helix, formed by two dsDNA molecules. Of course, this helix of the higher order is not so regular, since the molecules are not bound to one another. However, the force–extension behavior of the braids is very similar to the behavior of torsionally constrained dsDNA. The conformations of the stretched braid depend on the number of turns one DNA makes around the other (the linking number), n. At relatively low n and high forces the braids

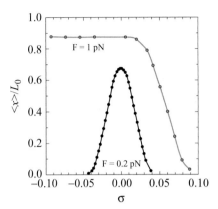

Figure 3.30 Extension of torsionally constrained DNA molecules. At low extending forces and sufficiently high torsional stress the DNA molecule forms numerous plectonemic branches, which causes strong contraction of the chain. At higher forces a larger torsional stress is needed to introduce writhe into stretched DNA. The symmetry between negative and positive supercoiling that is observed at lower forces disappears at high forces due to formation of alternative underwound DNA structures in the negatively supercoiled molecules. The data for the plot were taken from Fig. 3A of Strick *et al.* (1998), with permission.

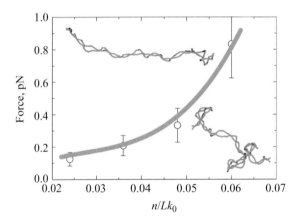

Figure 3.31 Phase diagram for the conformations of DNA braids. The braid density, plotted along the horizontal axis, is a ratio of the linking number between the two stretched DNA molecules, n, and the equilibrium linking number of the complementary strands in one molecule, Lk_0. The bold gray line divides the regions with and without plectonemic branches in the braids. The figure corresponds to near-physiological ionic conditions. The plot was reproduced from Fig. 8 of Charvin *et al.* (2005), with permission.

form extended helix-like conformations, while at higher n and lower stretching forces they form plectonemic branches of higher order (Fig. 3.31).

Supercoiled and braided DNA molecules not only contributed to our understanding of the phenomenon of DNA supercoiling, but they also turned out to be very useful in the studies of DNA topoisomerases (see below and Section 6.9).

3.4.5 Extension of ssDNA

ssDNA is many times more flexible than the double-stranded molecule. As a result, much larger forces are needed to extend the ssDNA, as it follows from Eq. (3.49). A typical force–extension curve of ssDNA is shown in Fig. 3.28. The extension, $\langle x \rangle / L_0$, does not follow Eq. (3.49) so well, since conformational properties of ssDNA are not described well by the WLC model. Since the Kuhn length of ssDNA is small, it forms a compact coil in solution. Therefore, the molecular conformations are strongly affected by the intersegment electrostatic interaction. Also, the molecular backbone is stretchable, probably due to the change of the sugar conformation from C3'-*endo* to C2'-*endo* (Smith *et al.* 1996). In addition, bent conformations of ssDNA are not as smooth as conformations of the double-stranded molecules. It was found that accounting for the contour length extension gives an accurate description of the force–extension curve for ssDNA by the freely jointed model (Smith *et al.* 1996). The resulting force–extension equation is similar to Eq. (3.48):

$$\frac{\langle x \rangle}{L_0} = \left[\coth \left(\frac{Fb}{k_\mathrm{B}T} \right) - \frac{k_\mathrm{B}T}{Fb} \right] \left(1 + \frac{F}{S} \right) \tag{3.50}$$

where S is the stretch modulus of ssDNA. Under physiological ionic conditions Eq. (3.50), with b of 1.5 nm, S of 800 pN and undisturbed contour length of 0.56 nm per nucleotide, fits the experimental data well (Smith *et al.* 1996). However, in the absence of the stretching force the value of b for ssDNA was found to be notably larger, between 3 and 6 nm, even in the absence of base stacking (Mills *et al.* 1999, Murphy *et al.* 2004, Chi *et al.* 2013).

3.4.6 DNA–protein interaction

Many proteins interacting with DNA are able to affect its force-induced extension. In such cases monitoring DNA extension could serve as a sensitive detection of DNA transformation/translocation by the enzymatic systems. The method allows the monitoring of the activity of individual proteins in real time. It became very popular over the last two decades and has been applied to studies of RNA polymerases (Yin *et al.* 1995, Forde *et al.* 2002, Abbondanzieri *et al.* 2005), DNA topoisomerases (Crisona *et al.* 2000, Strick *et al.* 2000, Dekker *et al.* 2002, Stone *et al.* 2003, Charvin *et al.* 2005, Gore *et al.* 2006), DNA helicases (Johnson *et al.* 2007, Lionnet *et al.* 2007) and translocases (Smith *et al.* 2001, Saleh *et al.* 2004, Pease *et al.* 2005). Two examples of application of the method to DNA topoisomerases and DNA helicases are considered below.

Helicases are enzymes that unwind dsDNA (Lohman & Bjornson 1996). They are capable of doing this by consuming the energy of ATP hydrolysis that is coupled to their translocation along DNA. They are required in many biological processes, such as replication, recombination and repair. They act by binding to a single DNA strand and preventing local base pairing with the complementary strand. Single-molecule force–extension experiments turned to be a very informative tool in the studies of helicase action. An example of such experiment with gp41 helicase from bacteriophage T4 is shown in Fig. 3.32 (Lionnet *et al.* 2007). The experiment allows determination of the rate

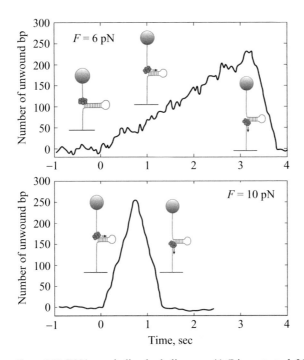

Figure 3.32 DNA unwinding by helicase gp41 (Lionnet *et al.* 2007). The DNA substrate is a 231 bp hairpin with extended single-stranded ends tethered to a magnetic bead and the glass surface. The magnetic field exerted a constant force of 6 pN or 10 pN on the bead. The experiment started by binding the enzyme with the single-stranded segment of DNA located above the hairpin in the diagram. Translocation of the helicase along the single-stranded segment does not change the total DNA extension (the initial flat part of each plot). Then the enzyme starts unwinding the helix, which increases the extension (converted to the number of unwound base pairs on the plots). The large extending force used in the experiment prevents DNA reassociation behind the helicase. Comparison of the two panels shows that the rate of unwinding increases with increasing extending force. When the hairpin unwinding is complete, the protein continues translocation along the ssDNA, allowing the reassociation of the two strands behind the helicase starting from the hairpin loop. The slope of the corresponding segment of the plot does not depend on the extending force, showing that protein translocation along ssDNA is force independent. Moving the helicase along the lower ssDNA end corresponds to the last horizontal segment of the plots. The plots were reproduced from Fig. 1 of Lionnet *et al.* (2007), with permission.

of DNA unwinding by the protein and the dependence of this rate on the extending force. It is worth to note that this experimental setting with a sufficiently large extending force causes helix unwinding without any helicase, and was used to estimate thermodynamic stability of base pairs (Huguet *et al.* 2010).

If extended DNA is supercoiled, it represents a good substrate for type II DNA topoisomerases, which catalyze strand-passing of one dsDNA segment through another (see Section 6.9). The relaxation of such a substrate causes an increase of the molecular extension that can be used to monitor the progress of the reaction. The sensitivity of the method is sufficient to detect individual acts of strand-passing catalyzed by the enzymes. Crisona *et al.* used this approach to compare relaxation of negatively ($-$)

and positively (+) supercoiled DNA molecules by topoisomerase topo IV from *E. coli* (Crisona *et al.* 2000). First, they found in the bulk experiments that the relaxation of (+) supercoiled DNA proceeds approximately 20 times faster than the relaxation of (−) supercoiled DNA. The single-molecule experiments provided a detailed picture of the process. The researchers used very low enzyme concentration to ensure that only one active enzyme is bound with the DNA molecule. Under this condition the binding topo IV with a DNA segment (G segment) was a rate-limiting step in the relaxation of (+) supercoiled DNA. This step was followed by fast processive relaxation by multiple capture and strand-passing of many T segments (transferred segments) until all (+) supercoils were removed. In the relaxation of (−) supercoiled DNA the picture is quite different. Usually only a single strand-passing event occurs before the enzyme dissociation from the G segment. Each reaction cycle takes about 10 s due to slow capturing of a T segment. These data allowed the researchers to suggest a model of the process based on a different distribution of angles between juxtaposed segments in positively and negatively supercoiled DNA molecules (Crisona *et al.* 2000).

3.5   DNA nanotechnology

DNA nanotechnology is a way to construct molecular-sized objects by using DNA molecules as a building material. In recent years this field has expended at a remarkable pace. Two major properties of DNA are critically important for nanotechnology. The first property is the ability of DNA to form a very rigid double helix from single-stranded segments with complementary sequences of the nucleotides. The second property is the very high flexibility of ssDNA, allowing formation of various junctions at relatively low energy cost. Seeman was the first to realize the potential of DNA molecules as a building material (Seeman 1982). Of course, the current development of the field only became possible due to the remarkable accessibility of synthetic oligonucleotides with specific sequences.

The main structural element of DNA nanotechnology is the four-way junction. Although the junctions can be formed by any pair of duplexes with identical sequences, such junctions are unstable, since their free energy is much larger than the free energy of two separated duplexes (Lu *et al.* 1992). The branch point can migrate in such junctions, and this migration results in junction decay. The junction can be stable, however, if it is formed by oligonucleotides with specially designed sequences that cannot form perfect duplexes but can form four-way junctions with perfect base pairing in all branches (Fig. 3.33). The termini of the branches can have hanging single-stranded ends that allow them to be joined with complementary single-stranded ends of other structural blocks. Such sticky ends represent the second major structural element of DNA nanotechnology. By joining sticky ends the junctions can form 2D arrays. It was found, however, that the four-way junctions were not good structural elements for array formation, probably due to their conformational flexibility. As a result of this flexibility the growing arrays contain too many defects, which makes it difficult to obtain arrays of a notable size (Seeman 2010).

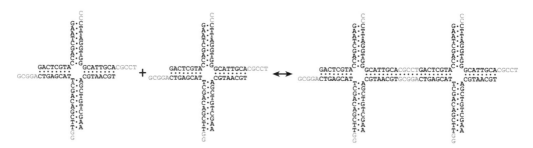

Figure 3.33 Association of two four-way junctions by joining their sticky ends. For convenience the sticky ends are shown in gray. The sequences of the junctions are designed in such a way that they cannot form duplexes with sufficient homology, and the branch point in the junction cannot migrate. Theoretically, this four-way junction can form an infinite 2D array, since in both directions the junction would be separated by a nearly integer number of helix turns (21 bp).

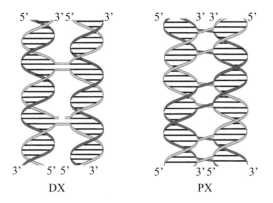

DX   PX

Figure 3.34 DNA structural motifs consisting of two parallel double helices with switching strands. The structures were first obtained by Seeman and co-workers, who named them DX and PX (Fu & Seeman 1993, Shen *et al.* 2004).

Of course, junctions assembled from synthetic oligonucleotides can have various numbers of branches. In particular, three-way junctions were used in the assembly of a DNA cube, probably the first object created by DNA nanotechnology (Chen & Seeman 1991).

2D arrays were obtained from more rigid elements than the simple four-way junctions. These elements are very interesting in themselves and deserve special attention. They are formed by two parallel double helices in which strands are often switching from one helix to another. A diagram of two such structural blocks is shown in Fig. 3.34. The sequences of DNA strands that form the structures diagramed in the figure were designed so that the strands could not form isolated duplexes with perfect base pairing. Although the PX structure can be formed by two identical duplexes, the equilibrium is strongly shifted to the duplex formation, so the PX structure cannot be observed in a mixture of identical duplexes. Estimation of the free energy corresponding to formation of the PX structure with five crossovers from two identical duplexes gives a value in the

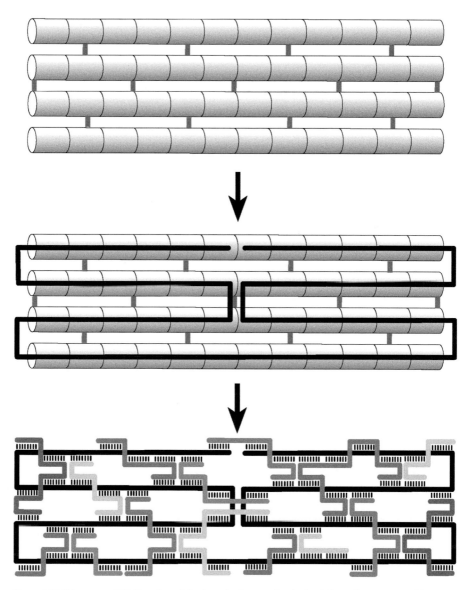

Figure 3.35 Diagram of DNA origami design. At the top the desired shape is represented by parallel double helices joined by crossovers. The scaffold ssDNA provides one of two complementary strands in each double-stranded segment, so the projected shape of the scaffold sets the desired shape of the origami (middle panel). Synthetic oligonucleotides, shown by different-colored lines at the bottom of the figure, serve as the complementary strands of the double-stranded segments. These oligonucleotides also form cross-links between the double-stranded segments by switching from one segment to another. The sequence of the oligonucleotides is uniquely defined by their location in the origami and by the sequence of M13 DNA. Picture adopted from Rothemund (2006). A black-and-white version of this figure will appear in some formats. For the color version, please refer to the plate section.

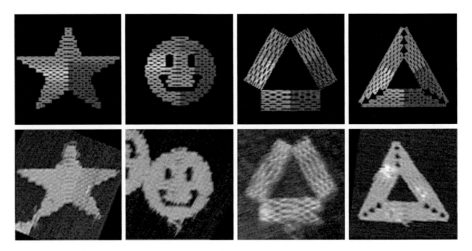

Figure 3.36 AFM images of different DNA origami shapes. The first row shows the projected shapes. The AFM images of the obtained origami are in the second row. This figure was reproduced from Fig. 2 of Rothemund (2006), with permission.

range of 60–75 kcal/mol at high Mg^{2+} concentration, and 35–50 kcal/mol at low Mg^{2+} (Spink *et al.* 2009). There are data, however, that suggest that the PX structure can be formed by two DNA segments with head-to-head homology in negatively supercoiled DNA, due to the supercoiling relaxation caused by the structure formation (Wang *et al.* 2010).

The fact that the structures with two parallel helices and frequent switches of strands between them can be formed is a clear manifestation of the remarkable conformational flexibility of ssDNA. Such strand switches between parallel helices are widely used in DNA nanotechnology. In particular, they are used in DNA origami, an elegant way to assemble relatively large objects from ssDNA molecules. In this approach one long DNA molecule forms a scaffold, which is hybridized with many shorter oligonucleotides to set a particular shape of a two- or three-dimensional (3D) construct. The key idea there was to use natural ssDNA from phage M13 (Rothemund 2006), although the approach had been used earlier (Yan *et al.* 2003, Shih *et al.* 2004). It is important to note in this context that chemical synthesis of oligonucleotides has a limited accuracy, so synthetic oligos larger than 100 nucleotides in length are not usually used in experiments. The length of M13 DNA is 7249 nucleotides, so it represents a different scale. Rothemund realized that the predefined sequence of a scaffold DNA is not a limitation, and a great variety of shapes can be created using a single scaffold DNA (Rothemund 2006). The main idea of this approach is illustrated in Fig. 3.35. Of course, during the origami design one has to take into account the helical periodicity of the double helix, so the cross-links can only be located at certain positions of the double-stranded segments. Atomic force microscopy (AFM) images of different shapes obtained in this way are shown in Fig. 3.36.

The principle outlined above has also been used to construct origami-based 3D objects (Andersen *et al.* 2009, Han *et al.* 2011). The key feature of DNA origami that has made it very popular in DNA nanotechnology is the large size of the origami-based unit. This unit can have numerous shapes and carry a huge number of different structural elements. It is almost impossible to make nonperiodic structures of this size from synthetic oligonucleotides only.

Over the years researchers have managed to make simple molecular devices from DNA oligos that perform a desired movement (Mao *et al.* 1999, Yurke *et al.* 2000) and controllable DNA walkers moving along DNA-based tracks were suggested (Sherman & Seeman 2004, Shin & Pierce 2004, Omabegho *et al.* 2009). The work in these and other new emerging directions is in progress. Although the impressive achievements of DNA nanotechnology have not brought practical applications so far, this may happen at any time.

References

Abbondanzieri, F. A., Greenleaf, W. J., Shaevitz, J. W., Landick, R. & Block, S. M. (2005). Direct observation of base-pair stepping by RNA polymerase. *Nature* 438, 460–5.

Allison, S. A., Herr, J. C. & Schurr, J. M. (1981). Structure of viral phi 29 DNA condensed by simple triamines: a light-scattering and electron-microscopy study. *Biopolymers* 20, 469–88.

Andersen, E. S., Dong, M., Nielsen, M. M., Jahn, K., Subramani, R., Mamdouh, W., Golas, M. M., Sander, B., Stark, H., Oliveira, C. L. P., Pedersen, J. S., Birkedal, V., Besenbacher, F., Gothelf, K. V. & Kjems, J. (2009). Self-assembly of a nanoscale DNA box with a controllable lid. *Nature* 459, 73–6.

Anderson, C. F. & Record, M. T. J. (1990). Ion distributions around DNA and other cylindrical polyions: theoretical description and physical implications. *Annu. Rev. Biophys. Biophys. Chem.* 19, 423–65.

Anderson, P. & Bauer, W. (1978). Supercoiling in closed circular DNA: dependence upon ion type and concentration. *Biochemistry* 17, 594–601.

Arscott, P. G., Li, A. Z. & Bloomfield, V. A. (1990). Condensation of DNA by trivalent cations. 1. Effects of DNA length and topology on the size and shape of condensed particles. *Biopolymers* 30, 619–30.

Balasubramanian, S., Xu, F. & Olson, W. K. (2009). DNA sequence-directed organization of chromatin: structure-based computational analysis of nucleosome-binding sequences. *Biophys. J.* 96, 2245–60.

Barkley, M. D. & Zimm, B. H. (1979). Theory of twisting and bending of chain macromolecules; analysis of the fluorescence depolarization of DNA. *J. Chem. Phys.* 70, 2991–3007.

Baumann, C. G., Bloomfield, V. A., Smith, S. B., Bustamante, C., Wang, M. D. & Block, S. M. (2000). Stretching of single collapsed DNA molecules. *Biophys. J.* 78, 1965–78.

Baumann, C. G., Smith, S. B., Bloomfield, V. A. & Bustamante, C. (1997). Ionic effects on the elasticity of single DNA molecules. *Proc. Natl. Acad. Sci. U. S. A.* 94, 6185–90.

Bednar, J., Furrer, P., Katritch, V., Stasiak, A. Z., Dubochet, J. & Stasiak, A. (1995). Determination of DNA persistence length by cryo-electron microscopy. Separation of the static and dynamic contributions to the apparent persistence length of DNA. *J. Mol. Biol.* 254, 579–94.

Benham, C. J. (1978). The statistics of superhelicity. *J. Mol. Biol.* 123, 361–70.

Bensimon, D., Dohmi, D. & Mezard, M. (1998). Stretching a heteropolymer. *Europhys. Lett.* 42, 97–102.

Binder, K. & Heermann, D. W. (1997). *Monte Carlo Simulations in Statistical Physics*. Berlin: Springer.

Bloomfield, V. A. (1996). DNA condensation. *Curr. Opin. Struct. Biol.* 6, 334–41.

Bosaeus, N., El-Sagheer, A. H., Brown, T., Smith, S. B., Akerman, B., Bustamante, C. & Norden, B. (2012). Tension induces a base-paired overstretched DNA conformation. *Proc. Natl. Acad. Sci. U. S. A.* 109, 15179–84.

Bresler, S. E. & Frenkel, Y. I. (1939). The character of thermal motion of long organic chains with reference to the elastic properties of rubber. *Zh. Eksp. Teor. Fiz.* 9, 1094–106.

Brian, A. A., Frisch, H. L. & Lerman, L. S. (1981). Thermodynamics and equilibrium sedimentation analysis of the close approach of DNA molecules and a molecular ordering transition. *Biopolymers* 20, 1305–28.

Bryant, Z., Stone, M. D., Gore, J., Smith, S. B., Cozzarelli, N. R. & Bustamante, C. (2003). Structural transitions and elasticity from torque measurements on DNA. *Nature* 424, 338–41.

Bueche, F. (1962). *Physical Properties of Polymers*. New York: Interscience.

Bustamante, C., Marko, J. F., Siggia, E. D. & Smith, S. (1994). Entropic elasticity of lambda-phage DNA. *Science* 265, 1599–600.

Cantor, C. R. & Schimmel, P. R. (1980). *Biophysical Chemistry*. New York: Freeman.

Chan, S. H., Stoddard, B. L. & Xu, S. Y. (2011). Natural and engineered nicking endonucleases – from cleavage mechanism to engineering of strand-specificity. *Nucleic Acids Res.* 39, 1–18.

Charvin, G., Strick, T. R., Bensimon, D. & Croquette, V. (2005). Topoisomerase IV bends and overtwists DNA upon binding. *Biophys. J.* 89, 384–92.

Chattoraj, D. K., Gosule, L. C. & Schellman, A. (1978). DNA condensation with polyamines. II. Electron microscopic studies. *J. Mol. Biol.* 121, 327–37.

Chen, J. H. & Seeman, N. C. (1991). Synthesis from DNA of a molecule with the connectivity of a cube. *Nature* 350, 631–3.

Chi, Q., Wang, G. & Jiang, J. (2013). The persistence length and length per base of single-stranded DNA obtained from fluorescence correlation spectroscopy measurements using mean field theory. *Physica A* 392, 1072–9.

Cloutier, T. E. & Widom, J. (2004). Spontaneous sharp bending of double-stranded DNA. *Mol. Cell* 14, 355–62.

Cluzel, P., Lebrun, A., Heller, C., Lavery, R., Viovy, J. L., Chatenay, D. & Caron, F. (1996). DNA: an extensible molecule. *Science* 271, 792–4.

Cocco, S., Yan, J., Leger, J. F., Chatenay, D. & Marko, J. F. (2004). Overstretching and force-driven strand separation of double-helix DNA. *Phys. Rev. E* 70, 011910.

Crick, F. H. & Klug, A. (1975). Kinky helix. *Nature* 255, 530–3.

Crisona, N. J., Strick, T. R., Bensimon, D., Croquette, V. & Cozzarelli, N. R. (2000). Preferential relaxation of positively supercoiled DNA by *E. coli* topoisomerase IV in single-molecule and ensemble measurements. *Genes Dev.* 14, 2881–92.

Crothers, D. M., Haran, T. E. & Nadeau, J. G. (1990). Intrinsically bent DNA. *J. Biol. Chem.* 265, 7093–6.

Dahlgren, P. R. & Lyubchenko, Y. L. (2002). Atomic force microscopy study of the effects of Mg2+ and other divalent cations on the end-to-end DNA interactions. *Biochemistry* 41, 11372–8.

Dekker, N. H., Rybenkov, V. V., Duguet, M., Crisona, N. J., Cozzarelli, N. R., Bensimon, D. & Croquette, V. (2002). The mechanism of type IA topoisomerases. *Proc. Natl. Acad. Sci. U. S. A.* 99, 12126–31.

Depew, R. E. & Wang, J. C. (1975). Conformational fluctuations of DNA helix. *Proc. Natl. Acad. Sci. U. S. A.* 72, 4275–9.

Diekmann, S. (1987). DNA curvature. In *Nucleic Acids and Molecular Biology*, eds. F. Eckstein & D. Lilley, 138–56. Berlin: Springer.

Doty, P. & Bunce, B. H. (1952). The molecular weight and shape of desoxypentose nucleic acid. *J. Am. Chem. Soc.* 74, 5029–34.

Du, Q., Kotlyar, A. & Vologodskii, A. (2008). Kinking the double helix by bending deformation. *Nucleic Acids Res.* 36, 1120–8.

Du, Q., Livshits, A., Kwiatek, A., Jayaram, M. & Vologodskii, A. (2007). Protein-induced local DNA bends regulate global topology of recombination products. *J. Mol. Biol.* 368, 170–82.

Du, Q., Smith, C., Shiffeldrim, N., Vologodskaia, M. & Vologodskii, A. (2005). Cyclization of short DNA fragments and bending fluctuations of the double helix. *Proc. Natl. Acad. Sci. U. S. A.* 102, 5397–402.

Duguet, M. (1993). The helical repeat of DNA at high temperature. *Nucleic Acids Res.* 21, 463–8.

Flory, P. J. (1953). *Principles of Polymer Chemistry*. Ithaca, NY: Cornell University Press.

Forde, N. R., Izhaky, D., Woodcock, G. R., Wuite, G. J. & Bustamante, C. (2002). Using mechanical force to probe the mechanism of pausing and arrest during continuous elongation by *Escherichia coli* RNA polymerase. *Proc. Natl. Acad. Sci. U. S. A.* 99, 11682–7.

Frank-Kamenetskii, M. D., Anshelevich, V. V. & Lukashin, A. V. (1987). Polyelectrolyte model of DNA. *Sov. Phys. Usp.* 30, 317–30.

Frank-Kamenetskii, M. D., Lukashin, A. V., Anshelevich, V. V. & Vologodskii, A. V. (1985). Torsional and bending rigidity of the double helix from data on small DNA rings. *J. Biomol. Struct. Dyn.* 2, 1005–12.

Frank-Kamenetskii, M. D., Lukashin, A. V. & Vologodskii, M. D. (1975). Statistical mechanics and topology of polymer chains. *Nature* 258, 398–402.

Fu, H., Chen, H., Marko, J. F. & Yan, J. (2010). Two distinct overstretched DNA states. *Nucleic Acids Res.* 38, 5594–600.

Fu, H., Chen, H., Zhang, X., Qu, Y., Marko, J. F. & Yan, J. (2011). Transition dynamics and selection of the distinct S-DNA and strand unpeeling modes of double helix overstretching. *Nucleic Acids Res.* 39, 3473–81.

Fu, T. J. & Seeman, N. C. (1993). DNA double-crossover molecules. *Biochemistry* 32, 3211–20.

Fujimoto, B. S. & Schurr, J. M. (1990). Dependence of the torsional rigidity of DNA on base composition. *Nature* 344, 175–7.

Garcia-Ramírez, M. & Subirana, J. A. (1994). Condensation of DNA by basic proteins does not depend on protein composition. *Biopolymers* 34, 285–92.

Gavryushov, S. (2009). Mediating role of multivalent cations in DNA electrostatics: an epsilon-modified Poisson–Boltzmann study of B-DNA–B-DNA interactions in mixture of NaCl and MgCl2 solutions. *J. Phys. Chem. B* 113, 2160–9.

Geggier, S., Kotlyar, A. & Vologodskii, A. (2011). Temperature dependence of DNA persistence length. *Nucleic Acids Res.* 39, 1419–26.

Geggier, S. & Vologodskii, A. (2010). Sequence dependence of DNA bending rigidity. *Proc. Natl. Acad. Sci. U. S. A.* 107, 15421–6.

Giambasu, G. M., Luchko, T., Herschlag, D., York, D. M. & Case, D. A. (2014). Ion counting from explicit-solvent simulations and 3D-RISM. *Biophys. J.* 106, 883–94.

Goldstein, R. F. & Benight, A. S. (1992). How many numbers are required to specify sequence-dependent properties of polynucleotides? *Biopolymers* 32, 1679–93.

Gore, J., Bryant, Z., Stone, M. D., Nollmann, M., Cozzarelli, N. R. & Bustamante, C. (2006). Mechanochemical analysis of DNA gyrase using rotor bead tracking. *Nature* 439, 100–4.

Gosse, C. & Croquette, V. (2002). Magnetic tweezers: micromanipulation and force measurement at the molecular level. *Biophys. J.* 82, 3314–29.

Gosule, L. C. & Schellman, J. A. (1976). Compact form of DNA induced by spermidine. *Nature* 259, 333–5.

Goulet, I., Zivanovic, Y. & Prunell, A. (1987). Helical repeat of DNA in solution. The V curve method. *Nucleic Acids Res.* 15, 2803–21.

Gray, D. M. & Tinoco, I. (1970). A new approach to the study of sequence-dependent properties of polynucleotides. *Biopolymers* 9, 223–44.

Grosberg, A. Y., Erukhimovitch, I. Y. & Shakhnovitch, E. I. (1982). On the theory of psi-condensation. *Biopolymers* 21, 2413–32.

Grosberg, A. Y. & Zhestkov, A. V. (1985). On the toroidal condensed state of closed circular DNA. *J. Biomol. Struct. Dyn.* 3, 515–20.

Gueron, M. & Weisbuch, G. (1980). Poly-electrolyte theory. 1. Counterion accumulation, site-binding, and their insensitivity to poly-electrolyte shape in solutions containing finite salt concentrations. *Biopolymers* 19, 353–82.

Guth, E. & Mark, H. (1934). Zur innermolekularen, Statistik, insbesondere bei Kettenmolekiilen I. *Mon. Chem. verwandte Teile anderer Wiss.* 63, 93–121.

Hagerman, P. J. (1981). Investigation of the flexibility of DNA using transient electric birefringence. *Biopolymers* 20, 1503–35.

 (1988). Flexibility of DNA. *Annu. Rev. Biophys. Biophys. Chem.* 17, 265–86.

 (1990). Sequence-directed curvature of DNA. *Annu. Rev. Biochem.* 59, 755–81.

Hagerman, P. J. & Zimm, B. H. (1981). Monte Carlo approach to the analysis of the rotational diffusion of wormlike chains. *Biopolymers* 20, 1481–502.

Han, D., Pal, S., Nangreave, J., Deng, Z., Liu, Y. & Yan, H. (2011). DNA origami with complex curvatures in three-dimensional space. *Science* 332, 342–6.

Horowitz, D. S. & Wang, J. C. (1984). Torsional rigidity of DNA and length dependence of the free energy of DNA supercoiling. *J. Mol. Biol.* 173, 75–91.

Hsiang, M. W. & Cole, R. D. (1977). Structure of histone H1–DNA complex: effect of histone H1 on DNA condensation. *Proc. Natl. Acad. Sci. U. S. A.* 74, 4852–6.

Hud, N. V. & Downing, K. H. (2001). Cryoelectron microscopy of lambda phage DNA condensates in vitreous ice: the fine structure of DNA toroids. *Proc. Natl. Acad. Sci. USA* 98, 14925–30.

Hud, N. V. & Vilfan, I. D. (2005). Toroidal DNA condensates: unraveling the fine structure and the role of nucleation in determining size. *Annu. Rev. Biophys. Biomol. Struct.* 34, 295–318.

Huguet, J. M., Bizarro, C. V., Forns, N., Smith, S. B., Bustamante, C. & Ritort, F. (2010). Single-molecule derivation of salt dependent base-pair free energies in DNA. *Proc. Natl. Acad. Sci. U. S. A.* 107, 15431–6.

Jacobson, H. & Stockmayer, W. H. (1950). Intramolecular reaction in polycondensation. I. Theory of linear systems. *J. Chem. Phys.* 18, 1600–6.

Johnson, D. S., Bai, L., Smith, B. Y., Patel, S. S. & Wang, M. D. (2007). Single-molecule studies reveal dynamics of DNA unwinding by the ring-shaped T7 helicase. *Cell* 129, 1299–309.

Katritch, V. & Vologodskii, A. (1997). The effect of intrinsic curvature on conformational properties of circular DNA. *Biophys. J.* 72, 1070–9.

Klenin, K. V., Vologodskii, A. V., Anshelevich, V. V., Dykhne, A. M. & Frank-Kamenetskii, M. D. (1988). Effect of excluded volume on topological properties of circular DNA. *J. Biomol. Struct. Dyn.* 5, 1173–85.

(1991). Computer simulation of DNA supercoiling. *J. Mol. Biol.* 217, 413–19.

Klenin, K. V., Vologodskii, A. V., Anshelevich, V. V., Klisko, V. Y., Dykhne, A. M. & Frank-Kamenetskii, M. D. (1989). Variance of writhe for wormlike DNA rings with excluded volume. *J. Biomol. Struct. Dyn.* 6, 707–14.

Korolev, N., Lyubartsev, A. P. & Nordenskiold, L. (1998). Application of polyelectrolyte theories for analysis of DNA melting in the presence of Na+ and Mg2+ ions. *Biophys. J.* 75, 3041–56.

Korolev, N., Lyubartsev, A. P., Rupprecht, A. & Nordenskiold, L. (1999). Competitive binding of Mg^{2+}, Ca^{2+}, Na^+, and K_+ ions to DNA in oriented DNA fibers: experimental and Monte Carlo simulation results. *Biophys. J.* 77, 2736–49.

Kosikov, K. M., Gorin, A. A., Zhurkin, V. B. & Olson, W. K. (1999). DNA stretching and compression: large-scale simulations of double helical structures. *J. Mol. Biol.* 289, 1301–26.

Kovacic, R. T. & van Holde, K. E. (1977). Sedimentation of homogeneous double-stranded DNA molecules. *Biochemistry* 1977, 1490–8.

Kratky, O. & Porod, G. (1949). Rontgenuntersuchung geloster fadenmolekule. *J. Royal Neth. Chem. Sci.* 68, 1106–22.

Kuhn, W. (1934). Concerning the shape of thread shapes molecules in solution. *Kolloid-Z.* 68, 2–15.

Laemmli, U. K. (1975). Characterization of DNA condensates induced by poly(ethylene oxide) and polylysine. *Proc. Natl. Acad. Sci. U. S. A.* 72, 4288–92.

Landau, L. & Lifshitz, E. (1951). *Statistical Physics.* Oxford: Pergamon.

Landau, L. D. & Lifshitz, E. M. 1986. Fluctuations in the curvature of long molecules. In *Theory of Elasticity*, 396–400. Oxford, Elsevier.

Lang, D. (1973). Regular superstructures of purified DNA in ethanolic solutions. *J. Mol. Biol.* 78, 247–54.

Le Bret, M. (1980). Monte Carlo computation of supercoiling energy, the sedimentation constant, and the radius of gyration of unknotted and knotted circular DNA. *Biopolymers* 19, 619–37.

Le Bret, M. & Zimm, B. H. (1984). Distribution of counterions around a cylindrical polyelectrolyte and Manning's condensation theory. *Biopolymers* 1984, 287–312.

Lebrun, A. & Lavery, R. (1996). Modelling extreme stretching of DNA. *Nucleic Acids Res.* 24, 2260 7.

Legerski, R. J. & Robberson, D. L. (1985). Analysis and optimization of recombinant DNA joining reactions. *J. Mol. Biol.* 181, 297–312.

Lerman, L. S. (1971). A transition to a compact form of DNA in polymer solutions. *Proc. Natl. Acad. Sci. U. S. A.* 68, 1886–90.

Levene, S. D. & Crothers, D. M. (1986). Ring closure probabilities for DNA fragments by Monte Carlo simulation *J. Mol. Biol.* 189, 61–72.

Li, W. F., Nordenskiold, L. & Mu, Y. G. (2011). Sequence-specific Mg2+–DNA interactions: a molecular dynamics simulation study. *J. Phys. Chem. B* 115, 14713–20.

Licinio, P. & Guerra, J. C. (2007). Irreducible representation for nucleotide sequence physical properties and self-consistency of nearest-neighbor dimer sets. *Biophys. J.* 92, 2000–6.

Lionnet, T., Spiering, M. M., Benkovic, S. J., Bensimon, D. & Croquette, V. (2007). Real-time observation of bacteriophage T4 gp41 helicase reveals an unwinding mechanism. *Proc. Natl. Acad. Sci. U. S. A.* 104, 19790–5.

Lipfert, J., Kerssemakers, J. W., Jager, T. & Dekker, N. H. (2010). Magnetic torque tweezers: measuring torsional stiffness in DNA and RecA–DNA filaments. *Nat. Methods* 7, 977–80.

Lohman, T. M. & Bjornson, K. P. (1996). Mechanisms of helicase-catalyzed DNA unwinding. *Annu. Rev. Biochem.* 65, 169–214.

Lu, M., Guo, Q., Marky, L. A., Seeman, N. C. & Kallenbach, N. R. (1992). Thermodynamics of DNA branching. *J. Mol. Biol.* 223, 781–9.

Manning, G. S. (1969). Limiting laws and counterion condensation in polyelectrolyte solutions. I. Colligative properties. *J. Chem. Phys.* 51, 924–33.

Mao, C. D., Sun, W. Q., Shen, Z. Y. & Seeman, N. C. (1999). A nanomechanical device based on the B–Z transition of DNA. *Nature* 397, 144–6.

Marko, J. F. & Siggia, E. D. (1995). Stretching DNA. *Macromolecules* 28, 8759–70.

McCammon, J. A. & Harvey, S. C. (1987). *Dynamics of Proteins and Nucleic Acids*. Cambridge: Cambridge University Press.

Metropolis, N., Rosenbluth, A. W., Rosenbluth, M. N., Teller, A. H. & Teller, E. (1953). Equation of state calculations by fast computing machines. *J. Chem. Phys.* 21, 1087–92.

Millar, D. P., Robbins, R. J. & Zewail, A. H. (1980). Direct observation of the torsional dynamics of DNA and RNA by picosecond spectroscopy. *Proc. Natl. Acad. Sci. U. S. A.* 77, 5593–7.

Mills, J. B., Vacano, E. & Hagerman, P. J. (1999). Flexibility of single-stranded DNA: use of gapped duplex helices to determine the persistence lengths of poly(dT) and poly(dA). *J. Mol. Biol.* 285, 245–57.

Moroz, J. D. & Nelson, P. (1998). Entropic elasticity of twist-storing polymers. *Macromolecules* 31, 6333–47.

Murphy, M. C., Rasnik, I., Cheng, W., Lohman, T. M. & Ha, T. (2004). Probing single-stranded DNA conformational flexibility using fluorescence spectroscopy. *Biophys. J.* 86, 2530–7.

Nelson, P. (1998). Sequence-disorder effects on DNA entropic elasticity. *Phys. Rev. Lett.* 80, 5810–12.

Odijk, T. (1977). Polyelectrolytes near the rod limit. *J. Polym. Sci. Polym. Phys. Ed.* 15, 477–83. (1995). Stiff chains and filaments under tension. *Macromolecules* 28, 7016–18.

Olson, W. K., Gorin, A. A., Lu, X. J., Hock, L. M. & Zhurkin, V. B. (1998). DNA sequence-dependent deformability deduced from protein–DNA crystal complexes. *Proc. Natl. Acad. Sci. U. S. A.* 95, 11163–8.

Omabegho, T., Sha, R. & Seeman, N. C. (2009). A bipedal DNA Brownian motor with coordinated legs. *Science* 324, 67–71.

Onsager, L. (1949). The effects of shape on the interaction of colloidal particles. *Ann. N. Y. Acad. Sci.* 51, 627–59.

Owczarzy, R., Moreira, B. G., You, Y., Behlke, M. A. & Walder, J. A. (2008). Predicting stability of DNA duplexes in solutions containing magnesium and monovalent cations. *Biochemistry* 47, 5336–53.

Owczarzy, R., You, Y., Moreira, B. G., Manthey, J. A., Huang, L., Behlke, M. A. & Walder, J. A. (2004). Effects of sodium ions on DNA duplex oligomers: improved predictions of melting temperatures. *Biochemistry* 43, 3537–54.

Paik, D. H. & Perkins, T. T. (2011). Overstretching DNA at 65 pN does not require peeling from free ends or nicks. *J. Am. Chem. Soc.* 133, 3219–21.

Pease, P. J., Levy, O., Cost, G. J., Gore, J., Ptacin, J. L., Sherratt, D., Bustamante, C. & Cozzarelli, N. R. (2005). Sequence-directed DNA translocation by purified FtsK. *Science* 307, 586–90.

Peck, L. J. & Wang, J. C. (1981). Sequence dependence of the helical repeat of DNA in solution. *Nature* 292, 375–8.

Peterlin, A. (1953). Light scattering by very stiff chain molecules. *Nature* 171, 259–60.

Peters, J. P. & Maher, L. J. I. (2010). DNA curvature and flexibility *in vitro* and *in vivo*. *Q. Rev. Biophys.* 43, 1–41.

Podtelezhnikov, A. A., Mao, C., Seeman, N. C. & Vologodskii, A. V. (2000). Multimerization-cyclization of DNA fragments as a method of conformational analysis. *Biophys. J.* 79, 2692–704.

Podtelezhnikov, A. A. & Vologodskii, A. V. (2000). Dynamics of small loops in DNA molecules. *Macromolecules* 33, 2767–71.

Post, C. B. & Zimm, B. H. (1979). Internal condensation of a single DNA molecule. *Biopolymers* 18, 1487–501.

Protozanova, E., Yakovchuk, P. & Frank-Kamenetskii, M. D. (2004). Stacked–unstacked equilibrium at the nick site of DNA. *J. Mol. Biol.* 342, 775–85.

Rau, D. C., Lee, B. & Parsegian, V. A. (1984). Measurement of the repulsive force between polyelectrolyte molecules in ionic solution: hydration forces between parallel DNA double helices. *Proc. Natl. Acad. Sci. U. S. A.* 81, 2621–5.

Revet, B. & Fourcade, A. (1998). Short unligated sticky ends enable the observation of circularised DNA by atomic force and electron microscopies. *Nucleic Acids Res.* 26, 2092–7.

Rief, M., Clausen-Schaumann, H. & Gaub, H. E. (1999). Sequence-dependent mechanics of single DNA molecules. *Nat. Struct. Biol.* 6, 346–9.

Rothemund, P. W. (2006). Folding DNA to create nanoscale shapes and patterns. *Nature* 440, 297–302.

Rouzina, I. & Bloomfield, V. A. (1996). Macroion attraction due to electrostatic correlation between screening counterions. 1. Mobile surface-adsorbed ions and diffuse ion cloud. *J. Phys. Chem.* 100, 9977–89.

  (2001). Force-induced melting of the DNA double helix 1. Thermodynamic analysis. *Biophys. J.* 80, 882–93.

Rybenkov, V. V., Cozzarelli, N. R. & Vologodskii, A. V. (1993). Probability of DNA knotting and the effective diameter of the DNA double helix. *Proc. Natl. Acad. Sci. U. S. A.* 90, 5307–11.

Rybenkov, V. V., Vologodskii, A. V. & Cozzarelli, N. R. (1997). The effect of ionic conditions on DNA helical repeat, effective diameter, and free energy of supercoiling. *Nucleic Acids Res.* 25, 1412–18.

Saleh, O. A., Perals, C., Barre, F. X. & Allemand, J. F. (2004). Fast, DNA-sequence independent translocation by FtsK in a single-molecule experiment. *EMBO J.* 23, 2430–9.

SantaLucia, J., Jr. (1998). A unified view of polymer, dumbbell, and oligonucleotide DNA nearest-neighbor thermodynamics. *Proc. Natl. Acad. Sci. U. S. A.* 95, 1460–5.

Schellman, J. A. (1974). Flexibility of DNA. *Biopolymers* 13, 217–26.

Schellman, J. A. & Harvey, S. C. (1995). Static contributions to the persistence length of DNA and dynamic contributions to DNA curvature. *Biophys. Chem.* 55, 95–114.

Schurr, J. M., Babcock, H. P. & Gebe, J. A. (1995). Effect of anisotropy of the bending rigidity on the supercoiling free energy of small circular DNAs. *Biopolymers* 36, 633–41.

Seeman, N. C. (1982). Nucleic acid junctions and lattices. *J. Theor. Biol.* 99, 237–47.

  (2010). Nanomaterials based on DNA. *Annu. Rev. Biochem.* 79, 65–87.

Shaw, S. Y. & Wang, J. C. (1993). Knotting of a DNA chain during ring closure. *Science* 260, 533–6.

Shen, Z., Yan, H., Wang, T. & Seeman, N. C. (2004). Paranemic crossover DNA: a generalized Holliday structure with applications in nanotechnology. *J. Am. Chem. Soc.* 126, 1666–74.

Sherman, W. B. & Seeman, N. C. (2004). A precisely controlled DNA biped walking device. *Nano Lett.* 4, 1203–7.

Shih, W. M., Quispe, J. D. & Joyce, G. F. (2004). A 1.7-kilobase single-stranded DNA that folds into a nanoscale octahedron. *Nature* 427, 618–21.

Shimada, J. & Yamakawa, H. (1984). Ring-closure probabilities for twisted wormlike chains. Application to DNA. *Macromolecules* 17, 689–98.

(1988). Moments for DNA topoisomers: the helical wormlike chain. *Biopolymers* 27, 657–73.

Shin, J. S. & Pierce, N. A. (2004). A synthetic DNA walker for molecular transport. *J. Am. Chem. Soc.* 126, 10834–5.

Shokri, L., McCauley, M. J., Rouzina, I. & Williams, M. C. (2008). DNA overstretching in the presence of glyoxal: structural evidence of force-induced DNA melting. *Biophys. J.* 95, 1248–55.

Shore, D. & Baldwin, R. L. (1983a). Energetics of DNA twisting. I. Relation between twist and cyclization probability. *J. Mol. Biol.* 170, 957–81.

(1983b). Energetics of DNA twisting. II. Topoisomer analysis. *J. Mol. Biol.* 170, 983–1007.

Shore, D., Langowski, J. & Baldwin, R. L. (1981). DNA flexibility studied by covalent closure of short fragments into circles. *Proc. Natl. Acad. Sci. U. S. A.* 78, 4833–7.

Shure, M. & Vinograd, J. (1976). The number of superhelical turns in native virion SV40 DNA and minicol DNA determined by the band counting method. *Cell* 8, 215–26.

Skolnick, J. & Fixman, M. (1977). Electrostatic persistence length of a wormlike polyelectrolyte. *Macromolecules* 10, 944–8.

Smith, D. E., Tans, S. J., Smith, S. B., Grimes, S., Anderson, D. L. & Bustamante, C. (2001). The bacteriophage straight phi29 portal motor can package DNA against a large internal force. *Nature* 413, 748–52.

Smith, S. B., Cui, Y. & Bustamante, C. (1996). Overstretching B-DNA: the elastic response of individual double-stranded and single-stranded DNA molecules. *Science* 271, 795–9.

Smith, S. B., Finzi, L. & Bustamante, C. (1992). Direct mechanical measurements of the elasticity of single DNA molecules by using magnetic beads. *Science* 258, 1122–6.

Spink, C. H., Ding, L., Yang, Q., Sheardy, R. D. & Seeman, N. C. (2009). Thermodynamics of forming a parallel DNA crossover. *Biophys. J.* 97, 528–38.

Stigter, D. (1977). Interactions of highly charged colloidal cylinders with applications to double-stranded DNA. *Biopolymers* 16, 1435–48.

(1978). Comparison of Mannings polyelectrolyte theory with cylindrical Gouy model. *J. Phys. Chem.* 82, 1603–6.

(1995). Evaluation of the counterion condensation theory of polyelectrolytes. *Biophys. J.* 69, 380–8.

Stone, M. D., Bryant, Z., Crisona, N. J., Smith, S. B., Vologodskii, A., Bustamante, C. & Cozzarelli, N. R. (2003). Chirality sensing by *Escherichia coli* topoisomerase IV and the mechanism of type II topoisomerases. *Proc. Natl. Acad. Sci. U. S. A.* 100, 8654–9.

Strauss, F., Gaillard, C. & Prunell, A. (1981). Helical periodicity of DNA, poly(dA).poly(dT) and poly(dA-dT). poly(dA-dT) in solution. *Eur. J. Biochem.* 118, 215–22.

Strick, T. R., Allemand, J. F., Bensimon, D., Bensimon, A. & Croquette, V. (1996). The elasticity of a single supercoiled DNA molecule. *Science* 271, 1835–7.

Strick, T. R., Allemand, J. F., Bensimon, D. & Croquette, V. (1998). Behavior of supercoiled DNA. *Biophys. J.* 74, 2016–28.

Strick, T. R., Croquette, V. & Bensimon, D. (2000). Single-molecule analysis of DNA uncoiling by a type II topoisomerase. *Nature* 404, 901–4.

Svoboda, K. & Block, S. M. (1994). Biological applications of optical forces. *Annu. Rev. Biophys. Biomol. Struct.* 23, 247–85.

Tagashira, H., Morita, M. & Ohyama, T. (2002). Multimerization of restriction fragments by magnesium-mediated stable base pairing between overhangs: a cause of electrophoretic mobility shift. *Biochemistry* 41, 12217–23.

Taylor, W. H. & Hagerman, P. J. (1990). Application of the method of phage T4 DNA ligase-catalyzed ring-closure to the study of DNA structure. II. NaCl-dependence of DNA flexibility and helical repeat. *J. Mol. Biol.* 212, 363–76.

Trifonov, E. N., Tan, R. K. Z. & Harvey, S. C. (1988). Static persistence length of DNA. In *DNA Bending and Curvature*, eds. W. K. Olson, M. H. Sarma, R. H. Sarma & M. Sundaralingam, 243–53. New York: Adenine.

Upholt, W. B., Gray, H. B., Jr. & Vinograd, J. (1971). Sedimentation velocity behavior of closed circular SV40 DNA as a function of superhelix density, ionic strength, counterion and temperature. *J. Mol. Biol.* 62, 21–38.

van den Broek, B., Noom, M. C., van Mameren, J., Battle, C., Mackintosh, F. C. & Wuite, G. J. (2010). Visualizing the formation and collapse of DNA toroids. *Biophys. J.* 98, 1902–10.

van Loenhout, M. T., de Grunt, M. V. & Dekker, C. (2012). Dynamics of DNA supercoils. *Science* 338, 94–7.

van Mameren, J., Gross, P., Farge, G., Hooijman, P., Modesti, M., Falkenberg, M., Wuite, G. J. & Peterman, E. J. (2009). Unraveling the structure of DNA during overstretching by using multicolor, single-molecule fluorescence imaging. *Proc. Natl. Acad. Sci. U. S. A.* 106, 18231–6.

Vologodskaia, M. & Vologodskii, A. (2002). Contribution of the intrinsic curvature to measured DNA persistence length. *J. Mol. Biol.* 317, 205–13.

Vologodskii, A. (2006). Simulation of equilibrium and dynamic properties of large DNA molecules. In *Computational Studies of DNA and RNA*, eds. F. Lankas & J. Sponer, 579–604. Dordrecht: Springer.

   (2012). Bridged DNA circles: a new model system to study DNA topology. *Macromolecules* 45, 4333–6.

Vologodskii, A. & Frank-Kamenetskii, M. (2013). Strong bending of the DNA double helix. *Nucleic Acids Res.* 41, 6785–92.

Vologodskii, A. & Rybenkov, V. V. (2009). Simulation of DNA catenanes. *Phys. Chem. Chem. Phys.* 11, 10543–52.

Vologodskii, A. V. (1994). DNA extension under the action of an external force. *Macromolecules* 27, 5623–5.

Vologodskii, A. V., Amirikyan, B. R., Lyubchenko, Y. L. & Frank-Kamenetskii, M. D. (1984). Allowance for heterogeneous stacking in the DNA helix–coil transition theory. *J. Biomol. Struct. Dyn.* 2, 131–48.

Vologodskii, A. V., Anshelevich, V. V., Lukashin, A. V. & Frank-Kamenetskii, M. D. (1979). Statistical mechanics of supercoils and the torsional stiffness of the DNA. *Nature* 280, 294–8.

Vologodskii, A. V. & Cozzarelli, N. R. (1995). Modeling of long-range electrostatic interactions in DNA. *Biopolymers* 35, 289–96.

Vologodskii, A. V., Lukashin, A. V. & Frank-Kamenetskii, M. D. (1975). Topological interaction between polymer chains. *Sov. Phys. JETP* 40, 932–6.

Vologodskii, A. V., Lukashin, A. V., Frank-Kamenetskii, M. D. & Anshelevich, V. V. (1974). Problem of knots in statistical mechanics of polymer chains. *Sov. Phys. JETP* 39, 1059–63.

Vologodskii, A. V. & Marko, J. F. (1997). Extension of torsionally stressed DNA by external force. *Biophys. J.* 73, 123–32.

Wahl, P., Paoletti, J. & Le Pecq, J. B. (1970). Decay of fluorescence emission anisotropy of the ethidium bromide–DNA complex. Evidence for an internal motion in DNA. *Proc. Natl. Acad. Sci. U. S. A.* 65, 417–21.

Wang, J. C. (1969). Variation of the average rotation angle of the DNA helix and the superhelical turns of covalently closed cyclic lambda DNA. *J. Mol. Biol.* 43, 25–39.

(1979). Helical repeat of DNA in solution. *Proc. Natl. Acad. Sci. U. S. A.* 76, 200–3.

Wang, J. C. & Davidson, N. (1966). Thermodynamic and kinetic studies on the interconversion between the linear and circular forms of phage lambda DNA. *J. Mol. Biol.* 15, 111–23.

(1968). Cyclization of phage DNAs. *Cold Spring Harbor Symp. Quant. Biol.* 33, 409–15.

Wang, M. D., Yin, H., Landick, R., Gelles, J. & Block, S. M. (1997). Stretching DNA with optical tweezers. *Biophys. J.* 72, 1335–46.

Wang, X., Zhang, X., Mao, C. & Seeman, N. C. (2010). Double-stranded DNA homology produces a physical signature. *Proc. Natl. Acad. Sci. U. S. A.* 107, 12547–52.

Watson, J. D. & Crick, F. H. C. (1953). The structure of DNA. *Nature* 171, 123–31.

Widom, J. & Baldwin, R. L. (1980). Cation-induced toroidal condensation of DNA – studies with $Co^{3+}(NH_3)_6$. *J. Mol. Biol.* 144, 431–53.

Wiggins, P. A., Phillips, R. & Nelson, P. C. (2005). Exact theory of kinkable elastic polymers. *Phys. Rev. E* 71, 021909.

Williams, M. C., Rouzina, I. & Bloomfield, V. A. (2002). Thermodynamics of DNA interactions from single molecule stretching experiments. *Acc. Chem. Res.* 35, 159–66.

Yakovchuk, P., Protozanova, E. & Frank-Kamenetskii, M. D. (2006). Base-stacking and base-pairing contributions into thermal stability of the DNA double helix. *Nucleic Acids Res.* 34, 564–74.

Yan, H., LaBean, T. H., Feng, L. & Reif, J. H. (2003). Directed nucleation assembly of DNA tile complexes for barcode-patterned lattices. *Proc. Natl. Acad. Sci. U. S. A.* 100, 8103–8.

Yan, J. & Marko, J. F. (2004). Localized single-stranded bubble mechanism for cyclization of short double helix DNA. *Phys. Rev. Lett.* 93, 108108.

Yin, H., Wang, M. D., Svoboda, K., Landick, R., Block, S. M. & Gelles, J. (1995). Transcription against an applied force. *Science* 270, 1653–7.

Yoo, J. & Aksimentiev, A. (2012). Competitive binding of cations to duplex DNA revealed through molecular dynamics simulations. *J. Phys. Chem. B* 116, 12946–54.

Yurke, B., Turberfield, A. J., Mills, A. P., Simmel, F. C. & Neumann, J. L. (2000). A DNA-fuelled molecular machine made of DNA. *Nature* 406, 605–8.

Zhang, X., Chen, H., Fu, H., Doyle, P. S. & Yan, J. (2012). Two distinct overstretched DNA structures revealed by single-molecule thermodynamics measurements. *Proc. Natl. Acad. Sci. U. S. A.* 109, 8103–8.

Zhang, X., Chen, H., Le, S., Rouzina, I., Doyle, P. S. & Yan, J. (2013). Revealing the competition between peeled ssDNA, melting bubbles, and S-DNA during DNA overstretching by single-molecule calorimetry. *Proc. Natl. Acad. Sci. U. S. A.* 110, 3865–70.

Zhurkin, V. B., Lysov, Y. P. & Ivanov, V. I. (1979). Anisotropic flexibility of DNA and the nucleosomal structure. *Nucleic Acids Res.* 6, 1081–96.

4 DNA dynamics

4.1 Large-scale conformational dynamics

4.1.1 Motion without inertia

In this section we consider the dynamics of large-scale molecular motion in solution. This motion has one key feature that distinguishes it from macroscopic motion: it has no inertia. This fact is so important that it should be considered in detail.

Let us analyze the motion of a bead in water. The motion is described by Newton's equation

$$m\frac{dv}{dt} = F,$$ (4.1)

where m and v are the bead mass and velocity, and F is the net force acting on the bead. Let us assume that the frictional force, $F = -fv$, where f is the frictional coefficient, is the only force acting on the bead. Under this condition the solution of the equation is

$$v = v_0 \exp(-t/\tau),$$ (4.2)

where $\tau = m/f$, and v_0 is the initial velocity of the bead. Equation (4.2) shows that the initial velocity decreases to zero, owing to the frictional force, with characteristic time τ. The value of f for a bead in a liquid is specified by the Stokes formula:

$$f = 6\pi \eta r,$$ (4.3)

where r is the bead radius and η is the liquid viscosity. If we also express m through the bead radius and density, ρ, we obtain that

$$\tau = \frac{2}{9}r^2\rho/\eta.$$ (4.4)

Equation (4.4) shows that τ is proportional to r^2, so it changes dramatically when we go from the scale of our macroscopic world to the molecular scale. For $r = 1$ nm, $\eta = 0.01$ poise (the viscosity of water) and $\rho = 2$ g/cm^3, τ is equal to 4.4×10^{-13} s $\cong$ 0.5 ps. For this time interval the bead moves out of its position, on average, by ≈ 0.02 nm only, assuming that v_0 corresponds to the average velocity of the bead thermal motion at room temperature. This is a short distance even for the molecular scale. Thus, to a very good approximation, we can say that the bead "forgets" nearly immediately about the direction of its movement in water, as if it simply does not have any inertia.

This is consequence of the fact that the viscosity of water is huge for the motion of molecules.

This conclusion is applied, of course, to the motion of the bead under the action of force. As long as the force is acting, it is equilibrated by the frictional force, so the velocity, v, remains constant. Its value is specified by the condition that the net force acting on the bead is equal to zero,

$$v = F/f. \tag{4.5}$$

The velocity relaxation time is specified in this case by the same time constant τ.

4.1.2   Translational motion of DNA molecules

The simplest and best-studied type of motion of DNA molecules is the translational motion of their centers of mass in solution. When a constant force, F, acts on a molecule in a solution, the molecule moves with a constant velocity, v, as we have just discussed.

In his theoretical study of diffusion, Einstein found that the mobility of a molecule, μ ($\mu = 1/f$), is proportional to the diffusion coefficient D:

$$\mu = \frac{D}{k_B T}. \tag{4.6}$$

The diffusion coefficient can be measured experimentally. One method is based on measuring the broadening of scattered light from a monochromatic laser (Berne & Pecora 1976). Correlation spectroscopy gives another method to measure D (van Holde et al. 1998, Krichevsky & Bonnet 2002).

The value of μ is also related to the sedimentation coefficient, s:

$$s = [M(1 - \bar{v}\rho)/N_A]\mu, \tag{4.7}$$

where M is the molecular weight of the DNA molecules, $\bar{v}$ is the partial specific volume and ρ is the density of the solution (van Holde et al. 1998); N_A is Avogadro's number. Measurements of s were widely used in early DNA studies and a large volume of data was accumulated for both linear and circular molecules (van Holde et al. 1998, Bloomfield et al. 1999). The measurements and their analysis are described in detail in a few textbooks (Cantor & Schimmel 1980, van Holde et al. 1998, Bloomfield et al. 1999).

From a theoretical point of view, the mobility (or frictional coefficient) of DNA molecules does not depend, to a very good approximation, on their internal dynamics, which occurs on a much smaller timescale. Thus, the mobility can be calculated by averaging the mobility of rigid conformations over the equilibrium conformational ensemble. A fairly accurate approach to calculation of f is based on the Kirkwood–Riseman method, where a polymer chain is modeled by a chain of N beads with radius r_0 (Kirkwood & Riseman 1948). The approximation accounts for the hydrodynamic

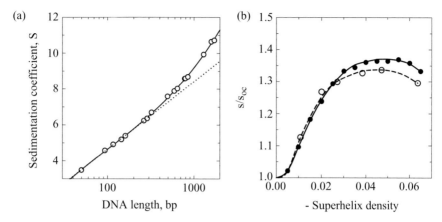

Figure 4.1 Comparison of experimental and theoretical sedimentation coefficients. All data correspond to [Na$^+$] of 0.2 M. (a) The experimental data for DNA restriction fragments ($\circ$) were fitted by the theoretical dependence for the WLC (solid line) to obtain the DNA persistence length (57.5 nm). The dotted line corresponds to the straight-rod model. The plot was reproduced from Fig. 4 of Kovacic and van Holde (1977), with permission. (b) The sedimentation of supercoiled DNA molecules. The measured sedimentation coefficients ($\circ$) are shown together with the computational results ($\bullet$) for the same 7 kb supercoiled DNA. There were no adjustable parameters in the calculations. The plot was reproduced from Fig. 2a of Rybenkov *et al.* (1997), with permission.

interaction of the beads. The value of f for a particular conformation of the polymer chain in this approximation is calculated as

$$ f = 6\pi r_0 \eta N \left/ \left[1 + \frac{r_0}{N} \sum_i \sum_{j \neq i} r_{ij}^{-1} \right] \right. , \tag{4.8}$$

where r_{ij} is the distance between centers of beads i and j. If we model dsDNA by a chain of touching beads, the value of r_0 should be equal to 1.59 nm (Hagerman 1981). For DNA molecules longer than three to five dozen base pairs in length the value of f has to be averaged over the equilibrium ensemble of the molecular conformations. This approach was used in both theoretical and computer calculations and showed a good agreement with the experimental measurements (Hearst & Stockmayer 1962, Yamakawa & Fujii 1973, Zimm 1980, Rybenkov *et al.* 1997).

Measurements of the sedimentation coefficients complemented by the theoretical analysis gave first estimations of DNA persistence length, a. The majority of the measurements were made on DNA samples with the broad distributions of the molecule lengths that were only available at that time. This definitely compromised the accuracy of the experiments. Also, the sensitivity of the sedimentation coefficient to the persistence length is not very high. The most accurate determination of a based on the sedimentation measurements was obtained when researchers used DNA restriction fragments (Kovacic & van Holde 1977). The experimental data and their theoretical fit are shown in Fig. 4.1(a).

The sedimentation measurements and their theoretical analysis were used to test the simulated conformational properties of supercoiled DNA molecules (Rybenkov *et al.* 1997). Although reasonably good agreement between the experimental and theoretical data was obtained in the study (Fig. 4.1(b)), the remaining discrepancy exceeded the statistical errors of the measurements and calculations, and it was difficult to understand the origin of this discrepancy.

4.1.3   Rotational motion

Rotational motion of macromolecules is characterized by a much stronger dependence on their size and shape. Although the theoretical analysis of the motion can be separated from the internal conformational changes in this case as well, it is more complex than the analysis of the translational motion. Rotational properties of a molecule are specified by three rotational diffusion coefficients, D_1, D_2 and D_3, for the rotation around the three principal axes of the molecular conformation (Cantor & Schimmel 1980). Correspondingly, to calculate the values of these diffusion coefficients averaged over the equilibrium conformational set, all three coefficients have to be calculated for each conformation of the set (see Hagerman and Zimm (1981) for an example of the computation). These diffusion coefficients specify the viscosity of the solution containing the macromolecules and their rotational orientation in the electric field or in a flow. The easiest experimental way to determine rotational characteristics of the molecules is measuring the relaxation of the rotational orientation due to thermal motion. The orientation is created by applying a short electric pulse to the solution, and relaxation is observed by monitoring an optical property of the solution. There are five time constants of the rotational relaxation, in general. These constants are expressed through linear combinations of D_i, and it is very difficult to obtain individual values of D_i from the measurements.

Hagerman and Zimm found that relaxation of relatively short DNA fragments, up to $5a$ in length, is specified, in a good approximation, by a single relaxation time (Hagerman & Zimm 1981). The measurements made for the corresponding restriction fragments, complemented by the computational analysis, allowed accurate estimations of a to be obtained (Fig. 4.2) (Hagerman 1981). Still, the approach based on orientation of the molecules by an intensive electric pulse can only be used in a solution of low ionic strength, to avoid its substantial heating.

4.1.4   Internal dynamics of the large-scale conformational changes

Let us consider a dilute solution of identical large DNA molecules, so there is no essential interaction between them. At any moment of time each molecule has a different conformation with some value of the end-to-end vector $\mathbf{R}(t)$. This vector is subject to random motion. Of course, the chain conformations and $\mathbf{R}(t)$, in particular, remain correlated over a certain period of time. The reduction of the correlation between the molecule conformations separated by time t can be specified by the correlation function $\langle \mathbf{R}(t)\mathbf{R}(0) \rangle$. This function has been a subject of extensive theoretical studies

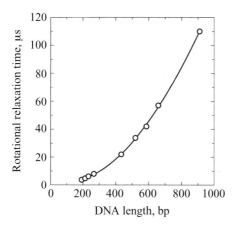

Figure 4.2 The rotational relaxation time for DNA fragments of different lengths. The experimental data (○) were fitted by the computational results (solid line) to obtain the DNA persistence length (the value of 50 ± 5 nm was obtained in the study). The data were obtained for [Na⁺] of 1 mM. The plot was reproduced from Fig. 9 of Hagerman (1981), with permission.

(see Grosberg and Khohlov (1994) for a review). It was found that, although the process is specified by a set of relaxation times, its main part is described by the longest relaxation time, τ. We can say that τ is the characteristic time over which a polymer chain remembers its previous conformations. The same relaxation time τ specifies the relaxation of an ensemble of disturbed polymer chains to their equilibrium conformational distribution.

The relaxation of isolated large DNA molecules has been studied in detail experimentally. Chu's group managed to observe the relaxation of λ DNA molecules from their fully extended conformation to the equilibrium random coils directly in optical microscope after staining them with a fluorescent dye (Perkins *et al.* 1994). The value of τ was also determined in flow experiments in bulk, and results of different approaches are in good agreement (Liu *et al.* 2007). For DNA molecules larger than 3 kb in length, the longest relaxation time in a water solution can be approximated by the empirical equation

$$\tau \approx L^{1.6} \text{ ms,} \tag{4.9}$$

where L is the DNA length measured in thousands of base pairs.

Another property that is important for understanding DNA functioning is the average time before the first collision of two chosen DNA sites. It is not clear how to address this question experimentally. We could try to detect the rate of bridging two sites by proteins, but the rate of such events is not diffusion limited, so many collisions precede the bridging. The same is true for joining sticky ends of DNA molecules. Therefore, all information about the rate of site juxtaposition is obtained from theoretical studies and Brownian dynamics (BD) simulations of DNA motion. The theoretical studies addressed dynamics of site juxtaposition in long polymer chains of many statistical segments (Wilemski & Fixman 1974a, Friedman & O'Shaughnessy 1994), while the simulation studied the process in shorter linear, circular and supercoiled molecules. Modern day

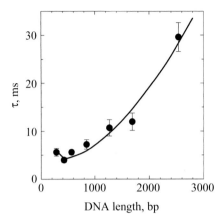

Figure 4.3 The average time of the first collision of DNA ends as a function of the molecule length. The data were obtained by the BD simulation performed by the present author. The chain ends were considered to be collided if the distance between them became smaller than 5 nm.

computers allow us to perform BD simulations with high efficiency. The information obtained in these simulations is quite reliable, since the computational results describe well other properties of the large-scale dynamics that can be measured experimentally. Thus, the BD simulation is a very important tool here. The major elements of this computational method are described in Box 4.1.

Both the theoretical analysis (Wilemski & Fixman 1974b) and BD simulation (Podtelezhnikov & Vologodskii 1997) show that to a good approximation the fraction of molecules that have not experienced collision of the ends up to the time moment t, $P(t)$, is described by a single-exponential decay:

$$P(t) = \exp(-t/\tau_c). \tag{4.10}$$

The values of the average time of the first collision of the fragment ends, τ_c, obtained by the BD simulation, are shown in Fig. 4.3. The simulation was performed for ionic conditions close to the physiological ones. We see from the figure that the values of τ_c for DNA fragments 0.3–3 kb in length are in the range of 5–30 ms. According to the theoretical results and computer simulations, τ_c should grow with DNA length L as $L^{2.2}$ for DNA molecules larger than 3 kb (Wilemski & Fixman 1974b, Podtelezhnikov & Vologodskii 1997).

It is interesting to compare τ_c with the rate of joining sticky ends of DNA fragments, τ_j. The only systematic study of this problem has been performed by Wang and Davidson for large phage DNA molecules (Wang & Davidson 1966, Wang & Davidson 1968). If we extrapolate their data to DNA molecules 3 kb in length, we find that for near-physiological ionic conditions τ_j should be close to 1 min. This estimation is in semi-quantitative agreement with the data on sticky-end joining in the work of Rybenkov *et al.* (1993), Geggier *et al.* (2011) and Vafabakhsh and Ha (2012). Thus, the values of τ_j are approximately three orders of magnitude larger than the values of τ_c. Clearly, the rate of DNA cyclization by joining the sticky ends is not limited by the rate of end collision.

Box 4.1 Brownian dynamics simulation of DNA dynamic properties

The large-scale dynamic properties of DNA molecules can be efficiently simulated by the BD method, which is based on the Langevin description of MD (Ermak & McCammon 1978). The traditional MD simulation of large molecules in solution accounts for thermal motion by placing solvent molecules explicitly or implicitly in the simulated system. Instead, the Langevin equations include the randomly fluctuating force exerted on the molecular subunits by the surrounding solvent. The equations of BD are essentially simpler for numerical integration than the Langevin equations, since the inertia term is skipped there, and therefore they are first-order differential equations (Ermak & McCammon 1978).

The fact that the inertia term can be skipped in the dynamic equations is due to the fast relaxation and the possibility of using a large integration time step, δt (see above). In general, the value of δt should be small enough that the forces acting on the molecular subunits do not change substantionally owing to the subunit displacements over time interval δt. The DNA model for large-scale dynamic simulation, outlined below, allows δt to be as large as 500 ps. In this case numerical solutions of the dynamic equations will not be affected by the subunit inertia since the momentum relaxation would occur over $\approx 1/1000$ of δt. A simple DNA model, a large time step and the possibility of skipping the inertia term allow BD simulations of DNA dynamics to be extended for time intervals up to a few seconds. These time intervals are sufficient to estimate characteristic rates for the majority of conformational rearrangements in DNA molecules a few kilobase pairs in length.

The model for the BD simulations is similar to the model used to simulate equilibrium properties of the double helix (see Section 3.3), although it has some specific features. It was originally developed by Allison et al. (Allison 1986, Allison et al. 1989, Allison et al. 1990) and later extended to the case of supercoiled DNA (Chirico & Langowski 1994, Chirico & Langowski 1996, Jian et al. 1998, Klenin et al. 1998, Huang et al. 2001). The main difference between the two models is that infinitely rigid stretching and electrostatic potentials in the model for equilibrium simulation are now replaced by the corresponding smooth potentials, since smaller derivatives of intersubunit potentials allow for a larger value of δt. The dynamic model also specifies hydrodynamic parameters of the subunits. The model is briefly described below.

A DNA molecule N base pairs in length is modeled as a chain consisting of m segments of equilibrium length l_0. The segments have no thickness. The chain energy consists of the following major terms.

1. The bending energy, E_b, specified by Eq. (3.37).
2. The stretching energy is computed as

$$E_s = \frac{h}{2l_0} \sum_{i=1}^{m} (l_i - l_0)^2, \tag{B4.1.1}$$

where l_i is the actual length of segment i, and h is the stretching rigidity constant. The stretching energy E_s is a computational device rather than an attempt to account for the actual stretching elasticity of the double helix. Smaller values of h allow larger timesteps in the BD simulations, but also imply larger departures from l_0. It is safe to choose $h = 100k_B T/l_0^2$, so that the variance of l_i is close to $l_0/100$. It was shown that a further increase of h does not change, within the simulation accuracy, the simulated dynamic properties of the model chain (Jian *et al.* 1997).

3. The energy of electrostatic intersegment interaction, E_e, is specified by the Debye–Hückel potential as a sum over all pairs of point charges located on the chain segments. The number of point charges placed on each segment, λ, is chosen to closely approximate continuous charges with the same linear density. The value of λ increases as the Debye length, $1/\kappa$, decreases (Jian *et al.* 1998, Huang *et al.* 2001). The energy E_e is specified as

$$E_e = \frac{v^2 l_0}{\lambda^2 D} \sum_{i<j}^{N} \frac{\exp(-\kappa r_{ij})}{r_{ij}}, \tag{B4.1.2}$$

where v is the effective linear charge density of the double helix, D is the dielectric constant of water, $N = kn\lambda$ is the total number of point charges and r_{ij} is the distance between point charges i and j. The value of v is chosen to approximate the solution of the Poisson–Boltzmann equation for DNA modeled as a charged cylinder by the Debye–Hückel potential (see Table 3.2).

4. The energy of torsional deformation of each segment, if needed for a considered problem, is specified explicitly in the model as described in detail by Chirico and Langowski (1994).

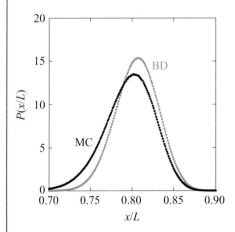

Figure B4.1.1 Comparison of DNA equilibrium properties computed by MC and BD simulations: the equilibrium distributions of DNA extension by the stretching force obtained by the two methods. The computations were performed for DNA 3000 bp in length, and the stretching force of 1 pN. The data were obtained by the present author.

It is assumed, to account for hydrodynamic interactions between different parts of the model chain, that beads of radius r_b are located at each vertex of the chain. The presence of the beads does not affect equilibrium properties of the model chain since there is no other interaction between the beads. The value of r_b specifies the frictional force from the solution exerted on each bead, and the hydrodynamic interaction between the beads. The interaction can be specified by Eq. (4.8) or the Rotne–Prager diffusion tensor (Rotne & Prager 1969).

Clearly, BD simulations can be used for simulation of DNA equilibrium properties, although the corresponding computations are about 100 times more time consuming than MC simulation. It is illustrated in Fig. B4.1.1 that the properties obtained by the two methods are really identical.

A more exotic type of internal motion in DNA was studied by Quake and co-workers (Bao *et al.* 2003). The researchers were able to tie different knots in a single DNA molecule by using optical tweezers and placing DNA in a flowing medium with high viscosity. Stretching the knotted molecules resulted in highly localized knots (Fig. 4.4).

Figure 4.4 Diffusion of a tight knot along a DNA molecule stretched by forces F applied to the beads attached to the molecule ends. The diagram shows the experimental design of the system. A simulated conformation of a DNA molecule 3000 bp in length is presented. Many times longer DNA molecules were used in the experiments (Bao *et al.* 2003).

Monitoring position of the knots versus time, the researchers measured the diffusion coefficients of different knots (Bao *et al.* 2003). This study provided valuable experimental data to test the accuracy of BD simulation of DNA internal motion. The test showed that measured and calculated diffusion coefficients of the knots did not differ by more than factor of two (Vologodskii 2006). This is definitely a good agreement, if we take into account that there were no adjustable parameters in the simulations. Therefore, the BD simulation allows reliable estimation of the characteristic times related to the large-scale dynamic properties of DNA molecules.

Relaxation of torsional orientation of the double helix is another DNA dynamic property that has attracted substantial attention over the years. The main approach in the study of the process is based on the depolarization of fluorescence of ethidium bromide intercalated into the double helix (Wahl *et al.* 1970, Millar *et al.* 1980, Thomas *et al.* 1980). This relaxation process, which involves different types of motion, occurs on the timescale of 10^{-10}–10^{-5} s (Shibata *et al.* 1985). Researchers tried to estimate DNA torsional rigidity from the relaxation dynamics (Barkley & Zimm 1979, Shibata *et al.* 1985, Fujimoto & Schurr 1990). The accurate analysis of the relaxation process is quite complicated, however (Shibata *et al.* 1985).

Figure 4.5 Diagram of a polymer chain in a concentrated solution of other chains. The chosen chain (shown in gray) is confined in a "tube" (shown in light gray) formed by other chains. The shape of the "tube" is specified by the conformation of the chosen chain.

4.1.5   Relaxation in dense solutions of large DNA molecules

A very different picture of polymer dynamics emerges in a highly concentrated solution of chain molecules. At equilibrium, the polymer coils penetrate one another, since such mixing increases the system entropy. This penetration is an extremely slow process, however, due to the inability of the chain segments to pass one through another. Each individual chain in the concentrated solution is surrounded by segments of other chains that form a "tube," and the chain can move only along this tube. The tube is defined by the conformation of the chosen chain (Fig. 4.5). Of course, the front end of the chain has freedom to choose any direction for the movement ahead, slowly resetting the tube. This model of polymer motion in concentrated solutions (or melts), *the reptation model*, was introduced into polymer physics by de Gennes (1979). Analysis shows that in the reptation motion the diffusion coefficient of the chain center of mass is proportional to L^{-2}, where L is the chain length (de Gennes 1979, Grosberg & Khohlov 1994). Therefore, for long DNA molecules in a concentrated solution the diffusion coefficient is extremely small. Correspondingly, the relaxation of the solution to its equilibrium state proceeds extremely slowly, and dense solutions of very long DNA molecules can never reach equilibrium.

The issue of relaxation of dense polymer solutions is related to chromosome territories in the cell nucleus. It was shown experimentally that during regular chromosome activity (interphase) chromosomes maintain their territories (Lieberman-Aiden *et al.* 2009). The dynamic of such initially segregated chromosomes was also studied by computer simulation (Rosa & Everaers 2008). The simulation assumed that each chromosome represents a random coil of the chromosome fiber, with diameter of 30 nm and persistence length of 150 nm. In the initial state of the solution the coils did not penetrate one another, forming well-defined boundaries between their territories. Thus, the initial state was very far from equilibrium. It was found that the fiber coils essentially preserve their territories over the cell lifetime. Thus, the fact that each chromosome maintains its own territory follows from the general properties of dense polymer solutions and does not require any assistance from cellular machines.

Figure 4.6 Diagram of DNA duplex formation by two complementary oligonucleotides. The first, rate-limiting process is the duplex nucleation, which requires formation of two to four consecutive base pairs. Its rate is specified by the rate constant k_1. This process is followed by fast extension of the helical region. The rate constants k_f and k_b correspond to the extension and reduction of a helical region by one base pair, respectively.

4.2   Kinetics of DNA melting and renaturation

4.2.1   Oligonucleotide duplexes

Let us consider first the kinetics of the duplex formation by two complimentary single-stranded chains, which is referred as association, renaturation or annealing. It consists of two distinctive processes, nucleation of a double-stranded helical segment and its extension by adding base pairs to the existing helical region (Fig. 4.6) (Ross & Sturtevant 1960, Craig *et al.* 1971, Pörschke & Eigen 1971). The former process is very slow because it requires two complementary single-stranded segments to be brought into close proximity. It is especially slow in solutions of low ionic strength due to the stronger electrostatic repulsion between the negatively charged DNA strands. Formation of a single base pair is not associated with the stacking interaction that stabilizes the double helix, and the majority of such pairs decay. Only when a few consecutive base pairs are formed does the helix nucleation become relatively stable. The second process of adding base pairs to an existing helical region, called zippering, is very fast. The rate constants of extension and reduction of a helical region by one base pair, k_f and k_b, are related to the equilibrium constant of the helix elongation, s:

$$k_f/k_b = s. \tag{4.11}$$

In general, the borders of a helical region move in both directions, extending or reducing the region. However, if the duplex is formed by short oligonucleotides, it is stable only at temperatures where $s \gg 1$ (see Section 2.2). Under this condition $k_f \gg k_b$, and the zippering is nearly a unidirectional process. Of course, this is not always the case for long helical regions, which are thermodynamically stable at s close to 1.

It was found that formation of two to four consecutive base pairs is needed for stable helix nucleation that initiates the zippering process. The rate constant of the nucleation, k_1, is close to 10^7/M/s at [Na$^+$] of 1 M (Nelson & Tinoco 1982, Freier *et al.* 1983, Williams *et al.* 1989). Since the usual concentration of oligonucleotides in the experiments is around 10 μM, the relaxation time of the process of duplex formation/dissociation is in the range of 10 ms. These relaxation times are easily accessible by the temperature-jump technique. The rate of nucleation is many times smaller than the rate of the first collision of the oligos, so the nucleation rate is not limited by diffusion. The value of k_1 depends weakly on the temperature (Nelson & Tinoco 1982). It also depends on ionic conditions, since the electrostatic repulsion between the complementary strands diminishes at higher ion concentration owing to the screening of DNA charges by small ions. Correspondingly, the mutual repulsion of the oligonucleotides increases at lower ionic strength. For association of hexanucleotides the value of k_1 decreases by a factor of 10 when the concentration of the sodium ions changes from 1 M to 0.01 M (Williams *et al.* 1989). Old data obtained for polynucleotides and high-molecular-weight DNA showed that the observed rate of their association changes by five orders of magnitude over the same range of [Na$^+$], suggesting that the effect can depend strongly on the oligonucleotide length (Ross & Sturtevant 1960, Blake & Fresco 1966, Studier 1969). There are no direct data, however, showing how the dependence of k_1 on ionic conditions changes with the oligonucleotide length.

A fundamental constant obtained in the studies of the oligonucleotide denaturation/renaturation kinetics is the rate constant for adding one base pair to the existing helix, k_f. Although its value, 10^6–10^7/s, was determined only for AU base pairs of RNA (Craig *et al.* 1971, Pörschke & Eigen 1971), it should be approximately the same for DNA base pairs (Nelson & Tinoco 1982). The value of k_f is very important for analysis of the transition kinetics. It allows one to estimate the denaturation rate, k_d, of a duplex of N base pairs using the equation

$$k_d \cong 2k_f N / s^N, \tag{4.12}$$

where s is the average equilibrium constant for the helix elongation (for an oligonucleotide of both AT and GC base pairs, s^N has to be substituted by $\prod_{i=1}^{N} s_i$) (Craig *et al.* 1971, Anshelevich *et al.* 1984). It follows from the equation that at room temperature the lifetime of a duplex of ten base pairs, five AT and five GC, at [Na$^+$] of 0.2 M, is equal to a few days. This time seems surprisingly large, since the random displacement of the borders of a helical region occurs so fast, approximately once per microsecond ($\tau = 1/k_f$). The explanation is very simple, however. The random process of the border displacement has very strong bias to extending the helical region rather than shrinking it, since $\frac{k_f}{k_b} = s \gg 1$ in the temperature range where a short duplex is stable. Therefore, an enormous number of border displacements occur before the helical region is shrunk to the helix nucleation size, two to four base pairs, and disappears.

It should be noted that Eq. (4.12) is only valid for cases when the average value of s is much larger than unity, which is always the case for short duplexes in the range of their stability but not necessary true for DNA molecules over $\approx$30 base pairs in length. However, the more general equation for the duplex dissociation,

$$k_d = 2k_f(s-1)^2(N(s-1)-2)/s^{N+3}, \tag{4.13}$$

maintains the main term of Eq. (4.12), s^N (Anshelevich *et al.* 1984). Therefore, the activation energy for the DNA duplex dissociation rate increases nearly linearly with the duplex length N and becomes extremely large for molecules of 100 or more base pairs.

4.2.2   Hairpins

Formation of DNA hairpins from single-stranded oligonucleotides (see Fig. 4.8 later) is similar to the duplex formation. The process consists of a slow nucleation step and a fast zippering process. The entire process is mach faster in this case, however. Its typical rate constant, k_1, is in the range of 10^4–10^5/s (Gralla & Crothers 1973, Pörschke 1974, Bonnet *et al.* 1998, Kuznetsov *et al.* 2008, Kuznetsov & Ansari 2012). The helix nucleation process should be very similar in this case to nucleation during the duplex formation and, therefore, should not be limited by the rate of diffusion of the complementary segments to one another. Indeed, k_1 is approximately three orders of magnitude lower than the estimated rate of the first collision (Kuznetsov & Ansari 2012). The value of k_1 for the hairpin formation is faster than for the duplex formation simply because of the large effective concentration of the complementary segments near each other in the short oligonucleotides. A simple quantitative estimation of k_1 based on this assumption is in agreement with experimentally observed values (Cantor & Schimmel 1980).

4.2.3   Long DNA molecules

Kinetics of melting of long DNA molecules, consisting of many thousands of base pairs, is a very slow process. It occurs on the timescale of minutes or even hours, depending on the experimental conditions (Freese & Freese 1963, Spatz & Crothers 1969, Record & Zimm 1972). A new factor contributing to this kinetics is the hydrodynamic friction of DNA unwinding. Indeed, complete denaturation of a DNA molecule N base pairs in length means that one strand of DNA must make $N/10.5$ turns relative the other strand. If the double helix is intact, DNA rotates like a speedometer cable, and its torsional relaxation should be relatively fast (Levinthal & Crane 1956, Wada & Netz 2009). However, when a long DNA molecule is partially melted, it consists of alternating double-stranded and denatured segments (Fig. 4.7). The melted regions form bulky coils, which create high friction for rotation, and this slows down the denaturation dramatically. Correspondingly, the kinetics of melting of long DNA molecules depends on the solution viscosity.

There is another slow step in the melting process: melting of helical regions surrounded by already melted segments (see Fig. 2.15) (Spatz & Crothers 1969, Hoff & Roos 1972, Perelroyzen *et al.* 1981). Complete melting of such regions is associated with large entropic gain, since one melted loop disappears. Therefore, the melting occurs under conditions when the base-pair stability constants, s_i, averaged over the region, are larger than unity. As a result, random displacement of the region borders is biased to its extension, on average. Therefore, similar to the melting of short duplexes, a huge number of border displacement steps are needed to complete the region melting. The process ends only when the region size is shrunk to few base pairs corresponding to the helix nucleation. However, the activation energy of the process is huge, and a temperature increase by a few tenths of a degree enormously accelerates the region denaturation. This was described in detail in Section 2.2.10.

Figure 4.7 Diagram of partially melted DNA molecule. The double-stranded regions (bold lines) alternate with compact melted regions (thin lines) that form bulky coils and dramatically slow down DNA rotation, which accompanies the melting.

An interesting version of DNA melting is observed after large temperature jumps, beyond the melting interval (Spatz & Crothers 1969). During the first stage a certain number of melted regions form very quickly, without unwinding of the denatured regions. This stage resembles the melting of circular DNA molecules, where unwinding of complementary strands is impossible owing to topological restrictions (see Section 6.5.3.3). Such melting without strand unwinding decreases the entropy of the denatured state. Therefore, it becomes possible only at temperatures that are much higher than the equilibrium melting temperatures. During the second, much slower stage of the process, shrinking of remaining helical regions follows unwinding of the melted regions.

Renaturation of long melted DNA molecules has even more obstacles. First, the concentration of the complementary strands is low and it slows down the helix nucleation. Second, when the temperature is decreased, a significant number of intramolecular hairpins form quickly. Although the free energy of conformations with these hairpins is much higher than the free energy of the perfect duplexes, they represent a kinetic obstacle for the duplex formation, especially at lower temperatures. Third, zipping of the complementary strands requires their rewinding, which is a slow process as well owing to the friction.

Since formation of intramolecular hairpins precedes the zippering, they have to be denatured during the zippering stage of the process. Clearly, the hairpin denaturation is slower at lower temperature, and it slows down the entire process. On the other hand, a certain stability of regular double-stranded segments is needed for efficient zippering. As a result, the renaturation rate reaches its maximum at temperatures that are below T_m by 20–30 °C (Fig. 4.8) (Marmur & Doty 1961, Wetmur & Davidson 1968). It was found that after fast cooling of DNA solution renaturation of long DNA molecules becomes nearly impossible (Marmur & Doty 1961).

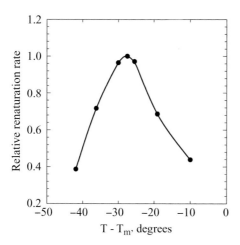

Figure 4.8 Temperature dependence of the renaturation rate of large DNA molecules. The rate approaches its maximum value at a temperature ≈30 °C below the DNA melting temperature. The data were obtained for [Na⁺] of 0.345 M. The plot was reproduced from Fig. 4 of Marmur and Doty (1961), with permission.

Of course, the renaturation rate depends on the DNA length and the size of the genome (Marmur & Doty 1961, Wetmur & Davidson 1968). It was also found that the rate of renaturation strongly depends on the concentration of monovalent salt if the concentration is below 0.4 M (Marmur & Doty 1961, Wetmur & Davidson 1968).

4.3 Kinetics of B–A and B–Z conformational transitions

Both the B–A and B–Z transitions represent a transition between double-stranded forms of DNA helices, and as such have many similarities. In this respect they are different from DNA melting, which involves isolated single strands. The B–A and B–Z transitions do not involve large changes of conformational freedom of the nucleotides, and, correspondingly, the enthalpies of each form in the transition range are close one to another. As a result, both transitions weakly depend on the solution temperature. Both transitions exhibit cooperativity, so in the transition interval DNA represents a mixture of alternating segments of one or other form. There are, of course, essential differences between the transitions, so below we consider them separately.

4.3.1 B–A transition

Although the equilibrium transition between B and A forms of the double helix was studied in detail in the 1980s (Ivanov & Krylov 1992), it turned to be difficult to investigate the transition kinetics. Equilibrium B–A transition is observed in water–ethanol (trifluoroethanol) mixtures when the ethanol fraction increases. It turned out that the mixing time of ethanol and water in the stop-flow apparatus is too large to observe the transition kinetics. Eventually, Porschke and Jose found that the transition can be slightly shifted by a strong electric field applied to the water–ethanol solution, owing to the higher dipole moment of the B form. This finding allowed them to study the transition kinetics (Jose & Porschke 2004, Jose & Porschke 2005).

From a structural point of view the B and A forms of the double helix are not dramatically different, and X-ray data showed a nearly continuous spectrum of conformations between the forms. Therefore, one could expect very fast kinetics of the transition, on the timescale of stacking–unstacking relaxation in the ssDNA molecules. This expectation was supported by the MD simulations, which predicted that the transition should occur on the nanosecond timescale. It turned out that the transition is not so fast, however. The kinetic curves are well approximated by two-exponential kinetics with time constants in the ranges of 2–10 μs and 50–100 μs (Jose & Porschke 2004, Jose & Porschke 2005). It seems reasonable to suggest, that cooperativity of the transition slows down the kinetics, similar to how it occurs in the helix–coil transition. Specific stages of the transition associated with the two relaxation constants remain unknown.

4.3.2   B–Z transition

The B–Z transition, as well as the B–A transition, represents a conversion of one dsDNA form to another. However, in structural terms the Z form is very different from the B form, and conversion of a base pair from B to Z conformation involves a large local distortion of the DNA structure. Thus, one could expect that both the rate constants of nucleation, k_1 and k_{-1}, and the rate constants of propagation, k_f and k_b, are very small. Neither of these constants have been determined separately for the B–Z transition, however. The studies of the transition kinetics were performed on poly[d(CG)] or on relatively short segments of d(CG)$_n$ · d(CG)$_n$ inserted into plasmids. The transition in the polynucleotide can be monitored by measuring the changes in CD or absorption spectra, which makes the experimental study fairly straightforward. On the other hand, the polynucleotide does not form a homogeneous structure, and this complicates analysis of the data. The pioneering study by Pohl and Jovin on poly[d(CG)] showed that the transition occurs in the time range of 10^2–10^3 s, and its activation energy is around 22 kcal/mol (Pohl & Jovin 1972).

It is much more difficult to monitor the transition in short inserts of d(CG)$_n$ · d(CG)$_n$ in plasmids (see Section 6.5.2). It was found that at very low negative supercoiling the characteristic time of the salt-induced transition constitutes a few hours at 25 °C (Peck *et al.* 1982, Pohl 1986).

4.4   Dynamics of DNA junctions

4.4.1   Migration of the Holliday junction

Any two DNA duplexes with identical sequences can form a four-way junction (Fig. 4.9). The junction has been the subject of numerous structural studies (reviewed by Lilley and Clegg (1993a), Lilley (2000)). This junction, also known as the Holliday junction,

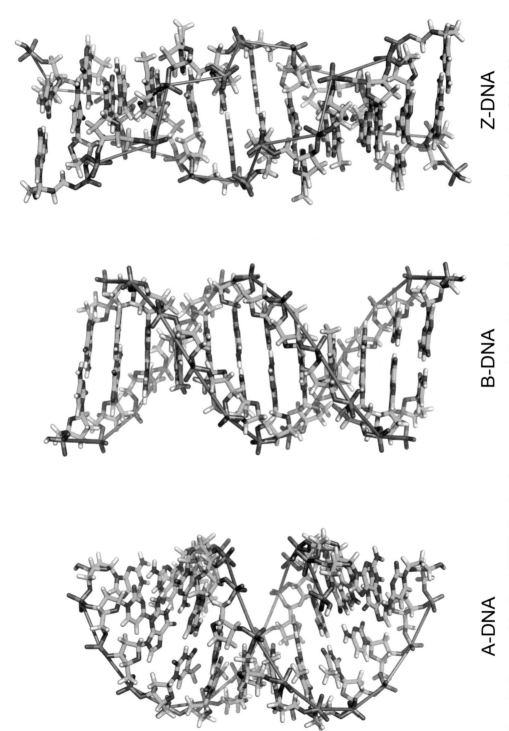

A-DNA   **B-DNA**   **Z-DNA**

Figure 1.5 Structure of A-, B- and Z-DNA. The phosphorus atoms are connected by a thin orange line to emphasize the geometry of the backbones. Image made by Richard Wheeler, published with permission.

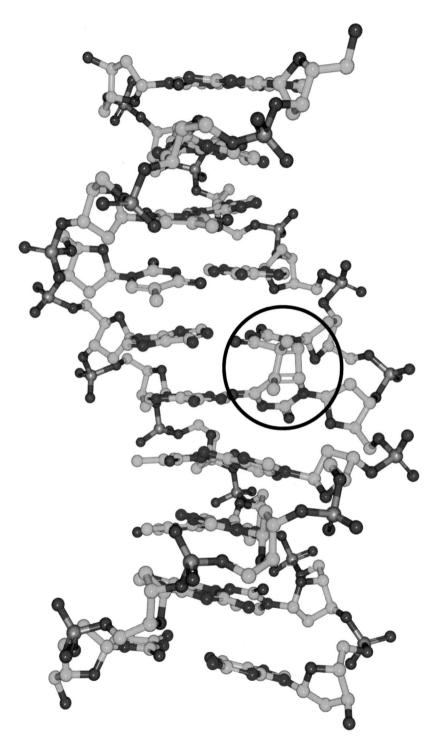

Figure 2.24 Structure of a DNA duplex containing a thymine dimer. The black circle shows the position of the dimer. The structure was obtained by X-ray analysis of the decamer crystals (Park *et al.* 2002).

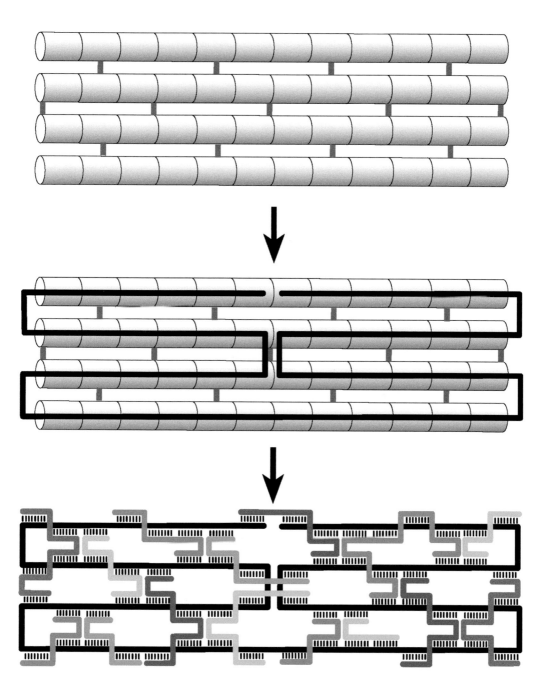

Figure 3.35 Diagram of DNA origami design. At the top the desired shape is represented by parallel double helices joined by crossovers. The scaffold ssDNA provides one of two complementary strands in each double-stranded segment, so the projected shape of the scaffold sets the desired shape of the origami (middle panel). Synthetic oligonucleotides, shown by different colored lines at the bottom of the figure, serve as the complementary strands of the double-stranded segments. These oligonucleotides also form cross-links between the double-stranded segments by switching from one segment to another. The sequence of the oligonucleotides is uniquely defined by their location in the origami and by the sequence of M13 DNA (adopted from Rothemund (2006)).

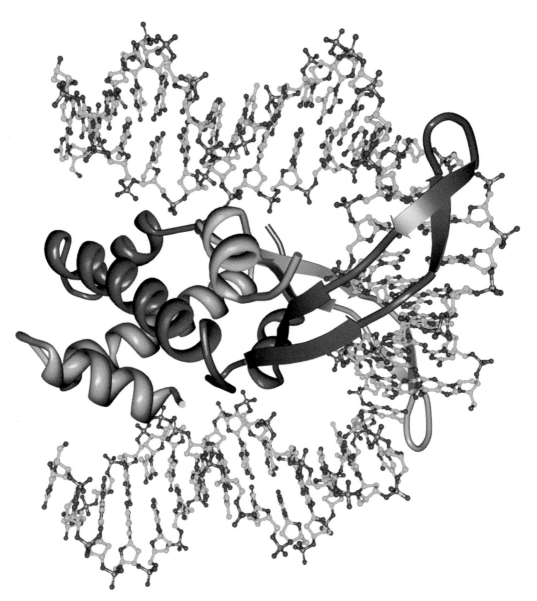

Figure 5.1 Diagram of DNA–IHF complex (Lynch *et al.* 2003). The DNA strands are shown in atom–bond presentation, while the protein is diagramed by the ribbons of different colors. Two prolines at the tips of *β*-hairpins are intercalated between the base pairs, and this causes strong bending of the double helix. The total bend angle is close to 160°. The image from RCSB PDB was obtained with Molecular Biology Toolkit (Moreland *et al.* 2005).

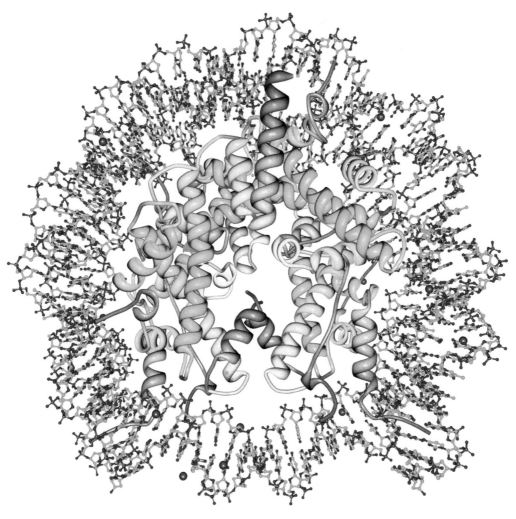

Figure 5.3 The structure of the nucleosome (Chua *et al.* 2012). The DNA strands are shown in atom–bond presentation. Different histone proteins are shown by ribbons of different colors. The 145 bp DNA fragment makes 1.75 helix turns around the histone core. The image from RCSB PDB was obtained with Molecular Biology Toolkit (Moreland *et al.* 2005).

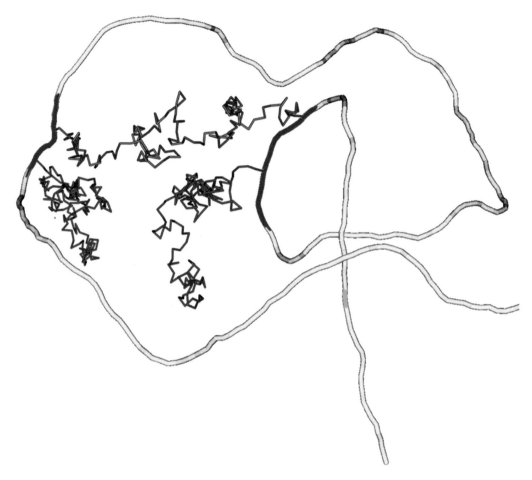

Figure 5.4 Diagram of the protein trajectory during the search for a specific binding site. The trajectory consists of segments of 3D diffusion (thin red lines) and random sliding along nonspecific segments of DNA (bold red lines, to emphasize that the sliding results in multiple visits of the same position along the DNA). DNA (bold yellow line) is assumed to have a random coil conformation in solution. There is a high probability that after dissociation from a double-helix segment the protein binds another segment of the same DNA molecule, as shown in the diagram. The specific binding site is shown in green.

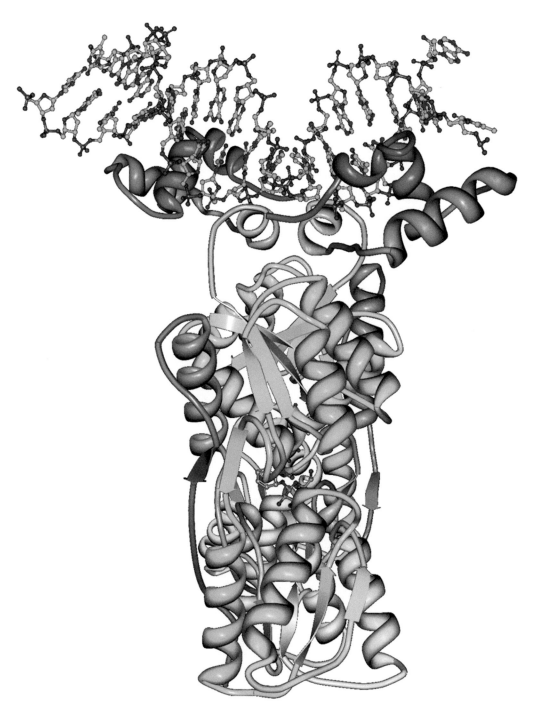

Figure 5.8 Structure of the complex of the *lac* repressor dimer with the palindromic operator (Lewis *et al.* 1996). The DNA-recognition domain consists of two HTH motifs separated from one another (shown as the dark-blue helices) and two α-helices, which interact with each other (the hinge region) and with the operator site (shown in light blue in the middle of the recognition domain). The image from RCSB PDB was obtained with Molecular Biology Toolkit (Moreland *et al.* 2005).

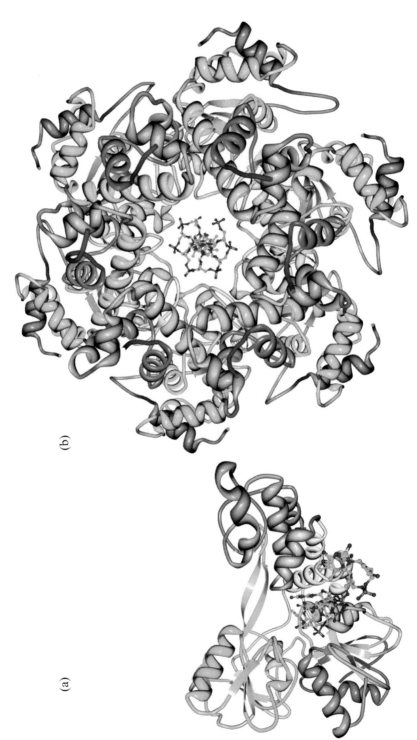

(b)

(a)

Figure 5.10 Two kinds of structure formed by DNA helicases. (a) NS3 helicase from the hepatitis C virus, which acts as a monomer (Gu & Rice 2010). (b) E1 helicase from papilloma virus, which forms a ring hexamer, with translocated ssDNA in the ring's central hole (Enemark & Joshua-Tor 2006). The images from RCSB PDB were obtained with Molecular Biology Toolkit (Moreland *et al.* 2005).

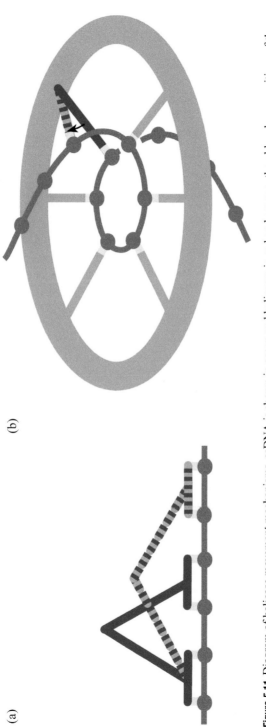

(b)

(a)

Figure 5.11 Diagram of helicase movement mechanisms. ssDNA is shown in green and helicases in red and orange; the old and new positions of the moving helicase parts are shown by red and red–orange lines, respectively. (a) The helicases of SF1 and SF2 superfamilies have two sites interacting with ssDNA, leading and lagging ones. They move by separating one of the sites and placing it ahead in the direction of movement, due to the conformational transition in the protein. (b) The hexameric helicases of superfamilies SF3, SF4, SF5 and SF6 force a spiral conformation of ssDNA inside their central hole, so that each subunit of the protein interacts with the DNA backbone. The helicases move along DNA by successively switching the position of each of their six DNA-interacting loops from lagging to leading.

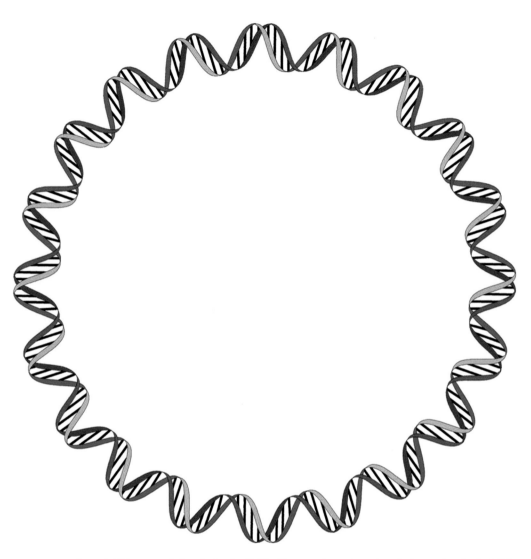

Figure 6.1 Diagram of closed circular DNA. The complementary strands form a topological link and cannot be separated from each other without breaking one of them. The topological characteristic of the link, the linking number, is equal to 20 in the diagram.

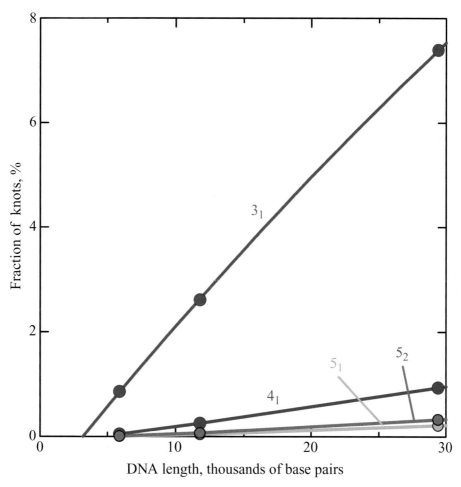

Figure 6.32 Equilibrium probability of the simplest knots in circular DNA. The results of computer simulations, shown by the symbols and solid lines, correspond to the physiological ionic conditions.

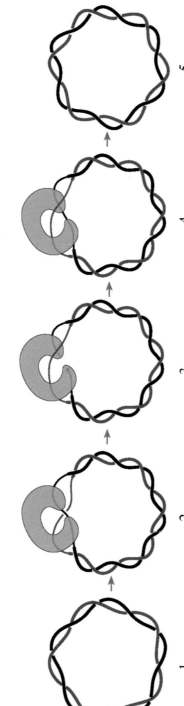

Figure 6.37 Diagram of supercoiling relaxation by type IA DNA topoisomerase. The enzyme is able to relax only negative supercoiling, and such a circular DNA is diagramed in panel 1. The enzyme binds only with a ssDNA segment (shown by thin lines), whose appearance is much more common in negatively supercoiled molecules, since base-pair opening accompanied by local unwinding reduces the torsional stress in the rest of the circular DNA (panel 2). An open region can accommodate either positive or negative interwinding of one strand around the other, but the left-handed interwinding is much more probable since it further reduces the torsional stress in the double-stranded region of negatively supercoiled DNA. After the binding the enzyme breaks the bound DNA strand, covalently attaches one end to itself, holds the other end of the broken strand, and creates a gap between the ends (panel 3). The segment of the complementary strand passes through the gap to a large cavity in the enzyme and the break is resealed after this (panel 4). The protein dissociates from the DNA, completing the catalytic cycle (panel 5). The reaction increases Lk of the complementary strands by 1, reducing negative supercoiling of the molecule.

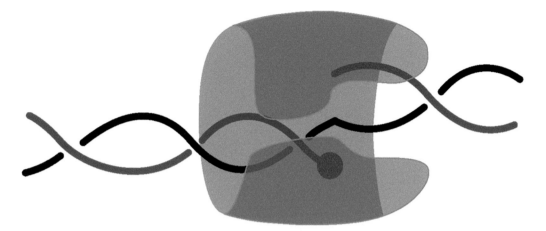

Figure 6.38 Diagram of type IB topoisomerase action. The enzyme binds a double-stranded segment and cuts one strand. One end of the cut strand is covalently attached to the protein while the other end is released for rotation around the intact strand. Eventually this end is caught by the enzyme and the broken strand is resealed. In this way the enzyme can relax both negative and positive supercoiling.

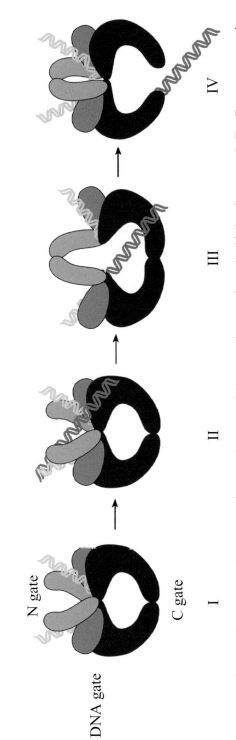

Figure 6.39 Diagram of type IIA topoisomerase action. In I, the enzyme in its open-clamp conformation binds a G segment. In II, a T segment enters the enzyme through the open clamp. The closure of the clamp upon ATP binding traps the T segment. The closure is coupled with breaking the G segment and opening the DNA gate. In III, the T segment reaches the central cavity of the enzyme, passing through the DNA gate. After this the DNA gate closes and the G segment is religated. In IV, closure of the DNA gate triggers the opening of the exit gate (C gate) and release of the T segment.

Figure 6.42 Diagram of the DNA pathway in the DNA–gyrase complex. Wrapping DNA around either the left or right C-terminal domain (shown in gray) creates a left-handed loop, so the strand-passing will reduce the value of Lk of the complementary strands.

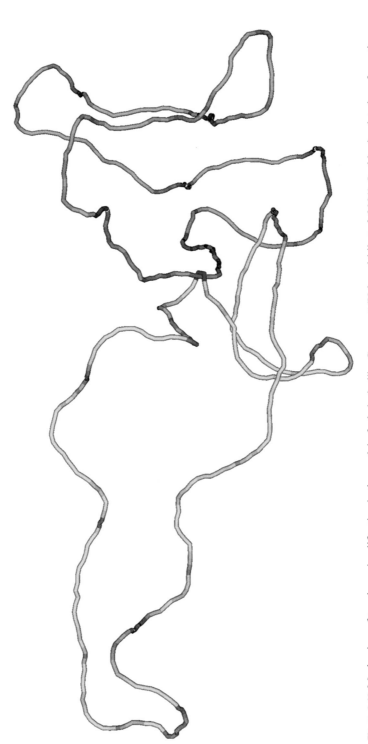

Figure 6.43 Mechanism of topology simplification in the model of a hairpin-like G segment (Vologodskii *et al.* 2001). In this simulated conformation of linked DNA molecules a potential T segment is located inside the hairpin made by the protein-bound G segment (the segment is shown in red, without bound protein). Such conformations make the link between the chains nearly local and increase conformational freedom for the rest of the linked chains. Therefore, the hairpin-like conformation of the G segment increases the probability that T and G segments will be properly juxtaposed in linked DNA molecules, accelerating unlinking.

(a) (b)

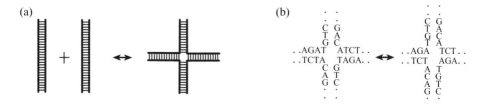

Figure 4.9 Four-way junctions. (a) Junctions with a mobile branch point can be formed by any two DNA duplexes with identical sequences. Under usual conditions in solution the equilibrium fraction of the four-strand complexes should be very small. (b) The single step of the junction displacement. In the example shown two AT base pairs in the horizontal branches are destroyed and replaced by two AT pairs in the vertical branches. Thus, the displacement of the branch point does not change the junction free energy.

plays a key role in general recombination. It is also a fundamental structural element in DNA nanotechnology, discussed in Section 3.5.

When formed by two identical DNA duplexes, the junctions have a mobile branch point, and this mobility is the subject of the current section. Let us note, first of all, that if the junction is displaced by extending two branches and shortening the others, the number of AT and GC base pairs is not changed. This means that, to a good approximation, the free energy of the four-strand complex does not depend on the junction position in the complex. Therefore, at each moment the branch point has equal probability to extend or shorten two opposite branches, so the junction performs a random walk. When a pair of branches shrinks to 2–3 bp, the complex dissociates to two duplexes. Thus, the rate of the branch migration is completely specified by a single rate constant, k, corresponding to the displacement of the junction by one base pair. This single step can involve disruption and reformation of two AT base pairs or two GC base pairs, and the value of k should depend on the type of base pair involved in a particular step. However, if we consider a junction displacement larger than a few dozen base pairs, only the average value of k matters.

Warner and co-workers were the first to try to measure the value of k (Thompson *et al.* 1976). Their experimental system and the method of measurement did not allow them to obtain accurate results, however. The problem was solved only18 years later in the elegant experiments of Panyutin and Hsieh (1994). They prepared two types of DNA duplex, A and B. Both duplexes had 20 unpaired nucleotides at the end of each strand (there was no complementarity between these ends). The single-stranded ends of duplex A were complementary to the ends of duplex B, so the duplexes could easily form a complex of four strands (Fig. 4.10). The long double-stranded parts of both duplexes were identical. Therefore, when the duplexes were mixed, they quickly formed complexes with the four-way junction that could migrate by reducing the long identical ends of the complex. Owing to these migrations the complexes eventually decayed to the duplexes. The kinetics of the complex decay was measured and analyzed in terms of the random-walk model. It was found that the value of k was equal to 3.5/s in a buffer containing 10 mM MgCl$_2$, but was 1000 times larger in solution that

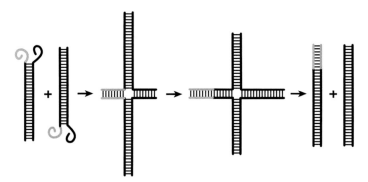

Figure 4.10 Diagram of a system prepared for the assay of the four-way junction migration (Panyutin & Hsieh 1994). Mixing the two duplexes with noncomplementary ends resulted in formation of the four-strand complex with all bases paired (strands shown in the same color are complementary to each other). The branch point in this complex could migrate only to long identical branches (shown in black). Eventually, the migration results in formation of two duplexes with fully complementary strands.

Figure 4.11 Displacement of one strand of DNA duplex by an identical single strand from the solution. The first step of the process is formation of the initial complex of three strands by hybridization of a single strand with a transiently melted end of the duplex. The formed branch point migrates randomly. Although the majority of these complexes dissociate to the original duplex and the single strand, a small fraction of them ends with strand replacement.

did not contain Mg^{2+} ions (Panyutin & Hsieh 1994). The effect of Mg^{2+} ions seems surprisingly large, although it correlates well with the finding that Mg^{2+} ions needed to preserve the stacking interaction between the branches at the junction (Lilley & Clegg 1993b).

The branch migration means that there should be spontaneous exchange of complementary strands between duplexes in solution. Indeed, the denaturation of the ends of two duplexes can be followed by formation of a four-way junction. Then the junction diffusion can result either in the formation of the duplexes with the same strands or with exchanged strands. Although for large duplexes the former outcome is much more probable, the latter outcome should occur from time to time as well. This exchange has been observed experimentally (Neschastnova *et al.* 2002). It was found that in a solution containing 7 mM $MgCl_2$, at 37 °C, the strand exchange between 170 bp duplexes takes about a day.

4.4.2   Strand exchange in presence of ssDNA

Isolated single strands can spontaneously replace the strands with identical sequence in DNA duplexes. If the duplex length exceeds dozen of base pairs and the solution temperature is not too high, the exchange occurs by strand displacement (Fig. 4.11). The

(a)   (b)

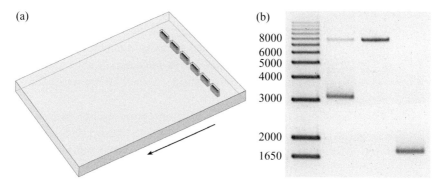

Figure 4.12 Separation of DNA molecules by agarose gel electrophoresis. (a) A flat gel with six wells, loaded with DNA samples. The gel is placed in the electrophoresis buffer so that DNA molecules run in the direction indicated by the arrow. (b) A picture of DNA separation in an agarose gel. The left-hand lane is a reference molecular-weight standard (1 kb Plus Ladder from Invitrogen). The size of the ladder bands is shown in base pairs on the left. After the electrophoresis DNA molecules in the gel were stained with Vistra Green fluorescent dye for visualization.

process starts by formation of a complex of three strands by hybridization of a single strand with the transiently melted end of the duplex. The new helical segment consists initially only of 2–4 bp and decays in most cases without any strand exchange. There is a chance, however, that the branch point formed by two identical strands will migrate to the position where the old duplex becomes unstable, resulting in strand displacement.

The process of strand exchange has been studied in detail for short oligonucleotide duplexes using labeled and unlabeled strands (Reynaldo *et al.* 2000). The strand displacement, described above, becomes a major route for the exchange for relatively long duplexes at low temperatures. The exchange in short duplexes at high temperatures occurs through complete dissociation of the duplex strands. The rate of random migration of this type of branch point was not determined in the study since it made a negligible contribution to the total time of the strand exchange. In accordance with this, the rate of the strand replacement was inversely proportional to the duplex length. A typical time for exchange was on the scale of hours even for a 12 bp duplex at 30 °C, and increased relatively fast with decreasing temperature.

4.5   Gel electrophoresis of DNA

Gel electrophoresis is the most important and widely used physical technique in molecular biology, and in DNA-related studies in particular. It allows separation of DNA molecules of different lengths and conformational properties with remarkable resolution. A picture of an agarose gel is shown in Fig. 4.12(a). DNA samples mixed with a colored buffer are loaded in the gel wells. The gel is placed into the camera with an electrophoresis buffer where electric field moves the negatively charged DNA in the direction parallel to the gel surface. DNA molecules of different sizes have different mobilities in the gels, so after running the electrophoresis for a certain time, typically a few hours, molecules of different sizes have different positions in the gels. The simplest way to locate positions of DNA in a gel is by soaking it with a fluorescent dye that binds with DNA. The stained gels are washed of excess staining dye and photographed. A typical pattern of separation of various DNA molecules is shown in Fig. 4.12(b). The second type of gel, based on polyacrylamide, is used for the separation of short DNA duplexes and ssDNA fragments.

Basic methods and their applications to various problems have been described in many books and reviews (see Rickwood and Hames (1982), Martin (1996), for example). Many specific applications of the technique have been developed, and some of them appear in various chapters of this book. Also, among such specific methods are the separation of DNA molecules in the gels with a temperature gradient (Fischer & Lerman 1979) and the method of localization of DNA bends in DNA fragments (Wu & Crothers 1984).

The theory of DNA gel electrophoresis is based on the reptation model of motion of a chain molecule in a polymer network (see Viovy (2000) for a review). This model is able to explain many experimental observations of the gel electrophoresis. Still, gel electrophoresis remains very much empirical, since the mobility of the molecules in gels cannot be calculated directly based on their sizes and conformational properties. Sometimes it is difficult to predict even relative mobility of molecules. For example, if in a particular electrophoresis conditions circular DNA molecules with topology A move faster than the molecules with topology B, in other conditions the molecules with topology B can move faster. Still, we have learned a lot about DNA physical properties using gel-electrophoresis separation of the molecules.

We will not go into any technical details of gel electrophoresis in this book, but will discuss one feature of the technique that is usually missed in the method description. Under proper conditions of the experiment DNA molecules separated by gel electrophoresis form very narrow bands in the gel. This is due to two factors. First, the velocities of the molecules along the electric field are relatively high compared with their diffusion rate in the gels. Second, the time of gel electrophoresis, t_{exp}, usually exceeds the time of conformational relaxation of the molecules, τ, by a few orders of magnitude, so the mobility of the molecules is very well averaged over the equilibrium ensemble of their conformations. There are cases, however, where the latter condition is not held. If τ is only a few times smaller than t_{exp}, poor conformational averaging results in the widening of the individual bands in the gel. The process of association and dissociation of short sticky ends of DNA fragments (made by restriction endonucleases) can belong to this category, as Fig. 4.13 illustrates. One can see from the figure that the fragment with more stable sticky ends forming four GC base pairs gives the broadest band in the gel, since the lifetime of the circular state is the largest in this case and conformational averaging requires more time than t_{exp}. The fragment in the middle lane of the gel has no sticky ends, and the conformational ensemble does not include the circular state. Therefore, the conformational averaging occurs very fast and the fragment forms a narrow band in the gel. The fragment on the right has sticky ends of two GC and two AT base pairs and forms less stable circles than the first fragment. Correspondingly, a better conformational averaging occurs in this case than for the first fragment, and the width of the corresponding band has an intermediate value.

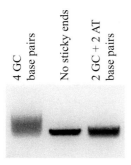

Figure 4.13 Separation of DNA fragments with sticky ends in an agarose gel. The fragment in the left-hand lane had sticky ends that form four GC base pairs, the fragment in the middle lane had blunt ends and the fragment in the right-hand lane had sticky ends that form two GC and two AT base pairs. In circular form each fragment had a length of 200 bp. Clearly, due to the long life of the circular form, the conformational averaging of the fragment mobilities is the worst for the fragment with the most stable sticky ends (left-hand lane). The electrophoresis was performed at room temperature. The picture is a gift from Q. Du.

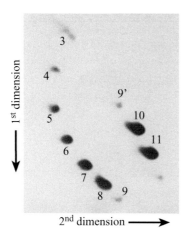

Figure 4.14 Separation of DNA topoisomers by 2D gel electrophoresis. Topoisomer 9 ($\Delta Lk = -9$) is split into two spots, 9 and 9′, corresponding to linear and cruciform states of the palindromic region of the circular DNA. The picture was reproduced from Fig. 1 of Lyamichev *et al.* (1983), with permission.

It is also possible that the value of τ exceeds t_{exp}. If the conformational ensemble consists of only two major groups, the DNA molecules form two separate bands in the gel. An example of such a case is a cruciform that can be formed in a palindromic region of supercoiled DNA. The transition between the linear form of the palindrome and the cruciform may be extremely slow, taking days or more (see Section 6.5.3.1). A 2D gel of supercoiled DNA with a palindromic region is shown in Fig. 4.14 (see Section 6.5.2 for a detailed explanation of 2D gel electrophoresis of supercoiled DNA). During the electrophoresis in the first dimension the mobility of DNA topoisomers (molecules with different values of ΔLk) increases monotonically with the increase of

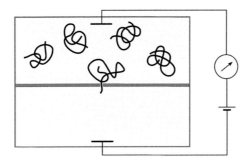

Figure 4.15 Diagram of DNA movement through a pore in a membrane that divides a minicell into two compartments. DNA molecules move through the pore due to the electric potential applied to the cell.

$|\Delta Lk|$ until the negative supercoiling causes the cruciform extrusion at a certain value of ΔLk. The transition abruptly reduces the absolute value of DNA writhe and, as a result, the DNA mobility. We can see from the figure that the transition occurs at $\Delta Lk = -9$ (topoisomer no 9 in the figure). The topoisomer forms two spots, marked 9 and 9′, which correspond to the linear and the cruciform states of the palindrome. Note that the total intensity of the two spots is much lower than the intensity of the surrounding spots, 8 and 10. This is due to the fact that a fraction of the molecules with ΔLk of -9 experienced the transition in one or other direction during the electrophoresis in the first direction, which lasted about 24 h (Lyamichev *et al.* 1983). The molecules that experienced the transition ended somewhere between spots 9 and 9′ and made no visible trace on the gel.

4.6   Moving DNA molecules through nanopores

Suppose that we have a small cell divided into two sections by a thin membrane with a tiny pore in it. DNA molecules are placed in one section of the cell. If electric potential is applied to the cell, DNA molecules can move through the pore. The translocation starts when a molecule end enters into the pore and proceeds relatively fast until the entire chain passes through (Fig. 4.15).

    The movement of a DNA molecule through the pore can be detected by measuring the electric current in the cell. This current exists due to the movement of small ions through the same pore. When DNA enters the pore, it essentially blocks the ion transport, if the pore diameter is close to the diameter of DNA molecules, so the current can be reduced dramatically.

    The movement of single-stranded nucleic acids through a nanopore was first investigated in 1996 (Kasianowicz *et al.* 1996). The researchers used α-hemolysin, a heptameric transmembrane pore secreted by *Staphylococcus aureus*. The pore was incorporated into a lipid bilayer that served as a membrane. This pore has a diameter of 2.6 nm (Song *et al.* 1996), which turned out to be sufficient for the translocation of ssDNA. It was found that the reduction of the ion current through the pore due to the translocation of a single-stranded plynucleotide can be different for each of the four DNA bases (Kasianowicz *et al.* 1996). This allowed the researchers to suggest that the translocation of ssDNA through a nanopore can provide direct, high-speed reading of the DNA sequence at the single-molecule level. Since then the area of DNA translocation through nanopores has become a very hot topic in biophysics (reviewed by Wanunu (2012)). A few protein-based wild-type and modified pores have been introduced over the years, as well as methods of making nanopores in synthetic materials. Moreover, different natural and synthetic membranes were tested in various experiments.

DNA translocation through nanopores has become a popular topic of theoretical studies. These studies have not brought complete understanding of the process, however. In particular, the electrical force acting on DNA when it migrates to the nanopore is not well understood. Comparison of the theoretical predictions with the experimental data shows that there is no simple model of the system that is able to explain all major experimental observations.

It has become clear, however, that the rate of passage of the DNA molecule through the channel is too high: the passing time of a single nucleotide is of the order of 1 μs (Kasianowicz *et al.* 1996). This time is too short for accurate measurement of the current through the minicell. The solution came from the enzymatic translocation of the DNA molecule through the nanopore (Chu *et al.* 2010, Cherf *et al.* 2012). The translocation of DNA by DNA polymerase, bound to the nanopore, can provide a long residence time of each base sampled in the pore (about 1 ms), and this allows more precise readout of the electric current. Also, the backward steps of the DNA are almost excluded in this case. At the same time, development of new genetically engineered nanopores allowed for a better identification of DNA bases (Stoddart *et al.* 2009).

In 2012 Oxford Nanopore Technologies announced success in DNA nanopore sequencing, with an initial error rate of 4% (Hayden 2012). Although this error rate is larger than the industry standard, the new sequencing technology should be able to cover tens of thousands of DNA bases in single reads. It may become the sequencing method of the very near future.

References

Allison, S., Austin, R. & Hogan, M. (1989). Bending and twisting dynamics of short DNAs: analysis of the triplet anisotropy decay of a 209 base pair fragment by Brownian simulation. *J. Chem. Phys.* 90, 3843–54.

Allison, S. A. (1986). Brownian dynamics simulation of wormlike chains: fluorescence depolarization and depolarized light scattering. *Macromolecules* 19, 118–24.

Allison, S. A., Sorlie, S. S. & Pecora, R. (1990). Brownian dynamics simulations of wormlike chains – dynamic light scattering from a 2311 base pair DNA fragment. *Macromolecules* 23, 1110–18.

Anshelevich, V. V., Vologodskii, A. V., Lukashin, A. V. & Frank-Kamenetskii, M. D. (1984). Slow relaxational processes in the melting of linear biopolymers: a theory and its application to nucleic acids. *Biopolymers* 23, 39–58.

Bao, X. R., Lee, H. J. & Quake, S. R. (2003). Behaviour of complex knots in single DNA molecules. *Phys. Rev. Lett.* 91, 265506.

Barkley, M. D. & Zimm, B. H. (1979). Theory of twisting and bending of chain macromolecules: analysis of the fluorescence depolarization of DNA. *J. Chem. Phys.* 70, 2991–3007.

Berne, B. J. & Pecora, R. (1976). *Dynamic Light Scattering: with Applications to Chemistry, Biology, and Physics*. New York: Wiley.

Blake, R. D. & Fresco, J. R. (1966). Polynucleotides. VII. Spectrophotometric study of the kinetics of formation of the two-stranded helical complex resulting from the interaction of polyriboadenylate and polyribouridylate. *J. Mol. Biol.* 19, 145–60.

Bloomfield, V. A., Crothers, D. M. & Tinoco, I., Jr. (1999). *Nucleic Acids: Structures, Properties, and Functions*. Sausalito, CA: University Science Books.

Bonnet, G., Krichevsky, O. & Libchaber, A. (1998). Kinetics of conformational fluctuations in DNA hairpin-loops. *Proc. Natl. Acad. Sci. U. S. A.* 95, 8602–6.

Cantor, C. R. & Schimmel, P. R. (1980). *Biophysical Chemistry*. New York: Freeman.

Cherf, G. M., Lieberman, K. R., Rashid, H., Lam, C. E., Karplus, K. & Akeson, M. (2012). Automated forward and reverse ratcheting of DNA in a nanopore at 5-Å precision. *Nat. Biotechnol.* 30, 344–8.

Chirico, G. & Langowski, J. (1994). Kinetics of DNA supercoiling studied by Brownian dynamics simulation. *Biopolymers* 34, 415–33.

  (1996). Brownian dynamic simulations of supercoiled DNA with bent sequences. *Biophys. J.* 71, 955–71.

Chu, J., Gonzalez-Lopez, M., Cockroft, S. L., Amorin, M. & Ghadiri, M. R. (2010). Real-time monitoring of DNA polymerase function and stepwise single-nucleotide DNA strand translocation through a protein nanopore. *Angew. Chem. Int. Ed.* 49, 10106–9.

Craig, M. E., Crothers, D. M. & Doty, P. (1971). Relaxation kinetics of dimer formation by self complementary oligonucleotides *J. Mol. Biol.* 62, 383–401.

de Gennes, P. G. (1979). *Scaling Concepts in Polymer Physics*. Ithaca, NY: Cornell University Press.

Ermak, D. L. & McCammon, J. A. (1978). Brownian dynamics with hydrodynamic interactions. *J. Chem. Phys.* 69, 1352–60.

Fischer, S. G. & Lerman, L. S. (1979). Length-independent separation of DNA restriction fragments in two-dimensional gel electrophoresis. *Cell* 16, 191–200.

Freese, E. B. & Freese, E. (1963). Rate of DNA strand separation. *Biochemistry* 2, 707–15.

Freier, S. M., Albergo, D. D. & Turner, D. H. (1983). Solvent effects on the dynamics of $(dG\text{-}dC)_3$. *Biopolymers* 22, 1107–31.

Friedman, B. & O'Shaughnessy, B. (1994). Scaling and universality in polymer reaction kinetics. *Int. J. Mod. Phys. B* 8, 2555–91.

Fujimoto, B. S. & Schurr, J. M. (1990). Dependence of the torsional rigidity of DNA on base composition. *Nature* 344, 175–8.

Geggier, S., Kotlyar, A. & Vologodskii, A. (2011). Temperature dependence of DNA persistence length. *Nucleic Acids Res.* 39, 1419–26.

Gralla, J. & Crothers, D. M. (1973). Free energy of imperfect nucleic acid helices. II. Small hairpin loops *J. Mol. Biol.* 73, 497–511.

Grosberg, A. Y. & Khohlov, A. R. (1994). *Statistical Physics of Macromolecules*. New York: AIP Press.

Hagerman, P. J. (1981). Investigation of the flexibility of DNA using transient electric birefringence. *Biopolymers* 20, 1503–35.

Hagerman, P. J. & Zimm, B. H. (1981). Monte Carlo approach to the analysis of the rotational diffusion of wormlike chains. *Biopolymers* 20, 1481–502.

Hayden, E. C. (2012). Nanopore genome sequencer makes its debut. *Nature* News and Comment. DOI: 10.1038/nature.2012.10051

Hearst, J. & Stockmayer, W. (1962). Sedimentation constants of broken chains and wormlike coils. *J. Chem. Phys.* 37, 1425–33.

Hoff, A. J. & Roos, A. L. (1972). Hysteresis of denaturation of DNA in the melting range. *Biopolymers* 11, 1289–94.

Huang, J., Schlick, T. & Vologodskii, T. (2001). Dynamics of site juxtaposition in supercoiled DNA. *Proc. Natl. Acad. Sci. U. S. A.* 98, 968–73.

Ivanov, V. I. & Krylov, D. (1992). A-DNA in solution as studied by diverse approaches. *Methods Enzymol.* 211, 111–27.

Jian, H., Schlick, T. & Vologodskii, A. (1998). Internal motion of supercoiled DNA: Brownian dynamics simulations of site juxtaposition. *J. Mol. Biol.* 284, 287–96.

Jian, H., Vologodskii, A. V. & Schlick, T. (1997). Combined wormlike-chain and bead model for dynamic simulations of long linear DNA. *J. Comput. Phys.* 136, 168–79.

Jose, D. & Porschke, D. (2004). Dynamics of the B–A transition of DNA double helices. *Nucleic Acids Res.* 32, 2251–8.

(2005). The dynamics of the B–A transition of natural DNA double helices. *J. Am. Chem. Soc.* 127, 16120–8.

Kasianowicz, J. J., Brandin, E., Branton, D. & Deamer, D. W. (1996). Characterization of individual polynucleotide molecules using a membrane channel. *Proc. Natl. Acad. Sci. U. S. A.* 93, 13770–3.

Kirkwood, J. G. & Riseman, J. (1948). The intrinsic viscosities and diffusion constants of flexible macromolecules on solution. *J. Chem. Phys.* 16, 565–73.

Klenin, K., Merlitz, H. & Langowski, J. (1998). A Brownian dynamics program for the simulation of linear and circular DNA and other wormlike chain polyelectrolytes. *Biophys. J.* 74, 780–8.

Kovacic, R. T. & van Holde, K. E. (1977). Sedimentation of homogeneous double-stranded DNA molecules. *Biochemistry* 1977, 1490–8.

Krichevsky, O. & Bonnet, G. (2002). Fluorescence correlation spectroscopy: the technique and its applications. *Rep. Prog. Phys.* 65, 251–97.

Kuznetsov, S. V. & Ansari, A. (2012). A kinetic zipper model with intrachain interactions applied to nucleic acid hairpin folding kinetics. *Biophys. J.* 102, 101–11.

Kuznetsov, S. V., Ren, C. C., Woodson, S. A. & Ansari, A. (2008). Loop dependence of the stability and dynamics of nucleic acid hairpins. *Nucleic Acids Res.* 36, 1098–112.

Levinthal, C. & Crane, H. (1956). On the unwinding of DNA. *Proc. Natl. Acad. Sci.* 42, 436–8.

Lieberman-Aiden, E., van Berkum, N. L., Williams, L., Imakaev, M., Ragoczy, T., Telling, A., Amit, I., Lajoie, B. R., Sabo, P. J., Dorschner, M. O., Sandstrom, R., Bernstein, B., Bender, M. A., Groudine, M., Gnirke, A., Stamatoyannopoulos, J., Mirny, L. A., Lander, E. S. & Dekker, J. (2009). Comprehensive mapping of long-range interactions reveals folding principles of the human genome. *Science* 326, 289–93.

Lilley, D. M. J. (2000). Structures of helical junctions in nucleic acids. *Q. Rev. Biophys.* 33, 109–59.

Lilley, D. M. J. & Clegg, R. M. (1993a). The structure of branched DNA species. *Q. Rev. Biophys.* 26, 131–75.

(1993b). The structure of the four-way junction in DNA. *Annu. Rev. Biophys. Biomol. Struct.* 22, 299–328.

Liu, Y. G., Jun, Y. G. & Steinberg, V. (2007). Longest relaxation times of double-stranded and single-stranded DNA. *Macromolecules* 40, 2172–6.

Lyamichev, V. I., Panyutin, I. G. & Frank-Kamenetskii, M. D. (1983). Evidence of cruciform structures in superhelical DNA provided by two-dimensional gel electrophoresis. *FEBS Lett.* 153, 298–302.

Marmur, J. & Doty, P. (1961). Thermal renaturation of deoxyribonucleic acids. *J. Mol. Biol.* 3, 585–94.

Martin, R. (1996). *Gel Electrophoresis: Nucleic Acids*. Oxford: BIOS.

Millar, D. P., Robbins, R. J. & Zewail, A. H. (1980). Direct observation of the torsional dynamics of DNA and RNA by picosecond spectroscopy. *Proc. Natl. Acad. Sci. U. S. A.* 77, 5593–7.

Nelson, J. W. & Tinoco, I., Jr. (1982). Comparison of the kinetics of ribooligonucleotide, deoxyribooligonucleotide, and hybrid oligonucleotide double-strand formation by temperature-jump kinetics. *Biochemistry* 21, 5289–95.

Neschastnova, A. A., Markina, V. K., Popenko, V. I., Danilova, O. A., Sidorov, R. A., Belitsky, G. A. & Yakubovskaya, M. G. (2002). Mechanism of spontaneous DNA–DNA interaction of homologous linear duplexes. *Biochemistry* 41, 7795–801.

Panyutin, I. G. & Hsieh, P. (1994). The kinetics of spontaneous DNA branch migration. *Proc. Natl. Acad. Sci. U. S. A.* 91, 2021–5.

Peck, L. J., Nordheim, A., Rich, A. & Wang, J. C. (1982). Flipping of cloned d(pCpG)$_n$ · d(pCpG)$_n$ DNA sequences from right- to left-handed helical structure by salt, Co(III), or negative supercoiling. *Proc. Natl. Acad. Sci. U. S. A.* 79, 4560–4.

Perelroyzen, M. P., Lyamichev, V. I., Kalambet, Y. A., Lyubchenko, Y. L. & Vologodskii, A. V. (1981). A study of the reversibility of helix–coil transition in DNA. *Nucleic Acids Res.* 9, 4043–59.

Perkins, T. T., Quake, S. R., Smith, D. E. & Chu, S. (1994). Relaxation of a single DNA molecule observed by optical microscopy. *Science* 264, 822–6.

Podtelezhnikov, A. & Vologodskii, A. (1997). Simulation of polymer cyclization by Brownian dynamics. *Macromolecules* 30, 6668–73.

Pohl, F. M. (1986). Dynamics of the B-to-Z transition in supercoiled DNA. *Proc. Natl. Acad. Sci. U. S. A.* 83, 4783–7.

Pohl, F. M. & Jovin, T. M. (1972). Salt-induced cooperative conformational change of a synthetic DNA – equilibrium and kinetic studies with poly(dG–dC). *J. Mol. Biol.* 67, 375–96.

Pörschke, D. (1974). Thermodynamic and kinetic parameters of an oligonucleotide hairpin helix. *Biophys Chem* 1, 381–6.

Pörschke, D. & Eigen, M. (1971). Co-operative non-enzymic base recognition 3 Kinetics of the helix–coil transition of the oligoribouridylic–oligoriboadenylic acid system and of oligoriboadenylic acid alone at acidic pH. *J. Mol. Biol.* 62, 361–81.

Record, M. T. & Zimm, B. H. (1972). Kinetics of helix–coil transition in DNA. *Biopolymers* 11, 1435–84.

Reynaldo, L. P., Vologodskii, A. V., Neri, B. P. & Lyamichev, V. I. (2000). The kinetics of oligonucleotide replacements. *J. Mol. Biol.* 297, 511–20.

Rickwood, D. & Hames, B. D. (1982). *Gel Electrophoresis of Nucleic Acids*. London: IRL Press.

Rosa, A. & Everaers, R. (2008). Structure and dynamics of interphase chromosomes. *PLoS Comput. Biol.* 4, e1000153.

Ross, P. D. & Sturtevant, J. M. (1960). The kinetics of double helix formation from polyriboadenylic acid and polyribouridylic acid. *Proc. Natl. Acad. Sci. U. S. A.* 46, 1360–5.

Rotne, J. & Prager, S. (1969). Variational treatment of hydrodynamic interaction in polymers. *J. Chem. Phys.* 50, 4831–7.

Rybenkov, V. V., Cozzarelli, N. R. & Vologodskii, A. V. (1993). Probability of DNA knotting and the effective diameter of the DNA double helix. *Proc. Natl. Acad. Sci. U. S. A.* 90, 5307–11.

Rybenkov, V. V., Vologodskii, A. V. & Cozzarelli, N. R. (1997). The effect of ionic conditions on the conformations of supercoiled DNA. I. Sedimentation analysis. *J. Mol. Biol.* 267, 299–311.

Shibata, J. M., Fujimoto, B. S. & Schurr, J. M. (1985). Rotational dynamics of DNA from 10^{-10} to 10^{-5} seconds: comparison of theory with optical experiments. *Biopolymers* 24, 1909–30.

Song, L. Z., Hobaugh, M. R., Shustak, C., Cheley, S., Bayley, H. & Gouaux, J. E. (1996). Structure of staphylococcal alpha-hemolysin, a heptameric transmembrane pore. *Science* 274, 1859–66.

Spatz, H. C. & Crothers, D. M. (1969). The rate of DNA unwinding *J. Mol. Biol.* 42, 191–219.

Stoddart, D., Heron, A. J., Mikhailova, E., Maglia, G. & Bayley, H. (2009). Single-nucleotide discrimination in immobilized DNA oligonucleotides with a biological nanopore. *Proc. Natl. Acad. Sci. U. S. A.* 106, 7702–7.

Studier, F. W. (1969). Effects of the conformation of single-stranded DNA on renaturation and aggregation. *J. Mol. Biol.* 41, 199–209.

Thomas, J. C., Allison, S. A., Appelof, C. J. & Schurr, J. M. (1980). Torsional dynamics and depolarization of fluorescence of linear macromolecules. II. Fluorescence polarization anisotropy measurements on a clean viral ϕ29 DNA. *Biophys. Chem.* 12, 177–88.

Thompson, B. J., Camien, M. N. & Warner, R. C. (1976). Kinetics of branch migration in double-stranded DNA. *Proc. Natl. Acad. Sci. U. S. A.* 73, 2299–303.

Vafabakhsh, R. & Ha, T. (2012). Extreme bendability of DNA less than 100 base pairs long revealed by single-molecule cyclization. *Science* 337, 1097–101.

van Holde, K. E., Johnson, W. C. & Ho, P. S. (1998). *Principles of Physical Biochemistry*. Upper Saddle River, NJ: Prentice Hall.

Viovy, J. L. (2000). Electrophoresis of DNA and other polyelectrolytespp: physical mechanisms. *Rev. Mod. Phys.* 72, 813–72.

Vologodskii, A. (2006). Brownian dynamics simulation of knot diffusion along a stretched DNA molecule. *Biophys. J.* 90, 1594–7.

Wada, H. & Netz, R. (2009). Rotational friction of a semiflexible polymer far from equilibrium. *Europhys. Lett.* 87, 38001.

Wahl, P., Paoletti, J. & Le Pecq, J. B. (1970). Decay of fluorescence emission anisotropy of the ethidium bromide–DNA complex: evidence for an internal motion in DNA. *Proc. Natl. Acad. Sci. U. S. A.* 65, 417–21.

Wang, J. C. & Davidson, N. (1966). Thermodynamic and kinetic studies on the interconversion between the linear and circular forms of phage lambda DNA. *J. Mol. Biol.* 15, 111–23.

(1968). Cyclization of phage DNAs. *Cold Spring Harbor Symp. Quant. Biol.* 33, 409–15.

Wanunu, M. (2012). Nanopores: a journey towards DNA sequencing. *Phys. Life Rev.* 9, 125–58.

Wetmur, J. G. & Davidson, N. (1968). Kinetics of renaturation of DNA. *J. Mol. Biol.* 31, 349–70.

Wilemski, G. & Fixman, M. (1974a). Diffusion-controlled intrachain reactions of polymers. I. Theory. *J. Chem. Phys.* 60, 866–77.

(1974b). Diffusion-controlled intrachain reactions of polymers. II. Results for a pair of terminal reactive groups. *J. Chem. Phys.* 60, 878–90.

Williams, A. P., Longfellow, C. E., Freier, S. M., Kierzek, R. & Turner, D. H. (1989). Laser temperature-jump, spectroscopic, and thermodynamic study of salt effects on duplex formation by dGCATGC. *Biochemistry* 28, 4283–91.

Wu, H. M. & Crothers, D. M. (1984). The locus of sequence-directed and protein-induced DNA bending. *Nature* 308, 509–13.

Yamakawa, H. & Fujii, M. (1973). Translational friction coefficient of wormlike chains. *Macromolecules* 6, 407–15.

Zimm, B. H. (1980). Chain molecule hydrodynamics by the Monte-Carlo method and the validity of the Kirkwood–Riseman approximation. *Macromolecules* 13, 592–602.

5 DNA–protein interaction

Interaction of DNA with various proteins is a key feature of its functioning inside the cell. This interaction can be either sequence specific or nonspecific. The basic features of these interactions will be considered in this chapter. We will start here from non-sequence-specific interaction.

5.1 Nonspecific DNA–protein binding

Many proteins interact with DNA without sequence specificity or with relatively low specificity. Also, proteins that bind specific DNA segments are able to bind the double helix in non-sequence-specific mode (see below). Affinity of these proteins to DNA is mainly due to the electrostatic interaction between the negatively charged double helix and positively charged protein interface responsible for the binding. In solution, small positive ions are bound (located in the close vicinity) with negatively charged DNA. Therefore, when a protein with a positively charged interface binds a DNA segment, it replaces the bound small ions. Therefore, the protein binding may give no overall reduction of the electrostatic energy of the solution, but results in an entropy gain, since a few released ions can now have independent locations anywhere in the solution. The protein binding corresponds to the reaction

$$\text{Pr} + \text{DNA} \leftrightarrow \text{Pr–DNA} + \Delta n \, \text{M}^+$$

where Δn is the difference in the number of bound small ions, M^+, between naked DNA and the DNA–protein complex, Pr–DNA. If one neglects the interaction between the adjacent bound ions, the equilibrium constant for this reaction, K, can be written as

$$K = \frac{[\text{Pr–DNA}]}{[\text{Pr}][\text{DNA}]}[\text{M}^+]^{\Delta n} = K_{\text{obs}}[\text{M}^+]^{\Delta n}, \tag{5.1}$$

where K_{obs} is the observed equilibrium constant for particular ionic conditions. By definition, K does not depend on $[\text{M}^+]$. Therefore,

$$K_{\text{obs}} = K[\text{M}^+]^{-\Delta n}. \tag{5.2}$$

Equation (5.2) allows one to determine Δn by measuring the dependence of K_{obs} on ion concentration. It follows from the equation that the value of K_{obs} reduces dramatically when the ion concentration in solution increases, if Δn is sufficiently large. It was found that Δn can be as large as 22 (for nonspecific binding of *E. coli* RNA polymerase (deHaseth *et al.* 1978)).

Of course, this treatment assumes that the same number of small ions is bound with DNA at any ion concentration in solution, which is definitely an oversimplification. A more elaborate analysis of the problem can be found in the work of Rouzina and Bloomfield (1997).

Electrostatic interaction between a bound protein and DNA can also involve protein dipoles. The dipoles are formed in the α-helices due to the hydrogen bonds in the backbone. The polar side chains of some residues can also be involved there.

In addition to electrostatic interaction, the proteins can form hydrogen bonds with the DNA backbone, which has both donors and acceptors of hydrogen bonds. Also, non-sequence-specific binding can be stabilized by what is called the hydrophobic interaction. This term means that energetically unfavorable contacts between water molecules and nonpolar groups of the DNA backbone and the protein are replaced by contacts between nonpolar groups of the backbone and protein. As a result of this replacement, the released water molecules form hydrogen bonds with other water molecules, reducing the total free energy associated with the protein binding.

Although the interactions listed above are able to provide sufficient binding free energy, sometimes non-sequence-specific binding also involves intercalation of aliphatic amino acids (isoleucine or proline) into the double helix. This intercalation causes strong bending of the double helix, which is usually related to the protein function. It is not clear to what extent if any it stabilizes the binding. Examples include IHF (Rice *et al.* 1996, Swinger & Rice 2004) and type IIA DNA topoisomerases (Dong & Berger 2007, Hardin *et al.* 2011). A diagram of the IHF–DNA complex is shown in Fig. 5.1.

Even if a protein interacts with the DNA backbone only, its affinity to the double helix has to have some sequence specificity, since DNA structure depends, at certain extent, on its sequence. It is known, for example, that the double-helix twist, measured in solution, depends on the sequence (see Table 3.3). The idea that the sequence-dependent DNA structure contributes to the specificity of the protein binding has become popular over the years (see Rohs *et al.* (2010), for example). This contribution of the sequence-dependent DNA structure to the specificity of protein binding is limited, however, due to relatively small sequence variations of DNA conformational parameters, and relatively large thermal fluctuations of these parameters. To illustrate this point we calculated the distribution of the average twist over all sequences of 16 bp in length and compared it with the twist distribution in a single segment of the same length due to thermal fluctuations (Fig. 5.2). The figure shows that the former distribution is substantially narrower than the latter one. This means that the distributions of twist for 16 bp segments with different sequences strongly overlap and therefore the magnitude of twist cannot serve as an essential selective marker of a specific sequence. We do not know the distributions of other conformational parameters of the double helix in solutions, but the data for crystals (Balasubramanian *et al.* 2009) allow us to assume that the same conclusion should be true for other DNA parameters as well. The energetic cost of larger disturbances that go beyond the elastic deformation, like DNA kinks, also has relatively low sequence dependence (Yakovchuk *et al.* 2006). Although a pair of kinks in the structure of a

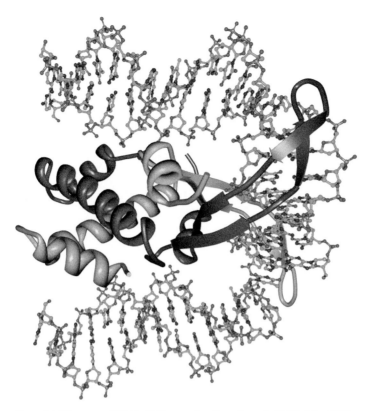

Figure 5.1 Diagram of DNA–IHF complex (Lynch *et al.* 2003). The DNA strands are shown in atom–bond presentation, while the protein is diagramed by the ribbons of different colors. Two prolines at the tips of β-hairpins are intercalated between the base pairs, and this causes strong bending of the double helix. The total bend angle is close to 160°. The image from RCSB PDB was obtained with Molecular Biology Toolkit (Moreland *et al.* 2005). A black-and-white version of this figure will appear in some formats. For the color version, please refer to the plate section.

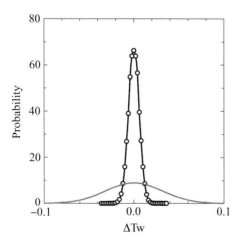

Figure 5.2 The distributions of the twist variations, ΔTw, for DNA segments 16 bp in length. The distribution shown by open circles was obtained by considering all possible sequences of the segment and using the values of twist for base-pair steps from Table 3.3. The gray line corresponds to the thermal fluctuation of twist in a single fragment 16 bp in length calculated for the DNA torsional rigidity of 3.0×10^{-19} erg·cm (see Eq. (B6.2.2)).

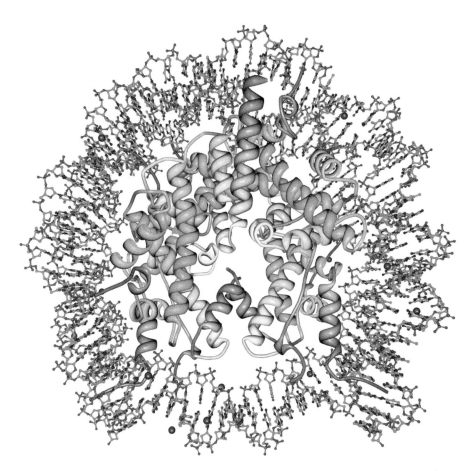

Figure 5.3 The structure of the nucleosome (Chua *et al.* 2012). The DNA strands are shown in atom–bond presentation. Different histone proteins are shown by ribbons of different colors. The 145 bp DNA fragment makes 1.75 helix turns around the histone core. The image from RCSB PDB was obtained with Molecular Biology Toolkit (Moreland *et al.* 2005). A black-and-white version of this figure will appear in some formats. For the color version, please refer to the plate section.

DNA–protein complex can provide some sequence specificity, their effect is limited. The binding of the nucleosome core particles with different DNA sequences, considered below, strongly supports this conclusion.

5.2   Selectivity of nucleosome core binding with DNA

Nucleosomes represent the basic structural units of eucaryotic chromosomes. They consist of a DNA segment 147 bp in length and a protein octamer (Luger *et al.* 1997, Richmond & Davey 2003). DNA makes nearly two complete turns around the protein core particle, and its structure is deformed substantially compared with the structure of the naked double helix (Fig. 5.3) (Chua *et al.* 2012). The decondensed chromatin isolated from eucaryotic cells looks like a string of nucleosomes that are bound by DNA linkers 5–80 bp in length. Although any DNA segment of appropriate length can bind the histone core to form a nucleosome, nucleosome positioning along DNA is not random either in vitro or inside the cells. The in vivo positioning is definitely influenced by other proteins tightly bound with DNA. Also, the nucleosome positioning in vivo is highly dynamic, as it changes according to the cell needs. The problem of nucleosome positioning in vivo is definitely beyond the scale of this book. Therefore, below we will touch on a simpler problem, the affinity of the nucleosome core to DNA segments with various sequences. This is a pure biophysical problem that makes a contribution to the nucleosome positioning in vivo.

The binding affinity of nucleosome core particles to DNA fragments with different sequences has received a lot of attention over the years (reviewed by Widom (2001), Struhl and Segal (2013)). Although it turned out to be difficult to measure the absolute values of the binding free energies, ΔG, the differences between these values for different sequences, $\Delta\Delta G$, have been determined for more than a hundred different sequences. It was found that $\Delta\Delta G$ spans over 4 kcal/mol (Widom 2001). Although this difference corresponds to a 1000-fold difference in the binding constants, it is a small difference for the interaction interface 147 bp in length. If we assume that sequence specificity of the affinity grows linearly with the length of the bound DNA segment, we obtain that an interface of 40 bp in length can provide an affinity span of about 1 kcal/mol. Clearly, this estimation is in accordance with the above statement that sequence-dependent DNA conformational properties can have very limited influence on the sequence specificity of the protein binding.

Much effort was put in over the years in attempts to predict DNA sequences that provide low and high affinity to the core particles. The majority of these attempts were concentrated on a search for empirical rules that connect the affinity with specific sequence patterns (Widom 2001, Trifonov 2011). A few studies tried, however, to predict the affinity by calculating the energy of DNA deformation to the experimentally determined nucleosome structure (Tolstorukov *et al.* 2007, Morozov *et al.* 2009, Balasubramanian *et al.* 2009). Of course, we do not know all parameters that specify the energy of DNA deformation in solution, so the researchers had to use the "statistical potentials" deduced from the database of DNA–protein structures in crystals (Olson *et al.* 1998, Balasubramanian *et al.* 2009). The approach brought some success and hope that we could learn how to predict the sequence specificity when better potentials describing the energetics of DNA deformation were known. This hope existed while the experimentally determined structures of nucleosomes included only DNA segments with very close sequences. The successive experimental studies eliminated this hope, however (Chua *et al.* 2012). The DNA structures in the nucleosomes formed by DNA segments with very different sequences turned to be essentially different as well (reviewed by Olson & Zhurkin (2011)). The finding that there is not a single DNA structure in the nucleosomes makes the problem of theoretical prediction of the sequence-dependent affinity of DNA to the core particles much more difficult. One has to find a way to evaluate the minimum energy of the nucleosome for a chosen DNA sequence by optimizing the entire nucleosome structure. This sounds like a very difficult task for the foreseeable future.

5.3   Search for specific DNA sites by proteins

Proteins participating in DNA replication and transcription with all its complex regulation, restriction endonucleases, methyltransferases and others, bind DNA sites with specific sequences. For some of these proteins there are only a few specific DNA sites over the entire genome, and they have to find the target sequences hidden in a huge molar excess of nonspecific sites. Various kinetic aspects of this very challenging problem of the sequence-specific protein binding have attracted a lot of attention of the biophysical community.

In 1970 Riggs *et al.* performed the first kinetic study of a site-specific DNA–protein association (Riggs *et al.* 1970). The researchers found that binding of *lac* repressor to *lac* operator, the specific binding site, is characterized by a rate constant of 7×10^9/M/s. This finding greatly influenced all subsequent studies of this kinetic problem because the obtained value exceeded the theoretical limit for a diffusion-controlled association of the molecules, 10^8/M/s. This limit corresponds to the rate of the first collision of two particles of a protein size due to thermal motion. The measured rate of association of *lac* repressor with *lac* operator was nearly two orders of magnitude larger. It became clear later, however, that there are a few specific sites in the *lac* operon used by Riggs *et al.* (1970), and the *lac* repressor has two separate binding surfaces for its cognate DNA sequence. Therefore, the measured rate of the specific binding did not exceed the theoretical limit by as much as was initially assumed.

Delbrück was probably the first to suggest that the search for a specific binding site may be facilitated by nonspecific binding of proteins with DNA and subsequent one-dimensional (1D) diffusion along the double helix (Adam & Delbrück 1968). This idea was elaborated in detail by Berg, Blomberg, Winter and von Hippel (Berg & Blomberg 1976, Berg & Blomberg 1977, Berg *et al.* 1981, von Hippel & Berg 1989). They found that the most efficient way for a protein to find a specific site consists of mixing of 1D and 3D diffusion. First, the protein binds DNA at a random nonspecific site and then diffuses until it finds a specific site or dissociates from the DNA molecule. In general, multiple DNA binding and dissociation events precede the binding of specific sites (Fig. 5.4). This general picture of the search has not been changed over the past 25 years, although its details have not been completely clarified (Halford & Marko 2004).

Numerous experimental studies of the search process confirmed that 1D diffusion along nonspecific sites makes an important contribution to the search process. It is more difficult, however, to determine the average DNA length searched by a protein before it dissociates from the DNA. The difficulty is due to the fact that there is an essential probability of repeating binding with the same DNA molecule (see Halford and Marko (2004), for example). Therefore, experimental data on the rate constant of specific binding for a particular system do not allow a unambiguous interpretation. More straightforward data on the sliding rate along nonspecific DNA, based on the direct visualization of fluorescently labeled proteins, have appeared only recently. The technique used in these studies requires strong stretching of the DNA, so its shape corresponds to a straight line, and strong nonspecific binding, that is, low concentration of ions in the solution (Gorman & Greene 2008). Overall, the majority of the studies

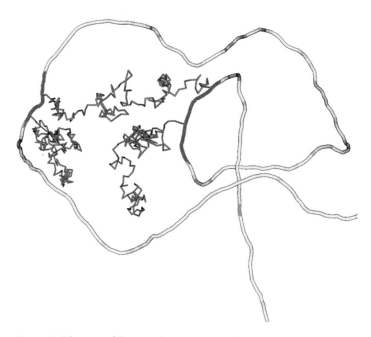

Figure 5.4 Diagram of the protein trajectory during the search for a specific binding site. The trajectory consists of segments of 3D diffusion (thin red lines) and random sliding along nonspecific segments of DNA (bold red lines, to emphasize that the sliding results in multiple visits of the same position along the DNA). DNA (bold yellow line) is assumed to have a random coil conformation in solution. There is a high probability that after dissociation from a double-helix segment the protein binds another segment of the same DNA molecule, as shown in the diagram. The specific binding site is shown in green. A black-and-white version of this figure will appear in some formats. For the color version, please refer to the plate section.

show that 1D diffusion coefficients are smaller by two to four orders of magnitude than the corresponding 3D coefficients (Gorman *et al.* 2007, Bonnet *et al.* 2008, Rau & Sidorova 2010, Dikic *et al.* 2012, Loth *et al.* 2013). Part of this reduction is due to the fact that during 1D diffusion a protein follows the helical path of the DNA backbone, which increases the total distance passed by the protein a fewfold (Schurr 1975). The current consensus is that many nonspecific bindings and dissociations occur during the search. Under ionic conditions close to physiological ones each of these bindings is followed by a 1D search that covers 50–100 bp of DNA (Gowers *et al.* 2005, Porecha & Stivers 2008, Rau & Sidorova 2010).

Although there is a very large amount of theoretical and experimental literature on the issue, nearly all these studies address the problem in vitro, where the setting is very different from the situation in vivo. Indeed, inside the cell DNA has a very compact conformation and is bound with many proteins, which should create obstacles for sliding, since the sliding has to follow the helical path of the DNA backbone. Recently, however, data on the search in bacteria started to appear (Elf *et al.* 2007, Hammar *et al.* 2012). According to the picture obtained in these studies, the search occurs as a combination of 1D diffusion along DNA and 3D diffusion in cytoplasm, as found in vitro. It was found that the *lac* repressor binds and dissociates from DNA about 10^5 times during the search (Elf *et al.* 2007). Correspondingly, the residence time of the protein bound with DNA nonspecifically is very short (<5 ms) compared with the total time of the search (>100 s). It was also estimated that the protein spends about 10% of the search time in the unbound state. Since 3D diffusion of the free protein inside the cell is very fast, a small fraction of the search time is sufficient to have uniform probability distribution of finding the protein at any point of the cytoplasm, regardless of the initial localization of the repressor. Thus, after very short time the protein "forgets" about the location of its appearance, and the distance between the protein gene and its specific binding site becomes irrelevant. The researchers concluded that the protein covers about 50 bp by sliding, on average, by a single binding event (Elf *et al.* 2007, Hammar *et al.* 2012). All these conclusions were indirect, however, and more studies on the search for the specific sites in vivo are needed.

5.4   Recognition of specific DNA sites by proteins

DNA bases are localized in the interior of the double helix. Therefore, opening the helical structure seems to be the simplest solution for the recognition of DNA segments with specific sequences. This opening is expensive energetically, however. Thus, it is important that DNA sequence can be read without base-pair opening, over specific chemical groups exposed in the double-helix grooves (Fig. 5.5). In the search for specific sites proteins use the latter method of sequence reading.

There was a hope, before the first structures of DNA–protein complexes were resolved, that there is a correspondence between DNA bases and amino-acid residues in the recognition interfaces. This hope evaporated quickly, however. It is now well established that the recognition takes place mainly by short and relatively rigid structural motifs of the proteins. The interfaces of these motifs consist mainly of α-helices and β-sheets. Many different amino-acid residues of these recognition motifs can interact with any of the four different bases. The structural motifs of the proteins can form surfaces that are complementary to the major groove of the recognition segments. Although ionic bonds and hydrophobic interactions contribute there, the hydrogen bonds between the proteins and DNA are the main elements of the recognition. It was found that certain structural motifs of the proteins, consisting of α-helices and β-sheets, appear especially often in the protein recognition sites. These motifs were carefully classified (reviewed by Luscombe *et al.* 2000, Rohs *et al.* 2010).

Due to its cylindrical surface only four to six consecutive base pairs can interact with a single α-helix or β-sheet of protein. Therefore, usually a pair of protein recognition regions is needed to provide sufficient specificity of the binding. One of the solutions here is the protein dimerization. Many sequence-specific DNA-binding proteins are homodimers that are arranged symmetrically, so they interact with specific palindromic or nearly palindromic sequences. In such cases each monomer interacts with a half of the specific site. Of course, this dimeric arrangement of the proteins greatly increases their affinity to the recognition sites.

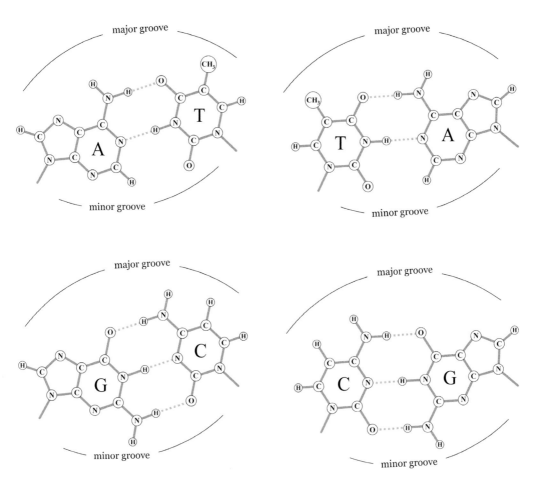

Figure 5.5 The chemical groups of the bases exposed to major and minor grooves of the double helix. Four possible configurations of base pairs in the double helix are shown. The difference between the groups exposed by different base pairs is much more pronounced in the major groove, and therefore the sequence-specific binding occurs mainly due to interactions with this groove of the double helix.

As discussed above, proteins that bind specific DNA sequences also bind the double helix in nonspecific mode. Although the latter binding is weaker, it also involves a substantial number of hydrogen bonds with phosphates (Kalodimos *et al.* 2004a). Therefore, due to the periodicity of the double helix, the stable positions of a nonspecifically bound protein along the DNA have to be separated by the free energy barriers at positions where formation of hydrogen bonds is not possible (Fig. 5.6). These barriers between the free-energy minima of the nonspecific binding have to be relatively low, to provide sufficiently fast diffusion of the protein along the DNA. The free energy at the barrier tops should also be lower than the free energy of the unbound proteins, so a step of random displacement of the protein along the DNA would have much higher probability than its dissociation from DNA. The dimeric structure of sequence-specific proteins can greatly facilitate solution of this problem. It was found that in some such dimeric

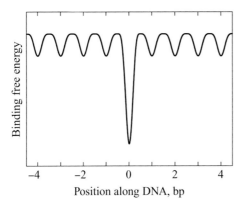

Figure 5.6 Diagram of the free-energy profile for a protein that binds DNA in both sequence-specific and non-sequence-specific modes. Even in the latter mode there are discrete positions of the protein along the double helix corresponding to free-energy minima. The position of specific binding corresponds to the deep free-energy minimum in the plot.

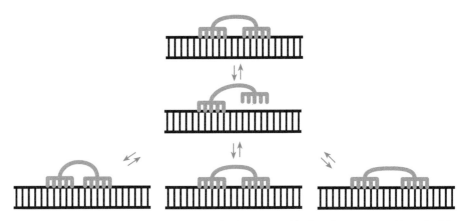

Figure 5.7 Diagram of displacement of a protein that has two binding sites connected by a flexible linker. The diagram shows a displacement of only one binding element, but displacement of the other can occur in the same way. This method of motion provides a faster protein displacement while keeping contact between the protein and the DNA.

proteins two relatively rigid recognition sites are connected by flexible linkers, so they can easily change their mutual position and orientation in certain limits. Therefore, it seems possible that the two protein recognition sites can dissociate from the DNA independently and then bind an adjacent DNA segment. Therefore, the protein can move along nonspecific DNA by detaching only one of two DNA-binding elements at each instant of time and rebinding it with a different DNA segment (Fig. 5.7). As we know from our everyday experience of walking, this is a more efficient way of moving around than jumping. The principle of movement by using two or more feet is also widely used in biological molecular motors, including DNA helicases that translocate along DNA (see below). Although it has not been shown that the proteins use this method to move along the DNA, the high flexibility of the connectors between the two DNA-binding segments makes this mechanism of the displacement very probable. This issue is well illustrated by the structures of *lac* repressor bound with specific and nonspecific DNA segments.

The *lac* repressor represents probably the best-studied system of gene regulation. The repressor acts as a homotetramer. The tetramer has two DNA-binding interfaces and can simultaneously bind two operator sites of the *lac* operon, which are well separated along the DNA contour. Thus, in the context of our current topic it is sufficient to consider a dimer that interacts with the specific and nonspecific DNA segments. Although the sequences of the operators are only approximately palindromic, in the specific complex they are bound with identical recognition motifs of two monomers. It was also found that the protein binds with 10 times greater affinity to a palindrome of the left half of the operator, so this palindrome has been widely used in biochemical and structural studies of the system. The X-ray structure of the complex of the *lac* repressor and the palindromic DNA segment is shown in Fig. 5.8. There are two pairs of protein recognition motifs in this structure. The helix–turn–helix (HTH) motifs, well separated in the structure, interact with DNA bases in the major grove. The two so-called hinge α-helices, located in the middle of the domain, also interact with DNA bases, making contributions to the binding specificity. The conformation of the HTH motifs remains nearly the same over all structures of DNA–*lac* repressor complexes obtained in various studies, although their orientation with respect to DNA binding site changes substantially (Kalodimos *et al.* 2004b). It is this change of orientation of the HTH motif that provides specific recognition of the natural, nonpalindromic operator by the dimer (Fig. 5.9). Amazingly, as a result of this reorientation the same α-helix specifically interacts with essentially different DNA sequences (Kalodimos *et al.* 2004b).

It was difficult to obtain the structure of *lac* repressor bound with a nonspecific DNA site, due to multiple positions of the protein on a DNA fragment. Kaptein and co-workers managed to overcome this problem by using a mutated dimer of the protein recognition domain and a specially chosen DNA duplex (Kalodimos *et al.* 2004a). They found that the same α-helix of the HTH motif can interact exclusively with DNA backbone just by changing its orientation relative to the double helix. They also found that the hinge regions of the protein subunits do not form α-helices observed in the specific complex and remain unstructured. Thus, in nonspecific complex each of the HTH motifs that interact with the DNA backbone is connected to the rest of the protein by the long flexible segment of the polypeptide chain. These linkers greatly increase the set of possible mutual positions of the HTH motifs, allowing them to bind DNA segments with varying separations along the DNA. Clearly, the structure of this complex supports the two-feet model of protein translocation along DNA.

5.5   Formation of protein-bridged DNA loops

Many proteins during their functioning inside the cells interact with two DNA sites simultaneously, and this results in formation of protein-assisted DNA loops. The looping

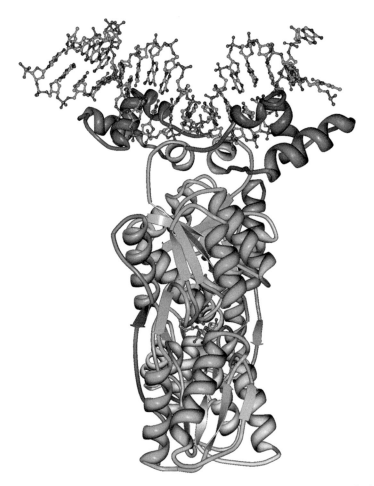

Figure 5.8 Structure of the complex of the *lac* repressor dimer with the palindromic operator (Lewis *et al.* 1996). The DNA-recognition domain consists of two HTH motifs separated from one another (shown as the dark-blue helices) and two α-helices, which interact with each other (the hinge region) and with the operator site (shown in light blue in the middle of the recognition domain). The image from RCSB PDB was obtained with Molecular Biology Toolkit (Moreland *et al.* 2005). A black-and-white version of this figure will appear in some formats. For the color version, please refer to the plate section.

appears in DNA replication, regulation of transcription, site-specific recombination, topological transformations by type II topoisomerases and DNA cleavage by type II restriction enzymes. The thermodynamics of protein-bridged formation of DNA loops has been a subject of many biophysical studies (see Schleif (1992), Halford *et al.* (2004), Allemand *et al.* (2006) for reviews). In terms of DNA conformational distribution, the problem of the looping–unlooping equilibrium is very close to the problem of juxtaposition of DNA ends, specified by the *j*-factor (see Figs. 3.23 and 3.24). Similar to the *j*-factor, the equilibrium constant of the looping exhibits oscillations with DNA length, if the loop size does not exceed 400–500 bp (Schleif 1992). The equilibrium is

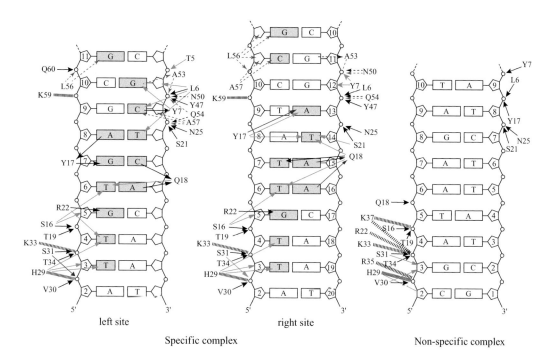

Figure 5.9 Diagram of interactions between *lac* repressor and specific and nonspecific DNA sites. The left and right halves of the recognition site are shown separately since their sequences are different in the natural operator. The bases that are specifically recognized by the protein are shaded. Dark and light arrows show hydrogen bonding and hydrophobic interactions, respectively; the dashed lines indicate electrostatic contacts. The two different interaction patterns with the halves of the specific site are achieved by changing conformations of Tyr7, Tyr17 and Gln18 in the α-helix and by slight rotation of the entire recognition motif. The same α-helix interacts with both specific and nonspecific DNA sites. Only one of two halves of the nonspecific site is shown. Figure reproduced from Fig. 7 of Kalodimos *et al.* (2004b), with permission.

also affected by DNA supercoiling, which can increase the probability of juxtaposition by up to 100-fold (Vologodskii & Cozzarelli 1996).

The kinetics of the protein-bridged looping has not been studied sufficiently, however. Considering this issue we, first of all, need to know if the rate of bridging is limited by the rate of the first collision of the DNA sites, or if many collisions, on average, precede the loop formation. The latter alternative would mean that the rate of bridging should be proportional to the equilibrium probability of site juxtaposition. So far only two studies have addressed this problem, for two different systems.

Halford and co-workers prepared DNA substrate with two recognition sites for the type II restriction enzyme FokI. The enzyme cleaves DNA efficiently only when two of its monomers bound to individual DNA binding sites form the dimer, creating a protein-bridged DNA loop. The researchers carefully dissected the multistep reaction pathway for this system and estimated the bridging time. The half-time of bridging was 0.23 s, while it should be in the range of milliseconds if it is diffusion limited (Huang *et al.* 2001, Klenin & Langowski 2001).

A system of transcription activation by a special DNA site called an enhancer was used in the other study addressing the issue (Polikanov *et al.* 2007). In this case the loop was made by the enhancer-bound protein interacting with RNA polymerase bound with the corresponding promoter. Formation of the loop activated the transcription complex, so the detection of the newly synthesized RNA molecules allowed the researchers to monitor the loop formation. The loop formation took minutes in this case, which itself served as a strong indication that the looping rate was not limited by diffusion. The researchers, however, used a different argument to show that the looping rate is specified by the equilibrium probability of the juxtaposition of the two sites, P. It was shown by computer simulations that the value of P is 100 times higher in supercoiled DNA than in the relaxed form. On the other hand, the average time of the first collision of two DNA sites is not affected, practically, by DNA supercoiling (Huang *et al.* 2001). It was found that the rate of transcription activation in the studied system is 50 times higher in supercoiled DNA compared with the relaxed form. This allowed the authors to conclude that the looping rate is proportional to P.

Thus, both studies showed that the rate of loop formation was not limited by the rate of diffusion of one DNA site to another. This does not mean, however, that this conclusion is general. Indeed, the question of how fast two juxtaposed sites are bridged by proteins depends on the interaction between the participating proteins. Sometimes only one protein participates in the bridging, and the interaction between the second protein binding site and the double helix play a decisive role in the looping. Opposite electric charges between the interacting molecules can strongly accelerate the binding. Clearly, more studies of the issue are needed to obtain a better picture of the bridging kinetics.

5.6   DNA helicases

DNA helicases represent a broad class of enzymes that are capable of not just binding DNA but unidirectionally translocating along ss and sometimes dsDNA. Among this kind of protein are RNA polymerases and DNA polymerases. The main function of helicases is unwinding the double helix, which they do by moving along ssDNA. Of course, a unidirectional movement requires energy consumption, and the translocation of helicases is coupled with ATP hydrolysis. The rate of the translocation depends on the ATP concentration in the surrounding solution and is strongly affected by other proteins that interact with helicases. During their unwinding activity, helicases are localized in front of the DNA replication fork; they move in the direction of the fork separating transiently opened strands of the double helix. There are indications that, while some helicases wait for transient opening of base pairs in front of them and then move ahead along the bound DNA strand, others promote the opening for faster unwinding.

Helicases are divided into six superfamilies (SF1,..., SF6) on the basis of their sequence and tertiary structure (Singleton *et al.* 2007). Proteins of superfamilies SF1 and SF2 can work as monomers, while other helicases function in the form of hexameric

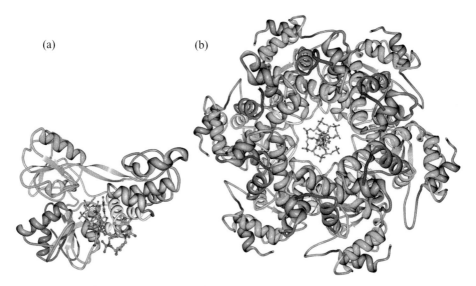

Figure 5.10 Two kinds of structure formed by DNA helicases. (a) NS3 helicase from the hepatitis C virus, which acts as a monomer (Gu & Rice 2010). (b) E1 helicase from papilloma virus, which forms a ring hexamer, with translocated ssDNA in the ring's central hole (Enemark & Joshua-Tor 2006). The images from RCSB PDB were obtained with Molecular Biology Toolkit (Moreland *et al.* 2005). A black-and-white version of this figure will appear in some formats. For the color version, please refer to the plate section.

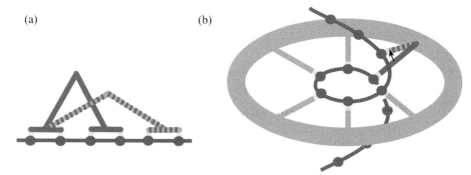

Figure 5.11 Diagram of helicase movement mechanisms. ssDNA is shown in green and helicases in red and orange; the old and new positions of the moving helicase parts are shown by red and red–orange lines, respectively. (a) The helicases of SF1 and SF2 superfamilies have two sites interacting with ssDNA, leading and lagging ones. They move by separating one of the sites and placing it ahead in the direction of movement, due to the conformational transition in the protein. (b) The hexameric helicases of superfamilies SF3, SF4, SF5 and SF6 force a spiral conformation of ssDNA inside their central hole, so that each subunit of the protein interacts with the DNA backbone. The helicases move along DNA by successively switching the position of each of their six DNA-interacting loops from lagging to leading. A black-and-white version of this figure will appear in some formats. For the color version, please refer to the plate section.

rings (in very rare cases the rings consist of four monomers). Examples of the two types of structure are shown in Fig. 5.10. All of them are able to interact nonspecifically with the backbone of ssDNA, although some of the members of SF1 and SF2 can bind and translocate dsDNA as well. Short, hardly structured loops of the proteins participate in the interaction with DNA. The members of SF1 and SF2 have two such loops per protein, while the hexameric rings have six interacting loops, one from each monomer. Thus, all helicases use the principle of foot walking for displacement. According to this principle only one of the protein's interacting loops can be detached from the DNA at any time. The loop reattaches to a new DNA segment, located in the direction of the translocation. The displacement is unidirectional due to the unidirectional conformational changes in the proteins coupled with the ATP hydrolysis. In the case of SF1 and SF2 the translocation of the proteins is provided by ATP-dependent change of the distance between two domains interacting with DNA (Fig. 5.11(a)) (Gu & Rice 2010). A more sophisticated mechanism of translocation is used by the hexameric helicases (Enemark & Joshua-Tor 2006). In this case the single step of displacement consists of switching the position of a single protein loop interacting with DNA from the lagging to the leading position (Fig. 5.11(b)).

References

Adam, G. & Delbrück, M. (1968). Reduction of dimensionality in biological diffusion processes. In *Structural Chemistry and Molecular Biology*, eds. A. Rich & N. Davidson, 198–215. San Francisco, CA: Freeman.

Allemand, J. F., Cocco, S., Douarche, N. & Lia, G. (2006). Loops in DNA: an overview of experimental and theoretical approaches. *Eur. Phys. J. E* 19, 293–302.

Balasubramanian, S., Xu, F. & Olson, W. K. (2009). DNA sequence-directed organization of chromatin: structure-based computational analysis of nucleosome-binding sequences. *Biophys. J.* 96, 2245–60.

Berg, O. G. & Blomberg, C. (1976). Association kinetics with coupled diffusional flows – special application to lac repressor–operator system. *Biophys. Chem.* 4, 367–81.

(1977). Association kinetics with coupled diffusion – extension to coiled-chain macromolecules applied to lac repressor-operator system. *Biophys. Chem.* 7, 33–9.

Berg, O. G., Winter, R. B. & von Hippel, P. H. (1981). Diffusion-driven mechanisms of protein translocation on nucleic acids. 1. Models and theory. *Biochemistry* 20, 6929–48.

Bonnet, I., Biebricher, A., Porte, P. L., Loverdo, C., Benichou, O., Voituriez, R., Escude, C., Wende, W., Pingoud, A. & Desbiolles, P. (2008). Sliding and jumping of single EcoRV restriction enzymes on non-cognate DNA. *Nucleic Acids Res.* 36, 4118–27.

Chua, E. Y., Vasudevan, D., Davey, G. E., Wu, B. & Davey, C. A. (2012). The mechanics behind DNA sequence-dependent properties of the nucleosome. *Nucleic Acids Res.* 40, 6338–52.

deHaseth, P. L., Lohman, T. M., Burgess, R. R. & Record, M. T., Jr. (1978). Nonspecific interactions of Escherichia coli RNA polymerase with native and denatured DNA: differences in the binding behavior of core and holoenzyme. *Biochemistry* 17, 1612–22.

Dikic, J., Menges, C., Clarke, S., Kokkinidis, M., Pingoud, A., Wende, W. & Desbiolles, P. (2012). The rotation-coupled sliding of EcoRV. *Nucleic Acids Res.* 40, 4064–70.

Dong, K. C. & Berger, J. M. (2007). Structural basis for gate–DNA recognition and bending by type IIA topoisomerases. *Nature* 450, 1201–5.

Elf, J., Li, G. W. & Xie, X. S. (2007). Probing transcription factor dynamics at the single-molecule level in a living cell. *Science* 316, 1191–4.

Enemark, E. J. & Joshua-Tor, L. (2006). Mechanism of DNA translocation in a replicative hexameric helicase. *Nature* 442, 270–5.

Gorman, J., Chowdhury, A., Surtees, J. A., Shimada, J., Reichman, D. R., Alani, E. & Greene, E. C. (2007). Dynamic basis for one-dimensional DNA scanning by the mismatch repair complex Msh2–Msh6. *Mol. Cell* 28, 359–70.

Gorman, J. & Greene, E. C. (2008). Visualizing one-dimensional diffusion of proteins along DNA. *Nat. Struct. Mol. Biol.* 15, 768–74.

Gowers, D. M., Wilson, G. G. & Halford, S. E. (2005). Measurement of the contributions of 1D and 3D pathways to the translocation of a protein along DNA. *Proc. Natl. Acad. Sci. U. S. A.* 102, 15883–8.

Gu, M. & Rice, C. M. (2010). Three conformational snapshots of the hepatitis C virus NS3 helicase reveal a ratchet translocation mechanism. *Proc. Natl. Acad. Sci. U. S. A.* 107, 521–8.

Halford, S. E. & Marko, J. F. (2004). How do site-specific DNA-binding proteins find their targets? *Nucleic Acids Res.* 32, 3040–52.

Halford, S. E., Welsh, A. J. & Szczelkun, M. D. (2004). Enzyme-mediated DNA looping. *Annu. Rev. Biophys. Biomol. Struct.* 33, 1–24.

Hammar, P., Leroy, P., Mahmutovic, A., Marklund, E. G., Berg, O. G. & Elf, J. (2012). The lac repressor displays facilitated diffusion in living cells. *Science* 336, 1595–8.

Hardin, A. H., Sarkar, S. K., Yeonee Seol, Y., Liou, G. F., Osheroff, N. & Neuman, K. C. (2011). Direct measurement of DNA bending by type IIA topoisomerases: implications for nonequilibrium topology simplification. *Nucleic Acids Res.* 39, 5729–43.

Huang, J., Schlick, T. & Vologodskii, T. (2001). Dynamics of site juxtaposition in supercoiled DNA. *Proc. Natl. Acad. Sci. U. S. A.* 98, 968–73.

Kalodimos, C. G., Biris, N., Bonvin, A. M., Levandoski, M. M., Guennuegues, M., Boelens, R. & Kaptein, R. (2004a). Structure and flexibility adaptation in nonspecific and specific protein–DNA complexes. *Science* 305, 386–9.

Kalodimos, C. G., Boelens, R. & Kaptein, R. (2004b). Toward an integrated model of protein–DNA recognition as inferred from NMR studies on the Lac repressor system. *Chem. Rev.* 104, 3567–86.

Klenin, K. V. & Langowski, J. (2001). Diffusion-controlled intrachain reactions of supercoiled DNA: Brownian dynamics simulations. *Biophys. J.* 80, 69–74.

Lewis, M., Chang, G., Horton, N. C., Kercher, M. A., Pace, H. C., Schumacher, M. A., Brennan, R. G. & Lu, P. (1996). Crystal structure of the lactose operon repressor and its complexes with DNA and inducer. *Science* 271, 1247–54.

Loth, K., Gnida, M., Romanuka, J., Kaptein, R. & Boelens, R. (2013). Sliding and target location of DNA-binding proteins: an NMR view of the lac repressor system. *J. Biomol. NMR* 56, 41–9.

Luger, K., Mader, A. W., Richmond, R. K., Sargent, D. F. & Richmond, T. J. (1997). Crystal structure of the nucleosome core particle at 2.8 A resolution. *Nature* 389, 251–60.

Luscombe, N. M., Austin, S. E., Berman, H. M. & Thornton, J. M. (2000). An overview of the structures of protein-DNA complexes. *Genome Biol.* 1, 1–10.

Lynch, T. W., Read, E. K., Mattis, A. N., Gardner, J. F. & Rice, P. A. (2003). Integration host factor: putting a twist on protein–DNA recognition. *J. Mol. Biol.* 330, 493–502.

Moreland, J. L., Gramada, A., Buzko, O. V., Zhang, Q. & Bourne, P. E. (2005). The Molecular Biology Toolkit (MBT): a modular platform for developing molecular visualization applications. *BMC Bioinform.* 6, 21.

Morozov, A. V., Fortney, K., Gaykalova, D. A., Studitsky, V. M., Widom, J. & Siggia, E. D. (2009). Using DNA mechanics to predict in vitro nucleosome positions and formation energies. *Nucleic Acids Res.* 37, 4707–22.

Olson, W. K., Gorin, A. A., Lu, X. J., Hock, L. M. & Zhurkin, V. B. (1998). DNA sequence-dependent deformability deduced from protein–DNA crystal complexes. *Proc. Natl. Acad. Sci. U. S. A.* 95, 11163–8.

Olson, W. K. & Zhurkin, V. B. (2011). Working the kinks out of nucleosomal DNA. *Curr. Opin. Struct. Biol.* 21, 348–57.

Polikanov, Y. S., Bondarenko, V. A., Tchernaenko, V., Jiang, Y. I., Lutter, L. C., Vologodskii, A. & Studitsky, V. M. (2007). Probability of the site juxtaposition determines the rate of protein-mediated DNA looping. *Biophys. J.* 93, 2726–31.

Porecha, R. H. & Stivers, J. T. (2008). Uracil DNA glycosylase uses DNA hopping and short-range sliding to trap extrahelical uracils. *Proc. Natl. Acad. Sci. U. S. A.* 105, 10791–6.

Rau, D. C. & Sidorova, N. Y. (2010). Diffusion of the restriction nuclease EcoRI along DNA. *J. Mol. Biol.* 395, 408–16.

Rice, P. A., Yang, S., Mizuuchi, K. & Nash, H. A. (1996). Crystal structure of an IHF–DNA complex: a protein-induced DNA U-turn. *Cell* 87, 1295–306.

Richmond, T. J. & Davey, C. A. (2003). The structure of DNA in the nucleosome core. *Nature* 423, 145–50.

Riggs, A. D., Bourgeois, S. & Cohn, M. (1970). The lac repressor–operator interaction. 3. Kinetic studies. *J. Mol. Biol.* 53, 401–17.

Rohs, R., Jin, X., West, S. M., Joshi, R., Honig, B. & Mann, R. S. (2010). Origins of specificity in protein–DNA recognition. *Annu. Rev. Biochem.* 79, 233–69.

Rouzina, I. & Bloomfield, V. A. (1997). Competitive electrostatic binding of charged ligands to polyelectrolytes: practical approach using the non-linear Poisson–Boltzmann equation. *Biophys. Chem.* 64, 139–55.

Schleif, R. (1992). DNA looping. *Annu. Rev. Biochem.* 61, 199–223.

Schurr, J. M. (1975). The one-dimensional diffusion coefficient of proteins absorbed on DNA hydrodynamic considerations. *Biophys. Chem.* 9, 413–14.

Singleton, M. R., Dillingham, M. S. & Wigley, D. B. (2007). Structure and mechanism of helicases and nucleic acid translocases. *Annu. Rev. Biochem.* 76, 23–50.

Struhl, K. & Segal, E. (2013). Determinants of nucleosome positioning. *Nat. Struct. Mol. Biol.* 20, 267–73.

Swinger, K. K. & Rice, P. A. (2004). IHF and HU: flexible architects of bent DNA. *Curr. Opin. Struct. Biol.* 14, 28–35.

Tolstorukov, M. Y., Colasanti, A. V., McCandlish, D. M., Olson, W. K. & Zhurkin, V. B. (2007). A novel roll-and-slide mechanism of DNA folding in chromatin: implications for nucleosome positioning. *J. Mol. Biol.* 371, 725–38.

Trifonov, E. N. (2011). Cracking the chromatin code: precise rule of nucleosome positioning. *Phys. Life Rev.* 8, 39–50.

Vologodskii, A. V. & Cozzarelli, N. R. (1996). Effect of supercoiling on the juxtaposition and relative orientation of DNA sites. *Biophys. J.* 70, 2548–56.

von Hippel, P. H. & Berg, O. G. (1989). Facilitated target location in biological systems. *J. Biol. Chem.* 264, 675–8.

Widom, J. (2001). Role of DNA sequence in nucleosome stability and dynamics. *Q. Rev. Biophys.* 34, 269–324.

Yakovchuk, P., Protozanova, E. & Frank-Kamenetskii, M. D. (2006). Base-stacking and base-pairing contributions into thermal stability of the DNA double helix. *Nucleic Acids Res.* 34, 564–74.

6 Circular DNA

In 1963 Dulbecco and Vogt, and Weil and Vinograd, discovered that dsDNA of the polyoma virus exists in a closed circular form (Dulbecco & Vogt 1963, Weil & Vinograd 1963). It turned out that this form is typical of bacterial DNA and of cytoplasmic DNA in animals. The distinctive feature of closed circular molecules is that its topological state cannot be altered by any conformational rearrangement that does not involve breaking DNA strands. This topological constraint is the basis for the fascinating properties of circular DNA molecules. The physical properties of circular DNA molecules is a subject of the current chapter.

6.1   Linking number of complementary strands and DNA supercoiling

Two forms of circular DNA molecules are extracted from the cell; they were designated as form I and form II (Weil & Vinograd 1963). The more compact form I was found to turn into form II after a single-stranded break was introduced into one chain of the double helix. Subsequent studies performed by Vinograd and co-workers linked the compactness of form I, in which both DNA strands are intact, to supercoiling. Form I is called the closed circular form. In this form each of the two strands that make up the DNA molecule are closed in on themselves. A diagram of closed circular DNA is presented in Fig. 6.1. The two strands of the double helix in closed circular DNA are topologically linked. In topological terms, the links between two strands of the double helix belong to a particular class, called the torus class (see Section 6.7). The quantitative description of such links is called the *linking number*, Lk, which may be determined in the following way (Fig. 6.2). One of the strands defines the edge of an imaginary surface (any such surface gives the same result). Lk is the algebraic (i.e. sign-dependent) number of intersections of the surface by the other strand. By convention, the Lk of a closed circular DNA formed by a right-handed double helix is positive. Lk depends only on the topological state of the strands and hence is maintained through all conformational changes that occur in the absence of strand breakage. It follows from the definition that Lk is an integer. Quantitatively, Lk of DNA strands is close to N/γ, where N is the number of base pairs in the molecule and γ is the number of base pairs per helix turn (the helical repeat) in linear DNA under given solution conditions. However, Lk and N/γ are not exactly equal to one another, and

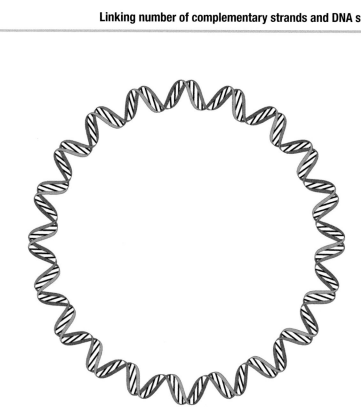

Figure 6.1 Diagram of closed circular DNA. The complementary strands form a topological link and cannot be separated from each other without breaking one of them. The topological characteristic of the link, the linking number, is equal to 20 in the diagram. A black-and-white version of this figure will appear in some formats. For the color version, please refer to the plate section.

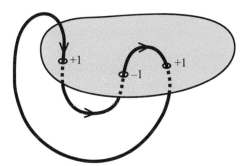

Figure 6.2 The definition of the linking number of two closed contours. One draws an imaginary surface on one contour, arbitrarily chooses the direction on the other contour and then calculates the algebraic number of intersections between this surface and the other contour. $Lk = 1$ for this example. Note that we could assign opposite signs to all the intersections, so Lk would equal -1. Thus, in this case the sign of Lk is not defined.

the difference between Lk and N/γ (which is also denoted as Lk_0) defines most of the properties of closed circular DNA. This difference is called the *linking number difference*, ΔLk:

$$\Delta Lk = Lk - N/\gamma. \qquad (6.1)$$

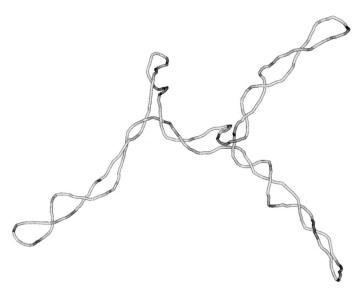

Figure 6.3 A typical conformation of supercoiled DNA. The yellow chain corresponds to dsDNA. The picture was obtained by computer simulation of a supercoiled molecule 4000 bp in length, $\sigma = -0.05$, at physiological ionic conditions.

Whenever $\Delta Lk \neq 0$, the entire double helix is stressed. This stress can either lead to a change in the actual number of base pairs per helix turn in closed circular DNA or cause a regular spatial deformation of the helix axis. It is this deformation of the helix axis that gave rise to the term "superhelicity" or "supercoiling," since the DNA axis forms in this case a helix of a higher order (Fig. 6.3).

There are two important inferences to be made from the definition of ΔLk.

1. The value of ΔLk is not a topological invariant. It depends on the solution conditions that determine γ. Even though γ itself changes very slightly with changing ambient conditions, these changes may substantially alter ΔLk, as the right-hand part of Eq (6.1) is the difference between two large quantities that are close in value.
2. The linking number is by definition an integer, whereas N/γ should not be an integer. Hence ΔLk is not an integer either. However, the values of ΔLk for a closed circular DNA with a particular sequence can differ by an integer only. This simply follows from the fact that in any particular conditions all variations in ΔLk can only be due to variations in Lk, since the value of N/γ is the same for all molecules. Molecules that have the same chemical structure and differ only with respect to Lk are called *topoisomers*.

It often proves more convenient to use the value of *superhelix density*, σ, which is ΔLk normalized to Lk_0:

$$\sigma = \Delta Lk/Lk_0 = \gamma \Delta Lk/N. \tag{6.2}$$

Circular DNA extracted from cells turns out to be always (or nearly always) negatively supercoiled, and has a value of σ between -0.03 and -0.09, but typically near the middle of this range (Bauer 1978).

6.2   Twist and writhe

6.2.1   The major theorem

Supercoiling can be structurally realized in two ways: by deforming the molecular axis and by altering the twist of the double helix. This can be ascertained by means of a simple experiment involving a rubber hose. To connect the ends of the hose we also need a short rod that can be pushed into the hose with some effort. The rod can be used to join together the ends of the hose and thus rule out their reciprocal rotation around the hose axis. If before joining the two ends we turn one of them several times around the axis, i.e. twist the hose, it will take a helical shape once the ends are joined. If one draws longitudinal stripes on the hose prior to the experiment, it will be clear that reciprocal twisting of the ends also causes the hose's torsional deformation.

There is very important quantitative relationship between the deformation of the DNA axis, its torsional deformation and Lk of the complementary strands of the double helix. The first mathematical treatment of the problem was presented in 1961 by Calugareanu, who found the basic relationship between the geometrical properties of a closed ribbon and topological properties of its edges (Calugareanu 1961). Ten years later Fuller suggested that Calugareanu's theorem can be very useful in the analysis of circular DNA (Fuller 1971). Two years earlier White had suggested a more general formulation of Calugareanu's theorem (White 1969), and sometimes the theorem is called the Calugareanu–White–Fuller theorem.

According to the theorem, the Lk of the edges of the ribbon is the sum of two values. One is *the twist* of the ribbon, Tw, a well known concept, and the second is a new concept, *writhe*, Wr. Thus,

$$Lk = Wr + Tw. \tag{6.3}$$

Tw is a measure of the number of times one of the edges of the ribbon spins about its axis. The Tw value of the entire ribbon is the sum of the Tw values of its parts. It is important that the Tw of each small segment of the ribbon is measured relative to the local axis of the segment. This axis changes its orientation in space when we move along the curve, and this is why the total Tw of the curve is not equal, in general, to Lk. The value of Wr is defined by the spatial course of the ribbon axis; i.e., it is a characteristic of a single closed curve, unlike Lk and Tw, which are properties of a closed ribbon. Thus Lk can be represented as a sum of two values that characterize the available degrees of freedom: torsional deformation around the ribbon axis and deformation of this axis. To apply the theorem to circular DNA, the two strands of the double helix should be considered as edges of a ribbon.

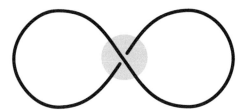

Figure 6.4 The nearly flat "figure eight". Only the gray-circled part of the contour juts out of plane. For such a curve Wr is very close to -1.

6.2.2   Properties of writhe

Let us consider the most important properties of Wr. The Wr of a DNA conformation is completely specified by the geometry of the DNA axis. It can be expressed through the Gauss integral (see Box 6.1). Wr can be thought of as a measure of a curve's net right-handed or left-handed asymmetry, i.e. its chirality, and is equal to zero for a planar curve. Unlike Lk, which can only be an integer, a curve's Wr can have any value. It changes continuously with the curve's deformation, which does not involve the intersection of segments. A curve's Wr does not change with a change in the curve's scale and depends solely on its shape.

Consider Wr of an almost flat figure-eight-shaped curve (Fig. 6.4). As it turns out, the Wr of such a curve is equal to -1 if the curve is a fragment of a right-handed helix (the case in Fig. 6.4), and $+1$ if the crossing corresponds to a left-handed helix. This result does not depend on the shape or size of the two loops of the "eight," or on the angle between the curve segments in the crossing area.

In general, for a nearly flat curve Wr can be easily and accurately estimated by analyzing the diagram of the curve projection on an arbitrary plane (Fuller 1971). For each crossing on the projection, breaks have to be drawn in the underpassing segments (Fig. 6.5). Then, one has to choose an arbitrary direction on the curve projection. The diagram prepared in such a way allows one to ascribe $(+1)$ or (-1) to each crossing according to the rule shown in Fig. 6.6. The Wr of the curve is close to the algebraic sum of the contributions from all "crossings." The angle between the crossing segments is not essential in this calculation. It should be emphasized, however, that this procedure cannot be used to estimate Wr of an arbitrary, nonflat, space curve.

Another important property of Wr holds not only for the "quasi-flat" curve but for an arbitrary curve as well. When the curve is deformed in such a way that one of its parts passes through another, the writhe value experiences a leap by 2 (or -2 for the opposite direction of the pass). If the strand-passing event occurs in dsDNA (as a result of type II topoisomerase action), when one double-stranded segment passes through another the value of Tw does not change. Thus, the strand-passing results in a change of Lk that is equal to the change of Wr (see Eq. (6.3)). This property helped greatly in the analysis of the action mechanisms of DNA topoisomerases (see Section 6.9).

The main result of the ribbon theory, expressed in Eq. (6.3), is that Lk can be structurally realized in two different ways. The first way consists in changing the twist

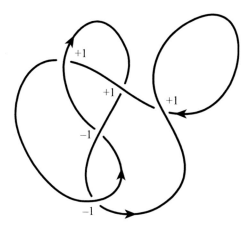

Figure 6.5 Estimation of Wr for "nearly flat" curve. The curve can be projected on an arbitrary plane and for each crossing the underpassing segments have to be shown with breaks. Then, one has to choose arbitrarily a direction on the curve projection (the result does not depend on this choice) and prescribe $(+1)$ or (-1) to each crossing according to the rule shown in Fig. 6.6. The curve Wr is equal to sum of these numbers over all crossings.

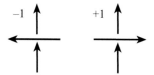

Figure 6.6 Two types of crossing on a projection of a contour. Each crossing makes $+1$ or -1 contribution to the curve's Wr.

of the double helix; the second lies in the deformation of the helix axis, giving rise to a certain writhe.

6.2.3   Application to DNA

As noted above, the properties of supercoiled DNA are specified by the value of ΔLk rather than Lk itself. Correspondingly, for applications to DNA Eq. (6.3) should be transformed by subtracting Lk_0 from both sides:

$$\Delta Lk = Wr + (Tw - Lk_0) = Wr + \Delta Tw. \tag{6.4}$$

The value of ΔTw, introduced here, specifies the deviation of DNA twist from its equilibrium value, $Lk_0 = N/\gamma$. It is important that the value of ΔLk does not change during conformational changes in the circular DNA at particular solution conditions, while the partitioning between ΔTw and Wr is different for each DNA conformation.

Fuller was the first to carry out a simple theoretical analysis of the shape of supercoiled DNA (Fuller 1971). He calculated the Wr of two possible geometries of the DNA axis, simple and interwound helices (Fig. 6.7). By neglecting the ends of the helices, he

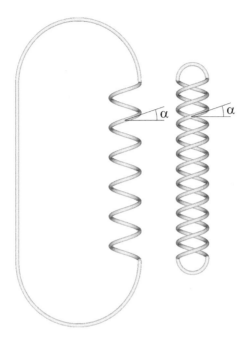

Figure 6.7 Diagram of simple (left) and interwound helices (right) closed by loops. The writhe of the helices depends on the winding angle α. The number of helical turns in the helices is equal to 6 and 12, respectively.

calculated that the Wr of a simple right-handed helix is given by

$$Wr = n(1 - \sin\alpha) \qquad (6.5)$$

and that the Wr of an interwound right-handed helix can be expressed as

$$Wr = -n\sin\alpha, \qquad (6.6)$$

where n is the number of helical turns and α is the winding angle. It follows from Eqs. (6.5) and (6.6) that, as α increases, the absolute value of Wr of a simple helix diminishes, whereas that of an interwound helix increases. On the other hand, if we increase the value of α, the curvature of the rod and, correspondingly, its bending energy decrease. Thus, the bending deformation, which corresponds to a particular value of Wr, can be made essentially smaller in the interwound form by increasing α and making the superhelix thinner. Therefore, the interwound superhelix should be favored over the simple one from the energetic point of view. Indeed, nearly all available experimental and theoretical data indicate that supercoiled DNA adopts interwound conformations (Vologodskii & Cozzarelli 1994).

The mathematical analysis of supercoiling described above suggests that supercoiling causes only elastic torsional and bending deformation of the double helix. However, sufficiently high negative supercoiling can also cause disruptions of the regular DNA structure (nonelastic deformation). Such disruptions emerge as opened base pairs, cruciforms and Z and H forms of DNA. They will be considered later in this chapter.

BOX 6.1 The Gauss integral

Lk can be also defined through the Gauss integral,

$$Lk = \frac{1}{4\pi} \oint_{C_1} \oint_{C_2} \frac{(d\mathbf{r}_1 \times d\mathbf{r}_2) \cdot \mathbf{r}_{12}}{r_{12}^3}, \qquad (B6.1.1)$$

where $\mathbf{r}_1$ and $\mathbf{r}_2$ are vectors whose ends run, upon integration, over the first and second contours, C_1 and C_2, respectively, $\mathbf{r}_{12} = \mathbf{r}_2 - \mathbf{r}_1$ (see Fig. B6.1.1).

If the integration in the Gauss integral is performed both times along a single circular contour, its value is equal to the contour Wr:

$$Wr = -\frac{1}{4\pi} \oint_{C} \oint_{C} \frac{(d\mathbf{r}_1 \times d\mathbf{r}_2) \cdot \mathbf{r}_{12}}{r_{12}^3}. \qquad (B6.1.2)$$

The sign before the integral is chosen so that it gives a negative value for a right-handed interwound superhelix (see Eq. (6.6)). Although the integral can be calculated analytically only for very simple closed curves, it is very useful for analyzing general properties of Wr. In particular, it follows from Eq. (B6.1.2) that Wr of a flat contour is equal to zero. Indeed, we can place the beginning of the coordinate system in the plane of the contour, so vectors $\mathbf{r}_1$, $\mathbf{r}_2$ and $\mathbf{r}_{12}$ lie in the same plane and their mixed product is equal to zero. It also follows from Eq. (B6.1.2) that segments of the contour that are close one to another make a major contribution to the value of Wr, since the term r_{12}^3 in the denominator of the Gauss integral reduces the contribution from remote segments.

Equation (B6.1.2) also helps to understand the expressions for Wr of simple and interwound helices (see Fig. B6.1.1). For a simple helix, only segments that are separated by a few helix turns (if α is not too small) make an essential contribution to the integral. In the case of an interwound helix an additional contribution to Wr comes from segments located opposite one another across the helix. The sign of the latter contribution is different from the first one because the tangent vectors have opposite orientations on the two halves of the interwound helix. The absolute value of this contribution is larger than the absolute value of the first one, and therefore the Wr of interwound and simple helices of the same handedness have opposite signs (see Eqs. (6.5) and (6.6)).

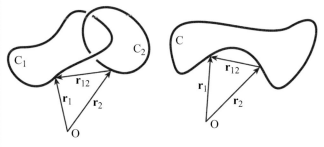

Figure B6.1.1 Notation of vectors in the calculation of the Gauss integral. The left-hand diagram corresponds to the integral for the linking number between two contours, C_1 and C_2; the right-hand one is related to the calculation of writhe of a single closed contour, C. O is the origin of the coordinate system. Note that the sign of Lk depends on the choice of directions on the circular contours, while the sign of Wr does not.

6.2.4   Surface topology

Surface topology was developed in the late 1980s to account for geometrical features of DNA located at a nonflat surface (White *et al.* 1988). The development was motivated by the need for a mathematical relationship between the DNA helical repeat in solution and the repeat in the nuclease digestion pattern of DNA bound to the surface of the nucleosome core (Klug & Lutter 1981). The concepts and results of the surface topology are definitely useful for such cases. There was also an attempt to apply the surface topology to free DNA molecules in solution, under the assumption that there is a "natural," topology-dependent choice of an imaginary surface where DNA is located. As soon as the surface is chosen, one can obtain a few relationships between DNA geometrical properties and its topology (White *et al.* 1988, Wasserman *et al.* 1988). However, large DNA molecules in solution adopt many random conformations with very different geometries, as we have seen in the pages of this book. Therefore, there is no "natural" surface for a molecule with a particular topology. Consequently, because the results of applying the equations of the surface topology depend on the choice of the imaginary surface, the treatment is inapplicable to free DNA molecules in solution.

6.3   Experimental studies of DNA supercoiling

It was shown in the previous section that there are at least two means of structural realization of supercoiling. This conclusion sets the following two questions.

1. How is DNA supercoiling distributed between different means of structural realization, change of DNA Tw and bending its axis, Wr?
2. How does this partitioning depend on σ and on solution conditions?

To address these and other questions related to conformational properties of supercoiled DNA we should consider how DNA supercoiling is studied experimentally. Over the years many methods have been developed that are specific for supercoiled DNA. These methods are very powerful and often open unique opportunities to study general properties of the double helix.

6.3.1   Measuring ΔLk

The value of DNA supercoiling is specified by ΔLk and thus experimental determination of ΔLk is the first problem to be solved. Two different experimental approaches to the problem are based on the fact that supercoiling makes DNA more compact, increasing its mobility in a gel or the sedimentation constant. The first method, developed in the pioneering studies of Vinograd and co-workers, uses the titration of supercoiling by an intercalating dye (Bauer & Vinograd 1968). This approach is applicable only to negatively supercoiled DNA. The molecules of some dyes intercalate between the base pairs upon binding with DNA, thus reducing the helical rotation angle between the adjacent base pairs. As an increasing number of these molecules bind to negatively

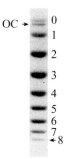

OC →

0
1
2
3
4
5
6
7
8

Figure 6.8 Gel electrophoretic separation of topoisomers of pUC19 DNA. The mixture of topoisomers covering the range of ΔLk from 0 to -8 was electrophoresed from a single well in 1% agarose from top to bottom. The topoisomer with $\Delta Lk = 0$ has the lowest mobility: it moves slightly more slowly than the opened (nicked) circular DNA (OC). The value of $(-\Delta Lk)$ for each topoisomer is shown. Illustration provided by M. Vologodskaia.

supercoiled DNA, the torsional tension within the DNA molecule is gradually reduced and conformations of the double-helix axis become less compact. This causes a decrease in the molecule's mobility, which can be monitored experimentally. The mobility of the molecules reaches the minimum value when the number of ligands per base pair, v, is given by

$$v = 360\Delta Lk/(\varphi N), \qquad (6.7)$$

where φ corresponds to the change of the angle (in degrees) between adjacent base pairs upon the intercalation of a ligand molecule between them. At this value of v there is no torsional stress in the circular DNA. At larger values of v the torsional stress in DNA starts growing again, causing compaction of the molecules and increasing their mobility. Having found the value of v at the mobility minimum, one can find ΔLk from Eq. (6.7), if the value of φ is known. For ethidium bromide, which is most often used for the titration of supercoiled DNA, $\varphi = -26°$ (Wang 1974a). The number of bound ligand molecules can be found with the help of spectroscopic methods. The described approach is not used now, however, since the sedimentation measurements are very laborious.

In 1975 Keller proposed the second approach to determining the linking number difference in closed circular DNA (Keller 1975). This approach is based on the fact that the values of ΔLk in any mixture of DNA topoisomers can differ by an integer only. The electrophoretic mobility of DNA is so sensitive to conformational changes that under appropriate experimental conditions molecules that differ in ΔLk by 1 can be separated. If a DNA sample contains all possible topoisomers with ΔLk starting from 0 and they are all well resolved in a gel, one can find the value of ΔLk corresponding to each band simply by band counting (Fig. 6.8). One can then scan the gel to find the relative amount of DNA in each band. The band that corresponds to $\Delta Lk \approx 0$ (since ΔLk is not an integer we should talk about ΔLk closest to 0) can be identified through a comparison with the band for the nicked circular form. An essential requirement of this approach is that the mixture of topoisomers has to have molecules with all possible

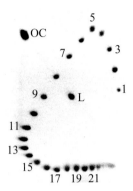

Figure 6.9 Separation of pUC19 DNA topoisomers by 2D gel electrophoresis. Topoisomers 1–4 have positive ΔLk; the rest have negative ΔLk. After electrophoresis was performed in the first direction, from top to bottom, the gel was saturated with chloroquine intercalating into the double helix. This shifted the supercoiling of each topoisomer. Upon electrophoresis in the second direction, from left to right, the 12th and 13th topoisomers turned out to be relaxed. The spot in the top left corner corresponds to the open circular form (OC); the spot in the middle of the gel corresponds to the linear DNA (L). Illustration provided by M. Vologodskaia.

values of ΔLk starting from 0, to perform the band counting. For samples of supercoiled DNA extracted from the cells $|\Delta Lk| \gg 1$, so to apply the method one has to prepare a reference mixture of topoisomers of that DNA with the complete set of possible values ΔLk between 0 and the maximum value in the sample of interest. One should also bear in mind the fact that topoisomer mobility is determined by the absolute value of ΔLk, so the presence of topoisomers with both negative and positive ΔLk can make interpreting the electrophoresis more difficult. Another restriction of the method is that the mobility increase with increasing $|\Delta Lk|$ saturates at relatively low values of $|\Delta Lk|$ (see Figs. 6.8 and 6.9).

A very elegant and effective way to overcome these shortcomings, 2D gel electrophoresis, was proposed by Lee *et al.* (1981). A mixture of DNA topoisomers is loaded into a well at the top left corner of a slab gel, and electrophoresed along the left-hand side of the gel. The bands corresponding to topoisomers with large absolute values of ΔLk merge into one spot. After this the gel is transferred to a buffer containing the intercalating ligand chloroquine and electrophoresed in the second, horizontal direction (Fig. 6.9). The mobility of topoisomers in the second direction is no longer determined by ΔLk, but by the value $(\Delta Lk - Nv\varphi/360)$, where v corresponds to the fraction of bound ligands. The latter value, $(\Delta Lk - Nv\varphi/360)$, can be regarded as the effective linking number difference. We can say that the distribution of ΔLk for the second direction is shifted to a positive value (since $\varphi < 0$). As a result, the mobilities of topoisomers that had ΔLk opposite but equal in absolute value in the first direction are different in the second direction. Also, the topoisomers with large negative ΔLk, which had identical mobility in the first direction, move with different speeds in the second direction since their supercoiling is reduced. The number of topoisomers that can be resolved almost doubles in the case of 2D gel electrophoresis.

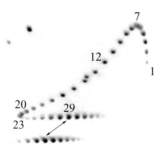

Figure 6.10 Determination of DNA superhelix density by 2D gel electrophoresis. The reference mixture of topoisomers (the numbered set of spots) and the experimental DNA were run in the first (from top to bottom) and second (from left to right) directions in the absence and presence of 1.5 μg/ml chloroquine, respectively. To avoid overlap of the samples the reference sample was loaded in a separate well with a time delay. The reference sample was obtained by adding the experimental sample to the mixture of topoisomers with lower supercoiling, so the intensity distribution for the higher topoisomers was identical for the two samples. This distribution was used to establish correspondence between the two sets of spots. Reproduced from Fig. 5C of Du *et al.* (2007), with permission.

Of course, to apply 2D gel electrophoresis to determine ΔLk of a particular sample one has to have a reference mixture of the topoisomers that contains molecules with all values of ΔLk between 0 (or even small positive values) and the maximum value in the sample. It is described in the next section how one can prepare such sample. It is convenient to load the reference sample and the sample with unknown ΔLk in a single 2D gel to perform the band counting (Figs. 6.8–6.10).

The separation of DNA topoisomers in a gel has proved to be one of the most potent techniques in research on circular DNA. This method and its 2D version have led to a whole series of remarkable experimental studies (see below).

It is important to keep in mind that ΔLk depends on the ambient conditions in solution. A temperature increase by 30 °C causes $|\sigma|$ to increase by approximately 0.01, since the double helix unwinds with rising temperature and Lk_0 reduces (Depew & Wang 1975). The increase constitutes about 20% of the characteristic σ for a closed circular DNA isolated from the cells (Bauer 1978). The double helix winds itself up with increasing ion concentration, and the change of σ resulting from changing ionic conditions can be as large as 0.01 (Bauer *et al.* 1980, Rybenkov *et al.* 1997b).

6.3.2   Obtaining DNA with a required ΔLk

To obtain a DNA with a preset superhelix density, closed circular DNA molecules are treated with an enzyme called topoisomerase IB. The action of this enzyme can alter the value of Lk in closed circular DNA as it introduces a single-stranded break into the double helix and then, after a while, reseals it. In between these two events the DNA may undergo changes in axial twist and shape, so that Lk gradually relaxes to its equilibrium value for the given solution conditions, i.e. the value corresponding to the minimum stress in closed circular DNA. This equilibrium value of Lk can be changed within a wide range by adding various ligands that alter the helical rotation angle of the double helix upon binding to it. Thus, by adding varying amounts of an intercalating dye, one can obtain, after enzyme treatment and removing both ligand and the enzyme from the solution, closed circular DNA with a desired negative supercoiling. Of course, since intercalating ligands unwind the double helix, in this way one could obtain only negatively supercoiled DNA. However, in recent years proteins were found that introduce either negative or positive writhe in supercoiled DNA upon binding, making it possible to prepare DNA molecules with practically any desired superhelix density (LaMarr *et al.* 1997, Dedon *et al.* 2009).

It is important that by such a method we always obtain a few DNA topoisomers with Lk distributed around a mean value rather than one topoisomer (if DNA is longer than 500 bp). This distribution is analyzed in detail in the next section.

6.3.3   Equilibrium distribution of topoisomers and the free energy of supercoiling

Closed circular DNA molecules with a non-zero ΔLk value have additional free energy, which is called supercoiling free energy, $G_s(\Delta Lk)$. Precise knowledge of $G_s(\Delta Lk)$ is very important for successful analysis of many problems related to DNA supercoiling. The simplest way to obtain $G_s(\Delta Lk)$ is based on the quantitation of the equilibrium distribution of topoisomers.

In 1975 two groups simultaneously managed to measure the distribution for a few circular DNA molecules (Depew & Wang 1975, Pulleyblank *et al.* 1975). The distribution was obtained by the same treatment of circular closed DNA with topoisomerase IB, in the absence of any intercalating ligands. As mentioned above, the enzyme can alter the value of Lk in closed circular DNA by making a nick in one strand in the double helix and then religating it. During this treatment the value of Lk changes towards the thermodynamic equilibrium. Due to thermal fluctuations large DNA molecules adopt a lot of different conformations in solution with comparable probability, and these conformations correspond to different values of Tw and Wr. After some time the distribution of $(Tw + Wr)$ in *nicked* circular DNA relaxes to the equilibrium distribution. Religation of the nick at random moments of time converts this distribution to the distribution of Lk in *closed* circular DNA. It is important to emphasize that the obtained distribution of topoisomers reflects the distribution of $(Tw + Wr)$ in the nicked molecules. This distribution contains in comparable amounts more than one topoisomer, if the DNA length exceeds 500 bp. The distributions were analyzed by gel electrophoresis. Figure 6.11 presents a typical result of such an experiment. The maximum of the equilibrium distribution always corresponds to $\Delta Lk = 0$. Since the molecules with close absolute values of ΔLk have close mobilities, they have to be close to each other in the gel. A distribution of the kind presented in Fig. 6.11, where molecules having positive and negative ΔLk values are separated, is the result of electrophoresis conditions that differ from those used for the topoisomerase reaction. The change in conditions means that γ in Eq. (6.1) needs to be replaced by a different value, γ', while Lk remains unchanged. As a result the entire distribution shifts by the value $\Delta Lk - \Delta Lk' = N(1/\gamma - 1/\gamma')$, and for a large enough value of this difference all the topoisomers have the same sign and are well separated.

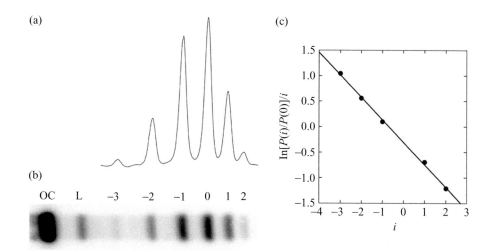

Figure 6.11 Equilibrium distribution of topoisomers in circular DNA pUC19 (2683 bp in length). The mixture of topoisomers forming the distribution was separated by gel electrophoresis (b). A scan of the gel is shown in (a). Adjacent peaks correspond to topoisomers that differ by 1 in values of Lk. For further analysis (c) the bands are numbered by index i, so that the zero value of i is assigned to the most intensive band. Since the distribution of topoisomers is the normal distribution, it can be presented as $P(i) = A \exp(-(i - \delta)^2/2\langle(\Delta Lk)^2\rangle)$, where δ is a parameter that specifies the distribution maximum. If this distribution is plotted as $\ln(P(i)/P(0))/i$ vs i, it gives a straight line (c). The slope of the line is equal to $-1/(2\langle(\Delta Lk)^2\rangle)$, and the intercept is equal to $\delta/(\langle(\Delta Lk)^2\rangle)$.

The experiments demonstrated that the resulting distribution is always the normal one:

$$p(\Delta Lk) = A \exp\left[-\frac{(\Delta Lk)^2}{2\langle(\Delta Lk)^2\rangle}\right] \tag{6.8}$$

where A is the normalization factor, and the brackets $\langle\,\rangle$ mean the average value over the equilibrium set of DNA conformations (Depew & Wang 1975, Pulleyblank et al. 1975, Horowitz & Wang 1984, Rybenkov et al. 1997b). Correspondingly, $\langle(\Delta Lk)^2\rangle$ is the distribution variance. It is known, on the other hand, that $P(\Delta Lk)$ has to be the Boltzmann distribution:

$$p(\Delta Lk) = A \exp\left[-\frac{G_s(\Delta Lk)}{RT}\right], \tag{6.9}$$

where $G_s(\Delta Lk)$ is the free energy of supercoiling. Comparing Eqs. (6.8) and (6.9) one concludes that

$$G_s(\Delta Lk) = \frac{RT(\Delta Lk^2)}{2\langle(\Delta Lk)^2\rangle} \tag{6.10}$$

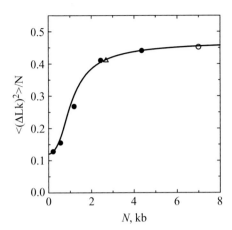

Figure 6.12 Dependence of $\langle(\Delta Lk)^2\rangle$ on DNA length, N, measured in thousands of base pairs. The values of $\langle(\Delta Lk)^2\rangle$ have been normalized by N. The shown experimental data are from Horowitz and Wang (1984) (●), Rybenkov *et al.* (1997b) (○) and Geggier *et al.* (2011) (△). These data were obtained with radioactive labeling of DNA, which provides better accuracy of the measurements. The line corresponds to the results of computer simulation (see below).

Experiments showed (Fig. 6.12) that $\langle(\Delta Lk)^2\rangle$ is proportional to DNA length, if $N > 2500$ (Depew & Wang 1975, Pulleyblank *et al.* 1975, Horowitz & Wang 1984):

$$\langle(\Delta Lk)^2\rangle = \frac{N}{2K}, \tag{6.11}$$

where K is a constant. Comparing Eqs. (6.10) and (6.11), one concludes that

$$G_s(\Delta Lk) = KRT(\Delta Lk)^2/N, \quad \text{if} \quad N > 2500. \tag{6.12}$$

Figure 6.12 shows that $\langle(\Delta Lk)^2\rangle$ diminishes faster when $N < 2500$.

The constant K that specifies $G_s(\Delta Lk)$ according to Eq. (6.12) depends on ionic conditions. This is understandable, since at lower salt concentration the electrostatic repulsion between DNA segments is stronger, and DNA conformations with close contacts between the segments are less probable. Correspondingly, at lower ion concentration thermal fluctuations of DNA Wr decrease and coefficient K increases. The majority of the measurements of $\langle(\Delta Lk)^2\rangle$ were made at ionic conditions close to physiological (≈ 0.2 M NaCl and/or 10 mM MgCl$_2$). For these ionic conditions $K = 1100$, if $N > 2500$ (Depew & Wang 1975, Pulleyblank *et al.* 1975, Horowitz & Wang 1984, Rybenkov *et al.* 1997b). However, the value of K becomes as large as 1600 in solution containing 1 mM NaCl and 0.5 mM MgCl$_2$, in agreement with theoretical prediction (Klenin *et al.* 1989, Rybenkov *et al.* 1997b). Many studies related to DNA supercoiling, in particular on noncanonical structures formed in supercoiled DNA, were performed in TBE buffer (90 mM Tris-borate, 1 mM EDTA) where the value of K is equal to 1400 (Rybenkov *et al.* 1997b). The value of K decreases with temperature, mainly due to reduction of DNA persistence length with the rise of temperature (Geggier *et al.* 2011).

Regardless of its intrinsic elegance, the method of obtaining $G_s(\Delta Lk)$ from the equilibrium distribution of topoisomers has a shortcoming. The method allows determination

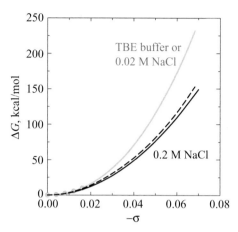

Figure 6.13 Dependence of the supercoiling free energy on the superhelix density, σ, for DNA molecules 5200 bp in length. The data for near-physiological ionic conditions (black solid line) correspond to Eq. (6.13) and were obtained both by ligand titration (Hsieh & Wang 1975) and by measuring the equilibrium distribution of topoisomers for small $|\sigma|$ (Depew & Wang 1975, Pulleyblank et al. 1975). All the cited data are in full agreement with one an other and with the results of the computer simulation (black dashed line) (Vologodskii & Cozzarelli 1994). The data for TBE buffer were obtained from the topoisomer distribution (gray circles) (Rybenkov et al. 1997b) and by computer simulation (gray line, which corresponds to Eq. (6.14)) (Vologodskii & Cozzarelli 1994).

of $G_s(\Delta Lk)$ only for very small values of $|\sigma|$, whereas the range of the greatest interest in physical and biological studies is between -0.06 and -0.03. Usually, the quadratic dependence of $G_s(\Delta Lk)$ on ΔLk is assumed to be valid for higher supercoiling as well, although in general this assumption is hardly justified. Another approach to $G_s(\Delta Lk)$ determination based on titration of supercoiling by an intercalating ligand, pioneered by Bauer and Vinograd (1970), showed that the quadratic dependence on ΔLk is retained until $|\sigma| < 0.06$ at physiological ionic conditions (Hsieh & Wang 1975). This is in agreement with the theoretical calculations of $G_s(\Delta Lk)$ (Vologodskii & Cozzarelli 1994). Thus, under near-physiological ionic conditions the free energy of supercoiling can be written as

$$G_s(\Delta Lk) = 1100RT(\Delta Lk)^2/N = 10RTN\sigma^2$$

$$\text{when} \quad |\sigma| \leq 0.06, \, N > 2500 \, \text{bp}. \tag{6.13}$$

The calculations showed (Vologodskii & Cozzarelli 1994), however, that Eq. (6.12) does not hold for larger values of $|\sigma|$ at lower concentration of ions (Fig. 6.13). In particular, in TBE buffer $G_s(\Delta Lk)$ is well approximated as

$$G_s(\Delta Lk) = 12RTN(\sigma^2 - 5.25\sigma^3) \quad \text{when} \quad -0.06 \leq \sigma \leq 0, \quad N > 2.5 \, \text{kb}. \tag{6.14}$$

Although the value of $\langle(\Delta Lk)^2\rangle$ that corresponds to Eq. (6.14) agrees well with the measured result for these ionic conditions (Rybenkov et al. 1997b), it is definitely desirable to confirm Eq. (6.14) for the entire range of σ.

Further analysis of the equilibrium distribution of topoisomers is given in Box 6.2.

Box 6.2 Contributions of Tw and Wr to the equilibrium distribution of topoisomers

The equilibrium distribution of topoisomers is an important subject in the field of circular DNA, so we are continuing its discussion. The experimentally observed equilibrium fluctuations in ΔLk in circular DNA are made up of fluctuations in Tw and Wr. The enzyme-assisted rejoining of the DNA strand at the break point, occurring at a random moment of time, fixes the momentary value of the sum $Tw + Wr$, i.e. Lk. In a circular DNA with a single-stranded break, torsional (changing Tw) and bending (changing Wr) fluctuations occur independently from each other (this is an assumption that should be considered as a good first approximation). Therefore, the variances, $\langle(\Delta Lk)^2\rangle$, $\langle(\Delta Tw)^2\rangle$ and $\langle(Wr)^2\rangle$, follow the equation

$$\langle(\Delta Lk)^2\rangle = \langle(\Delta Tw)^2\rangle + \langle(Wr)^2\rangle. \tag{B6.2.1}$$

Since the value of $\langle(Wr)^2\rangle$ has been found experimentally and $\langle(Wr)^2\rangle$ can be calculated (see below), Eq. (B6.2.1) makes it possible to determine $\langle(\Delta Tw)^2\rangle$. On the other hand, the fluctuations of ΔTw have to follow the Boltzmann distribution,

$$P(\Delta Tw) = A\exp[-C(2\pi \Delta Tw)^2/(2NlRT)], \tag{B6.2.2}$$

where $2\pi^2 C(\Delta Tw)^2/Nl$ is the energy of torsional deformation of DNA N bp in length, l is the separation between adjacent base pairs along the DNA axis, which is equal to 0.34 nm, and C is the torsional rigidity of the double helix. This is the normal distribution, and its variance, $\langle(\Delta Tw)^2\rangle$, for a chosen DNA and temperature depends only on the value of C:

$$\langle(\Delta Tw)^2\rangle = \frac{NlRT}{4\pi^2 C} = Nlk_{\mathrm{B}}T/(4\pi^2 C). \tag{B6.2.3}$$

k_{B} is the Boltzmann constant. In the last equation we switched to CGS units since constant C is expressed traditionally in these units. Thus, if we know $\langle(\Delta Tw)^2\rangle$, Eq. (B.6.2.3) allows us to estimate constant C.

It has been discussed in detail in Chapter 3 that DNA molecules a few hundred base pairs in length and larger adopt many different conformations due to thermal fluctuations. Each conformation of nicked circular DNA has its own value of Wr (Fig. B6.2.1). To evaluate $\langle(Wr)^2\rangle$ we can simulate the equilibrium ensemble of DNA conformations for a particular DNA length and calculate $\langle(Wr)^2\rangle$ over the simulated set of conformations (Benham 1978; Vologodskii et al. 1979a). Such simulation allows us to compute $\langle(Wr)^2\rangle$ for different DNA lengths with good accuracy (Klenin et al. 1989, Geggier et al. 2011). The simulation results showed that $\langle(Wr)^2\rangle$ is proportional to N if $N > 2.5$ kb, although it decreases faster for shorter DNA molecules, since their typical conformations become more and more flat as the length decreases (Fig. B6.2.2). The difference between the experimentally determined value of $\langle(\Delta Lk)^2\rangle$ and the calculated value of $\langle(Wr)^2\rangle$ is proportional to N over the whole range of N studied, $250 < N < 10\,000$ (Fig. B6.2.3), in full agreement with Eq. (B6.2.3). The values of $\langle(Wr)^2\rangle$ found in this way made it possible to determine the torsional rigidity of the double helix on the basis of Eq. (B6.2.3). The obtained value of C is equal to 3.1×10^{-19} erg·cm (it is a common practice to use these units

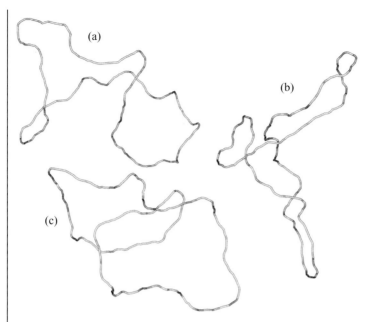

Figure B6.2.1 Typical conformations of nicked circular DNA 3000 bp in length. The illustration was obtained by computer simulation of the equilibrium conformational set as described in Chapter 3. The values of Wr for the simulated conformations are equal to -0.09 (a), -2.13 (b) and 1.30 (c).

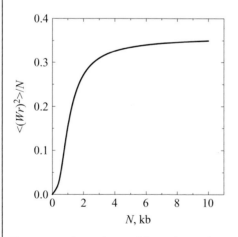

Figure B6.2.2 Dependence of the variance of Wr on DNA length. The line shows the results of computer simulations that correspond to a DNA persistence length of 45 nm (the value for 37 °C, see Section 3.2.2) and an effective diameter of the double helix of 5 nm.

for C) (Frank-Kamenetskii *et al.* 1985, Shimada & Yamakawa 1988, Klenin *et al.* 1989, Geggier *et al.* 2011). It is easy to show that correlation in torsional orientations of two base pairs decays exponentially when the separation between them along the DNA axis increases, so it is possible to introduce the torsional persistence length, $a_t = 2C/k_BT$. The above value of C corresponds to a_t of 150 nm.

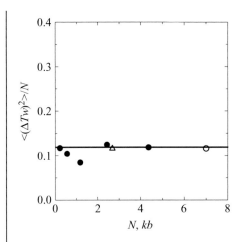

Figure B6.2.3 Dependence of the variance of ΔTw on DNA length. The data were obtained by subtracting computed $\langle(Wr)^2\rangle$ (Fig. B6.2.2) from measured $\langle(\Delta Lk)^2\rangle$ (Fig. 6.12). The line corresponds to Eq. (B6.2.3) with C of 3.1×10^{-19} erg·cm.

6.4   Conformations of supercoiled DNA

EM is the most straightforward way to study conformations of supercoiled DNA. This method has been used extensively since the discovery of DNA supercoiling by Vinograd and co-workers in 1965. They found that supercoiled DNA has a compact, interwound form. This conclusion was confirmed in many EM studies, which also brought quantitative information about DNA supercoiling (Laundon and Griffith, 1988; Adrian *et al.* 1990; Boles *et al.* 1990). The data obtained include the measurements of $\langle Wr\rangle/\Delta Lk$ and the average superhelix radius as a function of σ, and the branching frequency of the superhelix axis. It became clear, however, that labile DNA conformations may change during sample preparation for EM (see Vologodskii and Cozzarelli (1994) for details). A serious problem for interpreting EM results is that the ionic conditions on the grid are not specified. The DNA is exposed to several different solutions and dried before being viewed, and it is unknown when in the procedure the DNA is "fixed." Cryoelectron microscopy and AFM *in situ* are free from some of the disadvantages of more conventional EM, since in these methods DNA is viewed in solution or in a thin layer of vitrified water without shadowing or staining. It was shown by these methods that conformations of supercoiled DNA depend greatly on ionic conditions (Fig. 6.14) (Adrian *et al.* 1990, Bednar *et al.* 1994, Lyubchenko & Shlyakhtenko 1997). Still, independent solution studies are required to confirm conclusions by these methods about this very flexible object.

The solution methods, like hydrodynamic and optical ones, do not give direct, model-independent information about the 3D structure of supercoiled DNA. These methods do, however, measure structure-dependent features of the molecules for well-defined solution conditions while perturbing conformations only minimally. The methods turned

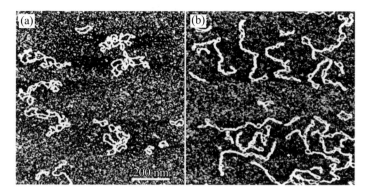

Figure 6.14 Images of supercoiled DNA obtained by AFM *in situ*. DNA were bound to the surface in buffer solutions without drying the samples. The samples were deposited in TE buffer (20 mM Tris·HCl, 1 mM EDTA) (a) and in TE buffer + 160 mM NaCl (b). The picture from the work of Lyubchenko and Shlyakhtenko (1997) is reproduced with permission.

to be very productive in the studies of DNA supercoiling when used in combination with the computer simulation of supercoiled molecules. The key element of the approach is comparison between simulated and measured properties of supercoiled DNA (Gebe *et al.* 1996, Rybenkov *et al.* 1997c, Rybenkov *et al.* 1997d, Hammermann *et al.* 1998). If the simulation and measured results are in agreement, one can assume that simulated and actual conformations look similar as well.

The efficiency of this approach originates from the fact that DNA conformational properties can be accurately described in terms of a simple model, as described in Chapter 3. The values of all three parameters of the model are known from numerous independent studies, so there are no adjustable parameters in the computation. Using this model one can simulate a random set of conformations that corresponds to the equilibrium conformational set for actual DNA molecules. The MC method is usually used to prepare the set (Vologodskii *et al.* 1992). This set allows one to determine mean values and distributions of many superhelix properties such as Wr, the number of branches or the sedimentation coefficient. There is an analogy between this computer approach and EM, which can also be considered as an MC method, because conclusions are drawn from a limited statistical sampling. Of course, a much larger set can be generated computationally.

Formation of DNA catenanes between supercoiled and nicked circular DNA gives a good example of this approach (Rybenkov *et al.* 1997d). In this work catenanes were formed by cyclizing linear DNA with long cohesive ends in the presence of supercoiled molecules (Fig. 6.15). The efficiency of the catenation depends on the distance between opposing segments of DNA in the interwound superhelix. Thus, the fraction of cyclizing molecules that becomes topologically linked with the supercoiled DNA is the product of the concentration of the supercoiled DNA and a proportionality constant, B, that depends on the conformations of the supercoiled DNA. The values of B were measured for different ionic conditions and various supercoilings. In parallel with the experiments, the same values were calculated using MC simulations of the equilibrium distribution

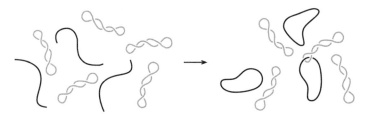

Figure 6.15 Probing conformational properties of supercoiled DNA by catenation (Rybenkov *et al.* 1997d). The diagram shows formation of catenanes between supercoiled and cyclizing linear molecules. The cyclization occurs via joining long cohesive ends of the linear molecules.

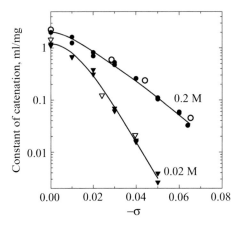

Figure 6.16 Probing conformations of supercoiled DNA by catenation. Measured and simulated equilibrium constants of catenation, B, are shown as a function of supercoiling at NaCl concentrations of 0.2 M (circles) and 0.02 M (triangles). The experimental values of B (open symbols) are shown together with calculated results (filled symbols). The solid lines are the best fits to the calculated data. The figure was replotted from Fig. 4 of Rybenkov *et al.* (1997d), with permission.

of DNA conformations. Very good agreement between measured and simulated values of B was found in this study over a broad range of σ and ionic conditions (Fig. 6.16).

Good agreement between experimental and simulated results was obtained in the other studies of supercoiled DNA based on the comparison of solution measurements and computer simulations. Thus, it seems reasonable to conclude that the simulations predict conformations of the supercoiled DNA with good accuracy and can be used to study properties of the supercoiled molecules that are hard to measure directly.

The data presented in Fig. 6.16 show that values of B strongly depend on ionic conditions. This dependence is due to the changes of the electrostatic repulsion between DNA segments, which is screened more efficiently at higher salt concentration (see Section 3.2). Figures 6.17 and 6.18 show typical simulated conformations of supercoiled molecules for two different ionic conditions, for σ of -0.05 (see Vologodskii *et al.* (1992) for the simulation details). Although the conformations for both ion concentrations correspond to branched interwound superhelices, there is a large difference between

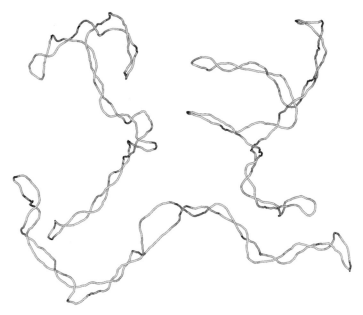

Figure 6.17 Typical simulated conformations of supercoiled DNA in solution containing 0.2 M NaCl. The conformations of the model chains correspond to DNA 4 kb in length and $\sigma = -0.05$.

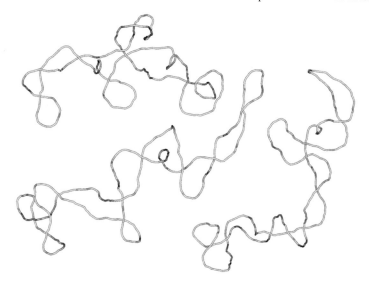

Figure 6.18 Typical simulated conformations of supercoiled DNA in solution containing 0.01 M NaCl. The conformations of the model chains correspond to DNA 4 kb in length and $\sigma = -0.05$.

them. Clearly, the simulated conformations are in good qualitative agreement with the AFM data shown in Fig. 6.14. The average conformational parameters of supercoiled DNA molecules estimated in the simulations are shown in Table 6.1.

One can see from the table that nearly three-quarters of the linking number difference is structurally realized by bending the DNA axis and creating corresponding value of Wr.

Table 6.1 Average conformational parameters of supercoiled DNA at ionic conditions close to physiological (0.2 M NaCl, or/and 10 mM MgCl$_2$). The data in the table were obtained by the computer simulations described by Vologodskii *et al.* (1992).

Superhelix density, σ	$Wr/\Delta Lk$	Superhelix diameter, nm	(Number of superhelix turns)/ΔLk	Number of superhelix branches
-0.03	0.73 ± 0.01	19 ± 1	0.9 ± 0.02	–
-0.06	0.73 ± 0.01	10 ± 1	0.9 ± 0.02	1 per 1500 bp

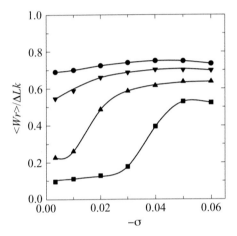

Figure 6.19 Effect of DNA length on the partitioning of ΔLk between Wr and ΔTw. The plotted values of $Wr/\Delta Lk$ were computed for DNA molecules of 300 bp (■), 600 bp (▲), 1200 bp (▼) and 3500 bp (●) in length by the author as described by Vologodskii *et al.* (1992).

This rule works well, however, only if the DNA length exceeds 2–3 kb. When the DNA length reduces, chiral conformations with notable values of Wr become more and more energetically expensive. Correspondingly, a larger fraction of ΔLk is structurally realized over the torsional deformation of the double helix, ΔTw (Fig. 6.19). In very small DNA circles, 200–300 bp in length, nonzero Wr appears at sufficiently high superhelix density, when they flip from nearly regular circles to figure-eight conformation. The latter transition has been studied in detail both theoretically and experimentally (Le Bret 1979, Gebe & Schurr 1996, Bednar *et al.* 1994).

6.5   Formation of alternative structures in supercoiled DNA

6.5.1   Negative supercoiling and alternative DNA structures

So far, we have considered properties of circular DNA that do not involve disruptions of the regular structure of the double helix. However, under sufficiently high negative supercoiling such disruptions become inevitable. In effect, the disruptions are a means of

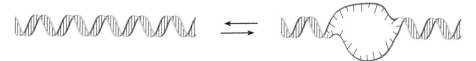

Figure 6.20 Formation of the opened region in dsDNA. In the absence of torsional stress the complementary strands are not interwoven, on average, in such a region.

(a)   (b)

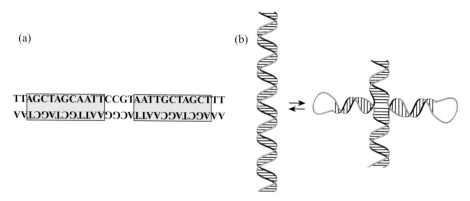

Figure 6.21 Formation of the cruciform in dsDNA. (a) An example of a palindromic sequence. The symmetrical parts of the sequence are shaded. (b) Diagram of the transition between B-DNA and the cruciform.

structural realization of supercoiling. In contrast to torsional and bending deformations, which are distributed along the whole circular DNA molecule, disruptions in the regular structure have a local character, while their type and location depend on the DNA sequence. All these alternative structures have a smaller (compared with the canonical B form of DNA) interwinding of one strand of the DNA relative to the other. Therefore, formation of such alternative structures reduces the deficit of interwinding in other parts of the negatively supercoiled DNA and can become thermodynamically favorable.

The simplest type of alternative structure (or conformation) induced by negative supercoiling is melted or opened regions where DNA strands are not interwoven in the relaxed state (Fig. 6.20). Such regions appear first of all in the DNA sections rich in AT base pairs, because melting of AT pairs requires less free energy than melting of GC pairs (see Section 2.2.2).

Negative supercoiling can induce formation of cruciform structures in palindromic regions of DNA (Fig. 6.21). Again, the complementary strands of DNA are not twisted relative to each other in the cruciform structures. At the same time, a large fraction of the nucleotides of the cruciforms are located in helical regions of the hairpins, which means that their free energy corresponds to the free energy of a regular linear double helix. This is why formation of long enough cruciform structures has to be preferable to the formation of opened conformations of the same size.

Formation of Z-DNA causes the largest release of negative torsional stress per base pair, since the strands are twisted there into a left-handed helix with a helical repeat of

Figure 6.22 Formation of the left-handed Z form inside a B-DNA segment.

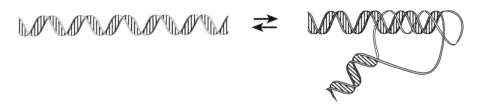

Figure 6.23 Formation of the H form inside a B-DNA segment. The pyrimidine strand and half of the purine strand make up a triple helix; the other half of the purine strand is free. Only Watson–Crick base pairs in the triple helix are shown. The CGC triplets are protonated in this structure.

12 bp per turn (Fig. 6.22). The Z form can predominantly appear in DNA regions with a regular alternation of purines and pyrimidines, first of all G and C.

The last known type of alternative structure occasioned by negative supercoiling is the H form, which may occur in DNA sections where one strand comprises only purine and the other only pyrimidine bases. The main element of the H form is the triple helix stabilized by Watson–Crick and Hoogsteen base pairs (Fig. 6.23). Topologically, the H form is equivalent to an open section or a cruciform structure: the complementary strands of the entire segment involved in the H form are not twisted relative to each other. Occurrence of the H form that involves CGC triplets is stimulated by reduction of the solution pH.

It is worth emphasizing once again that formation of all the above alternative structures is thermodynamically wasteful in linear DNA at near-physiological conditions. They can appear at such conditions only in negatively supercoiled DNA molecules.

Below we consider experimental approaches to studying the alternative structures, and thermodynamic treatment of the phenomena, and then analyze in some detail each of these structures.

6.5.2   Experimental detection of alternative structures

All the existing methods of studying alternative DNA structures occasioned by negative supercoiling can be divided into two groups. The methods of the first group are based on the localization of structural changes in DNA molecules. In principle, these methods could be equally used to register local structural changes in linear molecules, too. Successive analysis of DNA sequence in the disrupted regions helps to establish what kind of structure was formed there. The second group of methods is based on the measurement of global properties of DNA affected by local transitions. On formation of an alternative structure under the impact of negative supercoiling, the torsional stress itself decreases and so does the absolute value of Wr, which determines the electrophoretic mobility of DNA molecules in gels. This change in the mobility constitutes the core of the second approach. The method does not provide any information on where exactly the conformational change has happened, but allows one to obtain quantitative characteristics of the transition. The method offers a unique opportunity for reliable quantitative registration of conformational changes in a region that accounts for about 1% of the total length of the molecule. No other physical method registering integral properties of DNA (UV spectroscopy, circular dichroism, IR spectroscopy, NMR) makes it possible to monitor changes affecting such a small fraction of DNA base pairs. This is understandable since unwinding of 1% of the total length of negatively supercoiled DNA with σ of -0.05 reduces the supercoiling by 20%. Clearly, this change is substantial.

Most methods of localization of structural transitions are based on the breaking DNA in the area of an alternative structure with the subsequent mapping of these breaks. The breaks can be introduced by special enzymes, endonucleases, which specifically hydrolyze ssDNA (Beard *et al.* 1973, Wang 1974b, Gray *et al.* 1975). These endonucleases are sensitive even to minor irregularities in the helical structure of DNA. In cruciform structures they cause breaks in the loops of the hairpins (Lilley 1980). There are sites sensitive to these endonucleases in the areas of Z and H forms (Singleton *et al.* 1982, Lyamichev *et al.* 1986) and, naturally, in opened regions of DNA. In supercoiled DNAs there are also other areas of increased sensitivity to such endonucleases, though the nature of structural changes in such areas is not clear yet. Although after the application of the first single-stranded break alternative structures in DNA soon disappear as a result of the torsional relaxation, the enzymes often go on to cut the second strand of the DNA opposite to the first break. This is why a considerable fraction of the molecules end up in the linear form after treatment with endonucleases.

Another way of making breaks in the areas near the alternative structures is a chemical modification of DNA bases at the spots which in the regular B form are screened off by the double helix structure. Further chemical treatment of such selectively modified DNA leads to a break-up of the sugar-phosphate chains at the sites of the primary modification.

Restriction analysis is used for rough preliminary mapping of the breaks. DNA molecules converted into the linear form by nuclease or chemical cutting are treated with some restriction endonuclease to obtain a set of restriction fragments. The set of fragments, separated by gel electrophoresis, is compared with the set obtained by treatment of the original circular DNA with the same restriction endonuclease. If the cutting at the point of structural deformation occurred in one DNA region, two additional bands should appear in the electrophoresis of the first set of fragments, compared with the reference set. Besides, one of the lines of the reference set should be missing from the first set of fragments or become much less intense. By estimating the length of the additional fragments one can find two possible localizations of the structural transition. After the preliminary localization of the specific cleavage, one can map the cuts at the nucleotide level. Using this method Lilley and Panayotatos and Wells discovered the formation of cruciform structures in natural supercoiled DNAs (Lilley 1980, Panayotatos & Wells 1981).

A very elegant and effective method of analyzing local conformational changes in supercoiled DNA was first suggested by Wang and co-workers (Wang *et al.* 1983). The method allows one to obtain, as a result of a single experiment, the entire dependence of the probability of an alternative structure formation on ΔLk. Besides, it yields important information on the DNA unwinding associated with this transition. This method, described below, based on 2D gel electrophoresis, is the most informative and accurate method of quantitative analysis of local conformational transitions in supercoiled DNA.

To a good approximation we can assume that the electrophoretic mobility of a DNA topoisomer is determined by its value of $|Wr|$. A larger value of $|Wr|$ corresponds to more compact conformations of the molecule and a higher mobility. The electrophoretic mobility of DNA molecules increases monotonically with the growth of the absolute value of ΔLk, since it is accompanied by the growth of $|Wr|$. This, however, is true only as long as no conformational transitions occur in the molecule, since the transitions reduce the torsional stress and, consequently, $|Wr|$. DNA topoisomers that have undergone the transition will have lower mobility compared with what they would have had in the absence of the transition. As a result, the monotonic rise in the mobility of molecules with the growing $|\Delta Lk|$ may be disrupted. 1D electrophoresis of a broad mixture of topoisomers will show a pattern of irregular bands which are hard to assign to specific topoisomers. 2D electrophoresis helps to solve the identification problem. The electrophoresis in the first direction is performed under ordinary conditions, along one of the edges of a flat gel. This results in the distribution of topoisomers that was described above. Before the electrophoresis in the second direction the gel is soaked in a buffer containing an intercalating ligand, usually chloroquine. As a result of the ligand binding, the torsional stress in DNA is determined no longer by ΔLk but by the value $(\Delta Lk - v\varphi N/360)$, where v is the number of bound ligand molecules per base pair, φ is the angle of unwinding of the double helix upon the binding of a single ligand molecule ($\varphi < 0$) and N is the number of base pairs in the DNA. The concentration of the ligand is selected in such a way that the remaining torsional stress is insufficient for the formation of the alternative structure in the topoisomers of the set. In this case, the mobility of topoisomers monotonically depends on $|\Delta Lk - v\varphi N/360|$. This is why, after completion of the electrophoresis in the second direction, assigning the spots to specific topoisomers is not difficult at all (Fig. 6.24). Thus, the electrophoresis in the second direction identifies the mobility of each topoisomer during the electrophoresis in the first direction, and this gives important information on the alternative structure formation. One can see in the figure that the mobility of the 17th topoisomer is close to that of the 12th during the first direction of the electrophoresis. This means that these topoisomers have the same Wr and, consequently, the same torsional stress. Thus, the elastic torsional deformations for the topoisomers have to be the same as well. This is possible only if the difference in ΔLk for these topoisomers is equal to the magnitude of DNA unwinding in the region of the transition, δTw, which is equal to 5 in this case (for certain reasons for cruciform structures the observed value of δTw is often 1–1.5 turns larger than the expected value). The plot of δTw versus ΔLk corresponding to this 2D gel is shown in Fig. 6.25. Since formation of this small cruciform follows the two-state transition, values of δTw can be transformed into the probability of cruciform extrusion.

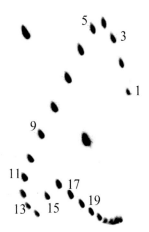

Figure 6.24 Cruciform extrusion in a palindromic region, d[CCC(AT)$_{16}$GGG], of circular DNA 2.7 kb in length. The mobility in the course of the first, vertical electrophoresis drops sharply between the 14th and 15th topoisomers, which becomes clear after the completion of electrophoresis in the second, horizontal direction, performed after soaking the gel by chloroquine. The mobility drop is due to the cruciform formation in the palindromic region. The picture, from Fig. 1 A of Vologodskaia and Vologodskii (1999), is reproduced with permission.

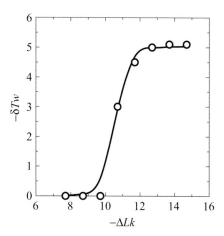

Figure 6.25 Changes of δTw versus ΔLk corresponding to the 2D gel shown in Fig. 6.24 (see the text for details). The values of ΔLk of the topoisomers in Fig. 6.24 were calculated as 4.3 − (topoisomer number).

6.5.3   Thermodynamic analysis

The analysis given below is based on the generally accepted assumption that the change of the free energy due to formation of an alternative structure can be presented as a sum of two terms (Hsieh & Wang 1975, Benham 1979, Vologodskii *et al.* 1979b, Wang *et al.* 1983). The first term corresponds to the transition-related change in the free energy of supercoiling, and the second term is the change in the free energy in the DNA segment where the conformational transition takes place. The second term does not depend on DNA supercoiling and corresponds to the free energy change in linear DNA due to the transition. As long as the DNA remains in the regular B form, its torsional stress and, consequently, the free energy of supercoiling is defined by ΔLk (see Eqs. (6.13) and (6.14)). This analysis assumes that even after formation of an alternative structure the free energy of supercoiling is specified only by DNA elastic deformation. The magnitude of elastic deformation after the structure formation is defined by $\Delta Lk - \delta Tw$ rather than by ΔLk, where δTw is the change of DNA twist due to the transition. The value of δTw can be expressed through the number of base pairs per helix turn in the B form of DNA, γ_B, and the number of base pairs per turn of *one strand around the other* in the alternative structure, γ_{alt}. If m base pairs adopt the alternative structure, then

$$\delta Tw = m(1/\gamma_{alt} - 1/\gamma_B). \tag{6.15}$$

Thereby, the free energy of supercoiling in the presence of the alternative structure can be expressed as $G_s(\Delta Lk + m\kappa/\gamma_B)$, where $\kappa = 1 - \gamma_B/\gamma_{alt}$. The value of κ specifies the change of DNA twist per base pair associated with a particular transition. Accordingly, the change in the free energy of supercoiling, δG_s, is given by

$$\delta G_s = G_s(\Delta Lk + m\kappa/\gamma_B) - G_s(\Delta Lk), \tag{6.16}$$

or

$$\delta G_s = G_s(\sigma + m\kappa/N) - G_s(\sigma). \tag{6.17}$$

Functions $G_s(\Delta Lk)$ and $G_s(\sigma)$ depend on ionic conditions and on DNA length (see Eqs. (6.13) and (6.14)). Equations (6.16) and (6.17) formally imply that the formation of any structure with a lesser interwinding of the complementary strands than in B-DNA reduces the energy of negative supercoiling and, consequently, must be stimulated by such supercoiling. This, of course, is true only under the condition that this structure forms in a comparatively short DNA fragment, so that the $m\kappa/\gamma_B$ value does not exceed $|\Delta Lk|$.

The change in the free energy in the DNA segment where the conformational transition has occurred can be calculated as $\sum_{i=1}^{m} \Delta G_i + 2G_j$, where ΔG_i is the free energy change associated with base pair i, and G_j is the free energy of each boundary between the B form and the alternative structure (see Chapter 3). At the transition point, σ_{tr}, the full change of the free energy is equal to zero,

$$\delta G_s + \sum_{i=1}^{m} \Delta G_i + 2G_j = 0. \tag{6.18}$$

Below we apply this analysis to various types of alternative structure. It should be noted however that, even if the transition occurs only in a single DNA segment, in some cases it is not necessarily converted to an alternative structure as a whole. If the length of the insert is sufficiently large and the DNA is not too long, the transition initially occurs only in a part of the insert. Such cases require more elaborated analysis (see Frank-Kamenetskii and Vologodskii (1984), for example).

6.5.3.1   Cruciform structures

Extrusion of cruciform structures under the action of negative supercoiling was first shown for very long palindromic regions in strongly negatively supercoiled DNA, where owing to their large size the cruciforms were observed directly by electron microscope (Gellert *et al.* 1979). Soon it was found that the cruciforms can appear in natural short palindromes 25–30 bp in length (Lilley 1980, Panayotatos & Wells 1981). Although the first studies of the cruciform extrusion used the approach based on the localization of structural transitions (see above), the quantitative information on the transition properties came from 2D gel electrophoresis (Lyamichev *et al.* 1983, Greaves *et al.* 1985, Haniford & Pulleyblank 1985, Panyutin *et al.* 1985, Naylor *et al.* 1986).

In the case of cruciforms the value of ΔG_i is equal to zero. Indeed, the cross grows in size, in a palindromic region, at the expense of elimination of two base pairs in the main helix and formation of two identical pairs in the hairpins. This should not change the free energy of the structure, because the number of paired bases and all boundary elements of the structure do not change. This means that the formation of cruciform structures must be characterized by a single energy parameter G_j (in this case $2G_j$ is sum of the free energy of the junction of four helices and the free energy of the hairpin loops). Thus, Eq. (6.18) is reduced to

$$\delta G_s(\sigma_{tr}) + 2G_j = 0. \tag{6.19}$$

Note that for the cruciform structures parameter $\kappa = 1$, since in cruciform structures the complementary strands are not twisted around one another ($\gamma_{alt} = \infty$). It follows from the above consideration that the absolute value of σ_{tr} for cruciforms decreases when m increases. Indeed, the free energy cost of the cruciform formation, $2G_j$, does not depend on m, while the gain in $\delta G_s(\sigma)$ increases with m. This can be shown explicitly if one substitutes $G_s(\sigma)$ from Eq. (6.13) or (6.14) into Eq. (6.19). Experimental studies show that cruciform extrusion at typical palindromic regions of 25–30 bp in length takes place at σ_{tr} of (-0.04)–(-0.06). It should be noted that nearly all the published data were obtained for TBE buffer, which is usually used for 2D gel electrophoresis of supercoiled DNA. Under these ionic conditions $G_s(\sigma)$ is specified by Eq. (6.14) rather than Eq. (6.13), since the latter was obtained for near-physiological ionic conditions. If we apply Eq. (6.14) to analyze the published experimental data, we find that $2G_j$ is between 20 and 24 kcal/mol (Greaves *et al.* 1985, Haniford & Pulleyblank 1985, Panyutin *et al.* 1985, Naylor *et al.* 1986, Vologodskaia & Vologodskii 1999). The only study where the transition was quantified in TBE buffer with an additional 10 mM MgCl$_2$ gave $2G_j$ of 18 kcal/mol (Vologodskaia & Vologodskii 1999). This value of $2G_j$ means that the cruciforms cannot be formed in palindromic regions shorter than two dozen base pairs in length, since torsional stress in supercoiled DNA nearly saturates at σ of -0.06 due to other local conformational transitions (see below).

An unusual feature of the cruciform extrusion in supercoiled DNA is its extremely slow kinetics. For long palindromic regions, where the equilibrium extrusion should occur at low $|\sigma|$, the transition is so slow that it is impossible to observe it experimentally at this σ at room temperature (Courey & Wang 1983, Gellert *et al.* 1983, Panyutin *et al.* 1984). The rate of the transition increases fast, however, with increase of σ (Vologodskii & Frank-Kamenetskii 1983, Panyutin *et al.* 1984,). Such slow kinetics of the extrusion is understandable, since cruciform nucleation goes through formation of an extended open region in the middle of the palindrome, which has extremely low probability in relaxed DNA. However, this probability increases fast when negative supercoiling increases. The detailed analysis of the extrusion kinetics can be found in the work of Vologodskii and Frank-Kamenetskii (1983).

6.5.3.2   Left-handed Z form

Formation of Z-DNA at near-physiological ionic conditions under negative torsional stress was discovered by Wells and co-workers by localization of the transition in a plasmid with a $d(CG)_n \cdot d(CG)_n$ insert (Singleton *et al.* 1982). Soon, the 2D gel electrophoresis became the main method in studies of the B–Z transition in supercoiled DNA. The method, as a way to investigate local conformational transitions in supercoiled DNA, was first introduced by Peck and Wang (Wang *et al.* 1983). It was described in detail in the previous section.

The transition of a DNA fragment from a right-handed B helix into a left-handed Z helix ($\gamma_Z = -12$) provides the maximum release of negative torsional stress compared with other alternative structures:

$$\kappa = 1 - 10.5/(-12) = 1.87.$$

Although initially formation of the Z form was observed for $d(CG)_n \cdot d(CG)_n$ inserts in supercoiled DNA (Singleton *et al.* 1982, Wang *et al.* 1983), it was soon found that negative supercoiling can also convert $d(AC)_n \cdot d(GT)_n$ inserts into the Z form (Haniford & Pulleyblank 1983a). This strongly suggested that all sufficiently long alternating purine–pyrimidine sequences can be converted into the Z form by negative torsional stress. Further studies showed that Z-DNA can be formed in many irregular sequences (Brahms *et al.* 1982, Pohl *et al.* 1982, Brahms *et al.* 1989). To describe the transition quantitatively, a thermodynamic model of the B–Z transition in DNA with arbitrary sequence was needed, and two similar models were suggested soon (Ho *et al.* 1986, Mirkin *et al.* 1987b). One of these models is described in Section 2.4. Study of the B–Z transition in the inserts with different deviations from $d(CG)_n \cdot d(CG)_n$ allowed researchers to determine all parameters of the models (Haniford & Pulleyblank 1983a, Ellison *et al.* 1985, Mirkin *et al.* 1987b). It should be noted, however, that the values of these parameters specify formation of the Z form only in TBE buffer (see Chapter 4). The complete set of parameters at ionic conditions close to physiological ones is not available at the present time.

Both the models of the B–Z transition and the experimental data show that any DNA sequence can be converted into the Z form under sufficient torsional stress. It has been found that in form V DNA (the form with unlinked circular complementary strands) about 30% of base pairs form segments of Z-DNA and about the same fraction of base pairs forms the right-handed B-DNA, whereas the remainder of the DNA stays in the denatured state (Pohl *et al.* 1982). The sequence of this DNA did not contain a notable share of purine–pyrimidine blocks, so this result means that the Z form appears in DNA segments with irregular sequences. This conclusion was confirmed by X-ray analysis of duplex $d(CGATCG) \cdot d(CGATCG)$, which formed a continuous Z helix (Wang *et al.* 1985).

Nearly all structural features of base pairs in B and Z forms of DNA are different (see Section 1.1), so it is not surprising that the rate of the transition between the two forms is very low. The characteristic time of the transition is on the scale of minutes or hours (Peck *et al.* 1986, Pohl 1986). Still, it is not slow enough to use 2D gel electrophoresis for kinetic studies of the transition. The fastest way to detect the transition in short inserts of plasmids found so far is based on binding antibodies specific to Z-DNA (Peck *et al.* 1986, Pohl 1986). The binding can reach its equilibrium level in approximately 50 s (Peck *et al.* 1986). Another problem in addressing the transition kinetics is the necessity of fast change of DNA supercoiling. It turned out that binding/dissociation of intercalating dyes with supercoiled molecules can be completed in 40 s (Peck *et al.* 1986), offering a good solution to the latter problem. It was found, by using these approaches, that at room temperature the transition in $d(CG)_n \cdot d(CG)_n$ inserts occurs over thousands of seconds, at DNA superhelix density around -0.05, but the transition rate increases fast with an increase of negative supercoiling (Peck *et al.* 1986, Pohl 1986).

6.5.3.3 Open regions and melting of supercoiled DNA

Formation of open regions or melting of closed circular DNA is more difficult to quantitatively describe than the B–Z transition and the cruciform extrusion, since the open regions do not have a specific structure. The unwinding of the double helix associated with the base-pair opening depends on the sign and the magnitude of the torsional stress.

The first experiments involving the melting of closed circular DNA were performed by Vinograd and co-workers (1968). The melting was registered by detecting the changes of the UV absorbance and buoyant density of supercoiled DNA. The researchers found that the melting of a negatively supercoiled DNA starts at considerably lower and ends at considerably higher temperatures than that of the corresponding linear DNA molecules. Such behavior is not surprising. Indeed, as long as the fraction of melted base pairs, θ, is less than $(-\sigma)$, the negative torsional stress promotes the melting. However, at $\theta > -\sigma$ the melted sections of DNA begin to accumulate a positive twist, so the topological constraint impedes further melting of the DNA. The decisive role of the topological constraint in the melting was clearly shown when researchers used a salt solution where melting temperatures of AT and GC base pairs coincide (Gagua *et al.* 1981). In this salt the melting interval of a linear DNA narrows down to several tenths of a degree. However, the melting of supercoiled DNA practically does not change, and the transition remains very broad, starting at 55 °C and ending at 110 °C (Fig. 6.26).

To obtain an accurate theoretical description of early DNA melting under negative torsional stress, we need to know how the free energy of open regions depends on their interwinding. The most elaborate attempt to solve this problem was undertaken by Bauer and Benham (1993). In their experiments early melting of supercoiled DNA was detected by 2D gel electrophoresis. The researchers suggested a detailed model of the process and optimized its parameters to obtain the best agreement with the experimental data. It is not clear, however, how accurate the obtained description of the process is, since it was obtained by simultaneous optimization of a few parameters and has not been tested quantitatively on melting of other supercoiled DNA molecules. Further progress in study of the melting of circular DNA molecules could be achieved by using specially designed plasmids with inserts that have much lower melting temperatures than the rest

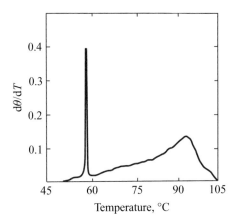

Figure 6.26 Melting of supercoiled DNA. The differential melting curve of the replicative form of phage ϕX174 DNA in the presence of 3 M tetraethylammonium bromide. In this solution melting temperatures of AT and GC base pairs are equal, so linear or nicked DNA melts in a very narrow temperature interval. A small fraction of nicked DNA in the sample is responsible for the sharp peak in the plot. The figure was redrawn from Fig. 1 a of Gagua *et al.* (1981).

of the circular DNA. In such plasmids the boundaries of the melted regions can be unambiguously determined, greatly facilitating theoretical analysis of the experiments.

6.5.3.4   H form

It was found in the early 1980s that homopurine–homopyrimidine sequences in negatively supercoiled DNA are hypersensitive to endonuclease S1, which specifically breaks down ssDNA (Hentschel 1982, Larsen & Weintraub 1982, Htun *et al.* 1984). This finding suggested that the segments form some alternative structure, causing a decrease of torsional stress in the molecules. Subsequent studies by Frank-Kamenetskii and coworkers based on 2D gel electrophoresis allowed them to design and prove the model of this structure (Lyamichev *et al.* 1985, Lyamichev *et al.* 1986, Lyamichev *et al.* 1987, Mirkin *et al.* 1987a). The main element of the structure is the triple helix formed by two homopyrimidine and one homopurine segments. In the topological sense, formation of this structure is equivalent to the complete unwinding of the double helix or to the transition of the region into a cruciform structure: the parameter $\kappa = 1$ for the H form. It was found that the superhelix density at which the transition takes place, σ_{tr}, strongly depends on the solution pH (Lyamichev *et al.* 1985):

$$\sigma_{tr} = (pH_0 - pH)/10\kappa r, \tag{6.20}$$

where r designates the number of base pairs corresponding to a single potential protonation site in the formed structure, and pH_0 is a constant (the CGC triplet is protonated in the structure, but not the TAT triplet).

The data based on the 2D gel electrophoresis were used to determine the energy parameters describing the H-form appearance for inserts $d(CT)_n \cdot d(AG)_n$ and $d(C)_n \cdot d(G)_n$ (Lyamichev *et al.* 1989). As expected, the free energy of the B–H junction proved to be close to the corresponding value for cruciform structures. It was also found that for the B–H transition the free energy of a base pair substantially depends not only on the type of the given pair, but also on the types of the adjacent ones. Detailed information on the H form in negatively supercoiled DNA can be found in the review of Mirkin and Frank-Kamenetskii (1994).

6.5.3.5   The influence of supercoiling on the B–A transition

Among the examined conformational changes which can be initiated by negative super-coiling, the B–A transition holds a special place. The influence of supercoiling on this transition is very small. The double helix is unwound in the A form relative to the B form but the unwinding is very small ($\gamma_A = 11$), so the value of κ is close to 0.05 (Krylov *et al.* 1990). Therefore, the possible gain in the supercoiling free energy from conversion of a single base pair from B to A form is also very small for any reasonable level of supercoiling. It is easy to estimate that it cannot exceed 0.05 kcal/mol per base pair. Such a small gain in G_s means that supercoiling cannot cause B–A transition in a DNA segment under conditions close to physiological. However, B–A transition that involves a large fraction of the DNA can greatly change the DNA supercoiling (Krylov *et al.* 1990).

6.5.3.5   Mutual influence and competition of different structural transitions

There is strong mutual influence between different transitions in closed circular DNA, caused by negative supercoiling. The primary reason for this is that any such transition reduces the torsional stress in the entire circular DNA, which is a driving force for other transitions. This mutual influence has to be taken into account in a quantitative analysis of conformational changes in supercoiled DNA. The phenomenon has been studied in detail both experimentally and theoretically (Benham 1983, Kelleher *et al.* 1986, Ellison *et al.* 1987, Aboul-ela *et al.* 1992). Here we only illustrate it by one striking example, B–Z transitions in a plasmid containing two d(CG)$_{12}$ · d(CG)$_{12}$ inserts separated by a certain distance. With the growth of negative supercoiling in this DNA, two consecutive cooperative transitions are observed, corresponding to B–Z transitions in the inserts (Fig. 6.27). Since the inserts are identical, it is impossible to say which of them undergoes the transition earlier and which later, but the transition in one of them reduces the torsional stress and holds back formation of the Z form in the other insert. Such a picture of conformational transitions was observed experimentally and obtained theoretically (Kelleher *et al.* 1986). This kind of behavior is absolutely impossible in linear DNA, where two such transitions occur independently and simultaneously.

6.6   How supercoiling can affect DNA biological functions

6.6.1   Supercoiling and DNA–protein interaction

DNA supercoiling can affect both equilibrium and dynamic aspects of DNA–protein binding. The affinity of a protein to DNA can be affected by supercoiling, if a protein unwinds the double helix locally upon binding with DNA. It is hardly possible that a protein can overtwist the double helix since the B form of DNA is already strongly

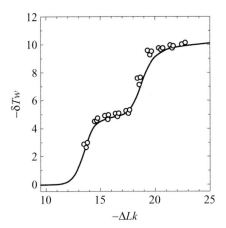

Figure 6.27 DNA unwinding due to B–Z transitions in two identical inserts, d(CG)$_{12}$ · d(CG)$_{12}$, within a 4400 bp plasmid. The theoretical curve is shown alongside the experimental data (O). The figure was redrawn from Fig. 3 of Kelleher *et al.* (1986), with permission.

wound. Therefore, the value of the local change of DNA twist, δtw, can only be negative. It is also possible that a protein introduces a local chirality into the bound DNA segment. This creates a local writhe, δwr, which can be either negative or positive. Thus, Lk of an unstressed double helix, Lk_0, is changed by some value, δlk, which is equal to $\delta tw + \delta wr$. This change results in a change of the supercoiling free energy, similar to the formation of local alternative structures. We can analyze this quantitatively. To a good approximation the free energy of supercoiling depends only on the DNA's elastic deformation. The magnitude of this deformation will be defined now by $Lk - (Lk_0 + \delta lk)$ rather than by $Lk - Lk_0$. Accordingly, under near-physiological ionic conditions the change in the free energy of supercoiling upon protein binding, δG_s, is given by (see Eq. (6.15))

$$\delta G_s = 1100RT/N[(Lk - Lk_0 - \delta lk)^2 - (Lk - Lk_0)^2] \approx -(2200RT/N)\Delta Lk\delta lk.$$

(6.21)

We can also express δG_s as a function of σ:

$$\delta G_s = -(2200RT\sigma/\gamma_B)\delta lk.$$

(6.22)

The value of δG_s affects the DNA–protein biding constant, which is changed by the multiplier

$$k_s = \exp[-\delta G_s/RT] = \exp[(2200\sigma/\gamma_B)\delta lk].$$

(6.23)

Since the sign of δG_s can be either positive or negative, negative supercoiling can either increase or decrease the binding constant. Equation (6.23) shows that the effect can very large. For example, if $\sigma = -0.06$ and $\delta lk = -1$, $k_s \cong 3 \times 10^5$.

6.6.1.1   Thermodynamics and kinetics of juxtaposition of remote DNA sites

The second way in which DNA supercoiling can influence enzymatic reactions is based on global changes of DNA conformations. These global changes increase the probability of juxtaposition of DNA sites separated along the chain contour. The probability of site

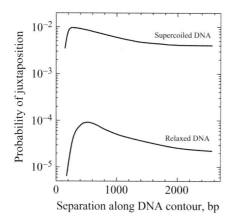

Figure 6.28 Probability of juxtaposition of two sites, separated along the DNA contour, for supercoiled and relaxed circular molecules. The data were obtained by the analysis of simulated ensembles of DNA conformations for molecules 5600 bp in length (Vologodskii & Cozzarelli 1996). The analysis assumed that two sites are juxtaposed if the distance between them does not exceed 10 nm.

juxtaposition is important for the processes where two or more sites participate in a DNA–protein complex. Numerous examples of such multi-site complexes are found in DNA replication, transcription, site-specific recombination and transposition.

The nature of the effect of supercoiling on the probability of site juxtaposition is easy to understand. If we take a look at a typical conformation of supercoiled DNA (Fig. 6.17), it becomes clear that supercoiling increases the probability of site juxtaposition. Indeed, there are many pairs of juxtaposed sites at each conformation of supercoiled DNA, while the juxtapositions are rare in typical conformations of the relaxed molecules (Fig. B6.2.1). Computer simulation shows that supercoiling magnifies the probability of juxtaposition by more than two orders of magnitude (Fig. 6.28) (Vologodskii & Cozzarelli 1996). Such an increase of the juxtaposition probability can greatly accelerate enzymatic processes that involve simultaneous interaction with two or more DNA sites. Supercoiling also changes preferable relative orientations of the juxtaposed sites, and this can be very important for the efficiency of some enzymatic reactions. In at least one case a strong influence of mutual orientation of the juxtaposed sites has been well documented (Stone *et al.* 2003).

Clearly, the probability of site juxtaposition and the distribution of their mutual orientations can affect the stability of DNA–protein complexes that involve DNA loops. It should affect the kinetics of the complex assembly as well. Indeed, as discussed in Section 5.4, the rate of assembly is proportional to the equilibrium probability of juxtaposition of DNA sites in a proper orientation, so supercoiling can both increase and decrease this rate.

6.6.1.2 **Changes of DNA helical repeat and protein binding**
Since supercoiling changes the DNA helical repeat, it can affect DNA binding with proteins. A simple analysis shows, however, that the effect should be negligible for proteins that interact with a single DNA site due to the relatively large amplitude of the

twist thermal fluctuations. Indeed, if the change of the average DNA twist in the binding site by supercoiling, δTw, is smaller than the amplitude of thermal fluctuations of Tw at this site, the value of δTw cannot essentially affect the protein binding. The amplitude of the twist thermal fluctuations, $\sqrt{\langle (\Delta Tw)^2 \rangle}$, in a segment m base pairs in length, is specified by Eq. (B6.2.3). Assuming that C is equal to 3×10^{-19} erg·cm one can obtain from the equation that

$$\sqrt{\langle (\Delta Tw)^2 \rangle} \cong 0.01\sqrt{m}. \tag{6.24}$$

The value of δTw can be estimated as

$$\delta Tw = 0.25\sigma m/\gamma \cong 0.0012m. \tag{6.25}$$

We assumed here that the change of DNA Tw is about $0.25\Delta Lk$ and that σ is close to -0.05. Comparing Eqs. (6.24) and (6.25), we conclude that the supercoiling-induced change of the site Tw cannot be important for the binding affinity if $m < 60$. There are very few proteins for which the binding site exceeds 60 bp, and only in these cases can the change of the average twist affect the binding properties.

There are proteins that interact simultaneously with two DNA sites separated along the contour of the double helix, creating DNA loops (Hochschild 1990, Halford *et al.* 2004). If the size of these loops is sufficiently small, there is a pronounced preferable torsional orientation of one end of the loop relative to the other end, when they are juxtaposed. This preferable orientation changes with the loop size, and it results in the oscillations of the protein binding affinity when the loop size alters (Hochschild & Ptashne 1986, Lee & Schleif 1989). DNA supercoiling can affect the protein binding in such cases, if the loop size is smaller than ≈ 300 bp. This effect has been observed experimentally (Kramer *et al.* 1988).

6.7   Knots and links in DNA molecules

As the ends of a linear DNA join, the molecule adopts a circular form with a specific topology. Until now we have examined different topological states associated with the linking number of the complementary strands of the double helix. The strands' linking number, however, is not the only topological characteristic of circular DNA. A circular DNA, as well as any circular polymer, may form an unknotted state or a knot of a certain type (an unknotted closed chain is considered as a trivial knot). In the case of dsDNA we should talk about knots formed by its axis. With all conformational rearrangements of the chains the type of the knot must remain unchanged. This factor must influence the conformational properties of circular DNA molecules, which depend, in general, on the topology of its axis.

Just as different types of knot may form through the cyclization of single chains, a closed pair or a larger number of polymeric chains may form links of different types, or catenanes. We have already considered a special type of link formed by the complementary strands of a double helix in a closed circular DNA. In this section, however, the

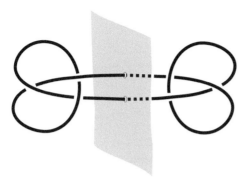

Figure 6.29 A composite knot. The shaded unlimited surface intersects this knot at two points only. By contracting these points into one along the surface, we obtain two simple knots.

3_1 (Trefoil) 4_1 5_1 5_2 6_1 6_2 6_3

Figure 6.30 Table of simple knots with fewer than seven intersections in the standard form.

double helix will be treated as a simple polymer chain. Knots and links often form in living cells and in the course of various laboratory manipulations with DNA. Different aspects of these issues are discussed in reviews (Wasserman & Cozzarelli 1986, Stark & Boocock 1995, Ullsperger *et al.* 1995)).

6.7.1   Types of knot and link

For the classification knots and links have to be deformed to the standard form of their projection on a plane. The standard form of a knot (link) projection is such an image thereof when the minimum number of intersections on the projection is achieved. Of course, in the course of a knot deformation self-intersections are not allowed. Deformations of this kind are known in topology as isotopic deformations. Two knots (or links) that can be transformed into each other by way of isotopic deformation belong to the same isotopic type.

Knots can be simple or composite. A knot is composite if there is an unlimited surface crossed by the knot at two points only that divides it into two nontrivial knots. This definition is illustrated in Fig. 6.29. There is an infinite number of topologically distinguishable types of knot and link. The simplest knot has three intersections in the standard form and is called a trefoil. Figure 6.30 shows the initial part of the table of simplest knots – namely, all knots with fewer than seven intersections in the standard form.

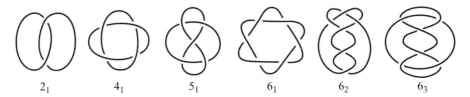

Figure 6.31 Table of links with fewer than seven intersections in the standard form.

A knot and its mirror image are considered to belong to the same type of knot, although they may belong to the same or different isotopic types. In particular, the trefoil and its mirror image cannot be transformed into each other by isotopic deformation, and therefore belong to different isotopic types. The figure-eight knot (4_1) and its mirror image belong to the same isotopic type. The table of knots is, in fact, a table of knot types, because it has only one representative of mirror pairs. With the growth of the number of intersections the number of types of simple knot grows very fast. As it turns out, there are 49 types of simple knot with nine intersections in the standard form, 165 with ten and about 552 types of knot with 11 intersections (Rolfsen 1976, Adams 1994).

The tables of two-contour links are based on the same principle as the table of knots (Rolfsen 1976). The initial part of the table of link types is shown in Fig. 6.31. The links formed by the complementary strands of the double helix in closed circular unknotted DNA belong to the torus class. The name of the class originates from the fact that the torus links can be placed on the surface of a torus without self-intersections. Three links in Fig. 6.31, 2_1, 4_1 and 6_1, belong to this class. However, the fraction of torus links among more complex links is very small.

6.7.2   Topological equilibrium

Since long DNA molecules may have many different topologies in solution, we can talk about the distribution of topological states in a sample of circular DNA. Among many possible distributions of topological states one is especially important, the distribution that corresponds to the thermodynamic equilibrium. As a result of thermal motion the distributions of nearly all conformational properties relax to their equilibrium form. Therefore, after some time we always have equilibrium distributions of conformational properties of the molecules in solution.

Distributions of topological states of circular molecules are different. These distributions do not change as long as the molecule backbones stay intact. The concept of the equilibrium distribution is very useful, however, in this case as well.

A formal definition of the equilibrium distribution of topological states of circular molecules is not different from the equilibrium distributions of properties of linear molecules. This is the most probable distribution, which is obtained by random exchange between different states in solution. The free energy of the molecules with this distribution has minimum possible value. We would obtain the equilibrium distribution of topological states if our circular molecules had phantom backbones, so their segments could pass through one another during the thermal motion. Thus, to calculate the distribution we should forget about forbidden exchange between different topological states and use usual rules of statistical mechanics.

Although exchange between different topological states is impossible in a solution of circular DNAs, we can obtain the topological equilibrium by slow cyclization of linear molecules. The cyclization can be performed by joining the sticky ends of DNA. If the sticky ends are long enough (about a dozen nucleotides or longer), a circular form is stable at room temperature; for shorter sticky ends ligation is required to seal at least one of the two single-stranded breaks.

Let us first consider a very diluted solution of opened circular DNA so that the equilibrium probability of forming a link between the molecules is negligible. In this case the equilibrium distribution of topological states is reduced to the distribution of unknotted circles and various knots. The first question is what the probability of any knot is for a circular DNA of a particular length. To solve the problem we have to learn how to distinguish between unknotted circular molecules and knots of various types.

6.7.3   Theoretical and experimental studies of knots and links in DNA

How can we determine the topology of a particular conformation of a closed chain? Let us imagine that we have a heavily entangled circular cord and we want to find out whether it is knotted or not. The fact that persistent attempts to untangle the cord have produced no result cannot serve as a proof that we are dealing with a nontrivial knot. What is needed to resolve this problem is an algorithm of verification of the topological identity of the conformation in question. The construction of such algorithms belongs to the realm of a branch of mathematics known as topology (Adams 1994). Topologists have developed invariants of the topological states, characteristics thereof that remain unchanged with any deformations of the chains performed without intersections of the chain segments (isotopic deformations). The simplest invariant of topological state is the Gauss integral, which defines the linking number of two chains (see Box 6.1). Of course, to classify the state of chains with a topological invariant, the latter must assume different values for different topological states. Not one topological invariant meets this requirement in full measure, but there are very powerful ones among them, which help identify many elementary types of knot (link) and distinguish them from more complex ones. The Gauss integral is a fairly weak topological invariant and is of no use for distinguishing many linked states of chains from the unlinked state. Still the integral is very useful for analysis of DNA supercoiling, since it distinguishes all links of the torus class. Another very useful invariant is the Alexander polynomial (Alexander 1928), which has been widely used in computer simulation (reviewed by Frank-Kamenetskii and Vologodskii (1981), Vologodskii (2007)). It is a polynomial of one variable in the case of knots and of two variables in the case of links of two circular chains.

Starting from 1974, many works have been devoted to computer calculations of the equilibrium probability of knots (reviewed by Vologodskii (2007)). The results show that the equilibrium probability of knotted circular molecules increases with their length. Figure 6.32 shows that for DNA molecules a thousand base pairs in length the equilibrium

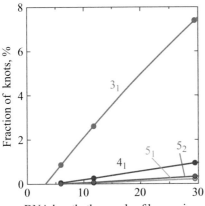

Figure 6.32 Equilibrium probability of the simplest knots in circular DNA. The results of computer simulations, shown by the symbols and solid lines, correspond to the physiological ionic conditions. A black-and-white version of this figure will appear in some formats. For the color version, please refer to the plate section.

fraction of knots is rather small, although the fraction of the simplest knots, trefoils, can be measured with good accuracy.

Knots and links of different types formed by circular DNA molecules have different electrophoretic mobilities in agarose gels, which makes it possible to separate them by electrophoresis (Fig. 6.33). The method requires special calibration, since it is impossible to predict the mobility of a particular topological form. The only way to identify the DNA topology in a particular band of the gel is to extract the molecules from this band and analyze their topology by EM. Covering the DNA with *E. coli* RecA protein greatly simplifies such analysis (Krasnow *et al.* 1983). In this way, the correspondence between the relative electrophoretic mobility and topology has been established for many of the simplest knots and links (Vologodskii *et al.* 1998). To separate knotted and linked DNA molecules by gel electrophoresis they must have single-stranded breaks, for otherwise the mobility will also depend on the double helix's linking number.

The computer calculations also showed that the probabilities of different nontrivial knots reduce dramatically when the thickness of the molecule increases (Le Bret 1980, Klenin *et al.* 1988). In the case of DNA, the effective thickness is defined not only by the geometrical diameter of the double helix, but also by electrostatic repulsion of the DNA segments (see Section 3.2.3). Therefore, the equilibrium fraction of trefoils strongly depends on ionic conditions (see Fig. 3.16) (Rybenkov *et al.* 1993, Shaw & Wang 1993, Rybenkov *et al.* 1997b).

If the concentration of DNA molecules during their random cyclization in solution is not too small, some fraction of the molecules will form links at thermodynamic equilibrium (in biology and chemistry, links are usually called catenanes). Let us suggest that the concentration is still small enough that most of the circular molecules are not linked with others at equilibrium; in this case, links of three and more molecules can be neglected. Thus, we can consider formation of catenanes as a bimolecular reaction,

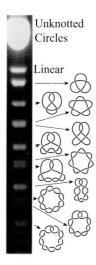

Figure 6.33 Separation of different knots by gel electrophoresis. All of the knots have the same molecular mass since they arise from the same unknotted substrate molecules. More complex knots form more compact conformations, on average, and move faster. Before gel electrophoresis, all DNA molecules were nicked to eliminate supercoiling of the molecules, so that the difference in the knot type was the only reason for the electrophoretic separation of knots into different bands. The DNA from each band was analyzed by EM and identified as the knot type drawn. The gel was reproduced from Fig. 7 of Crisona *et al.* (1994) with permission.

which is characterized by its equilibrium constant, B. The equilibrium concentration of catenanes, c_{cat}, can be found from the following equation, where c is the total concentration of circular molecules:

$$c_{cat} = B(c - 2c_{cat})^2 \approx Bc^2. \tag{6.26}$$

The value of B depends on DNA length and ionic conditions in solutions. The dependence of B on DNA length for physiological ionic conditions is shown in Fig. 6.34.

We can also consider topological equilibrium for a system that consists of two types of circular DNA (A and B). Such a system was used to probe conformations of supercoiled molecules (see Fig. 6.16) (Rybenkov *et al.* 1997d). It consisted of supercoiled molecules of one length (molecules A) and cyclizing molecules of another length (molecules B). The system was prepared in such a way that there was only partial topological equilibrium: there were no AA catenanes although concentration of molecules A was rather high, but there was equilibrium regarding AB and BB links. Since the goal was to measure the equilibrium constant for AB links, such partial equilibrium simplified the analysis of various products in this system. The same approach was used by Wang and Schwartz many years ago in their pioneering study of DNA topological properties (Wang & Schwartz 1967).

We have considered here different distributions of topological states: distributions of topoisomers of supercoiled DNA; distributions of knots and links. All these distributions have one very important advantage in comparison with distributions of other DNA

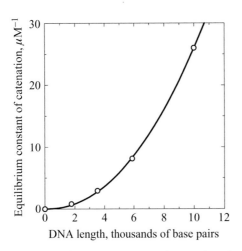

Figure 6.34 Equilibrium constant of catenation of two identical circular DNAs. The data were obtained by computer simulation for the physiological ionic conditions performed by the present author.

conformational properties. They are not changed during any manipulations of DNA that are required for their quantitative analysis. Conditions under which the distributions are analyzed are not important in this case. This contrasts the distributions of topological states from the distribution of all other conformational properties, which are so easy to disturb. This advantage of topological distributions has been widely used in studies of conformational properties DNA molecules and action of enzymes that are able to change DNA topology (see Wasserman and Cozzarelli (1986), Stark and Boocock (1995) for reviews of some biological applications).

6.8   DNA supercoiling inside living cells

6.8.1   DNA supercoiling tension in vivo

DNA molecules isolated from bacterial cells have a fairly high level of negative supercoiling. The superhelix density of plasmid DNA molecules extracted from bacteria is close to −0.06, although the precise value may depend on the plasmid, bacterium type and conditions of cell growth. The level of DNA supercoiling in bacterial cells results from competitive action of two topoisomerases, bacterial topoisomerase I and DNA gyrase (Menzel & Gellert 1987). This regulation is achieved by regulation of the corresponding genes by the level of supercoiling. According to this mechanism, reduction of the level of negative supercoiling increases transcription of DNA gyrase genes and, correspondingly, the activity of the enzyme that introduces negative supercoiling into bacterial DNA (Menzel & Gellert 1987). The level of the supercoiling tension is a subject of fine regulation: it does not change with temperature or other conditions that change the DNA helical repeat (Drlica 1992).

There are solid data, however, showing that the supercoiling tension in circular DNAs inside bacterial cells corresponds to only about one-half of its magnitude in purified DNA molecules extracted from the cells. This reduced supercoiling tension corresponds to the unconstrained supercoiling that is not associated with specific tertiary structures of DNA segments, imposed by DNA-bound proteins. Little is known about specific DNA–protein structures in bacterial cells (reviewed by Higgins and Vologodskii (2004), Browning *et al.* (2010)). It is important for the current issue that these structures have sufficient chirality to constrain Wr or Tw of (-3) per 100 turns of the double helix, on average. Various experimental approaches were used to measure the value of unconstrained supercoiling.

Pettijohn and co-workers were the first to address this problem (Pettijohn & Phenninger 1980, Sinden *et al.* 1980). They were able to show that there is unconstrained supercoiling in bacterial cells. Their main approach was based on measuring the number of DNA-bound molecules of trimethylpsoralen inside the cells, since the ligand can penetrate through the cell membrane. Upon binding, this ligand intercalates into the double helix and unwinds it. Therefore, binding of the ligand is more efficient with negatively supercoiled DNA than with nicked DNA. To measure amount of the ligand bound with DNA inside the cells they were exposed to 300 nm light, under which DNA-bound trimethylpsoralen forms photo cross-links. Thus, the radiation allowed the amount of the bound ligand to be measured after DNA extraction from the cells. To determine the effect of DNA supercoiling on the ligand binding, the researchers compared the amount of DNA-bound ligand in the intact cells and in the cells that had been exposed to γ-radiation. Since the radiation causes single-stranded breaks in the double helix, the radiated DNA had no supercoiling tension. Still, the approach had many assumptions and only provided a semiquantitative estimation of the unconstrained supercoiling. It was concluded that the supercoiling tension in *E. coli* cells corresponds to a σ of -0.05 ± 0.01.

No unconstrained supercoiling was detected in eukaryotic cells (Sinden *et al.* 1980). Although the latter conclusion was confirmed in subsequent studies, it should be noted that it is only applied to the supercoiling tension averaged over the entire genome (see below).

Another approach to measuring supercoiling tension within the bacterial cells was based on the topological analysis of the products of the site-specific recombination (Bliska & Cozzarelli 1987). The recombination in a specially constructed plasmid gave two double-stranded linked circular DNA molecules, and the linking number was proportional to the value of σ in the original plasmid. The approach had its own assumptions, and the obtained value of σ, -0.025, should be considered as the lower limit of the unconstrained supercoiling tension.

One more approach to this problem is based on formation of noncanonical DNA structures that appear under negative supercoiling. These structures have been detected in many plasmids extracted from bacterial cells (see Section 6.5), and it was suggested that formation of these structures plays a role in regulation of gene expression. Therefore, the question of formation of noncanonical structures inside the cells deserves more detailed analysis.

6.8.2   The existence of noncanonical structures inside the cells

Cruciform structures and segments of Z and H forms were detected in many plasmids isolated from bacterial cells. This, however, does not mean that the supercoiling tension in DNA inside the cells is sufficient for formation of these noncanonical structures. The issue required special investigation. It turned out that this investigation was easier to perform for the cruciform structures, owing to very slow kinetics of their formation and disappearance at temperatures around 0 °C. Under these conditions the relaxation times run into weeks or longer. Therefore, if DNA is isolated from the cells at these temperatures, the conformation of palindromic regions should remain unchanged. In this way Lyamichev *et al.* showed that the main palindrome of the pAO3 plasmid does not adopt the cruciform structure in *E. coli* cells (Lyamichev *et al.* 1984). The cruciform structure was extruded in the plasmids, however, after a short exposure of the extracted molecules to 40 °C. A similar result was obtained by Courey and Wang (1983), who examined the state of an artificial 68 bp palindrome inserted into a plasmid. Sinden *et al.* also found that there were no cruciform structures inside a cell by fixing the conformation of the palindrome with trimethylpsoralene interstrand cross-links and analyzing the structure of the palindromic region in the isolated plasmid (Sinden *et al.* 1983).

There could be two possible reasons that the palindromes do not adopt cruciform structures within the cells. First, the superhelix tension inside the cells could be insufficient for the extrusion. This is probably the case for the many natural palindromes, which adopt the cruciform conformation at a σ of -0.05 to -0.06. Second, at low superhelix tension the extrusion is so slow that the thermodynamic equilibrium is not reached over the cell lifetime. Indeed, the time of cruciform formation strongly depends on the superhelix density (see Section 6.5). In longer artificial palindromes the cruciform extrusion can occur at lower superhelix tension, but the process is too slow. The slow extrusion at low supercoiling tension is the reason why the cruciforms were not observed even for very long palindromes (Courey & Wang 1983, Borst *et al.* 1984).

Still, the cruciforms were detected inside the cells. They can be formed in sufficiently long segments of $d(AT)_n \cdot d(AT)_n$ where the cruciform extrusion occurs sufficiently fast at relatively low superhelix density. This was first shown by Haniford and Pulleyblank, who studied the distributions of topoisomers in the plasmid with and without $d(AT)_n \cdot d(AT)_n$ inserts (Haniford & Pulleyblank 1985). Their approach was based on the fact that the supercoiling tension inside the cells is maintained at a constant level by DNA topoisomerases. Formation of a cruciform (or a segment in Z form) causes a reduction of the negative supercoiling tension, compensated by DNA gyrase, which reduces the DNA linking number, Lk. The topoisomer distribution obtained in this kind of experiment is illustrated in Fig. 6.35. The figure shows that the structural transitions occurred only in a fraction of the plasmids carrying the insert. The transition was observed, however, only in cells with suppressed protein synthesis (Haniford & Pulleyblank 1983b, Haniford & Pulleyblank 1985). Earlier these researchers applied the same approach to the formation of Z form in a plasmid DNA in *E. coli* cells (Haniford & Pulleyblank 1983b).

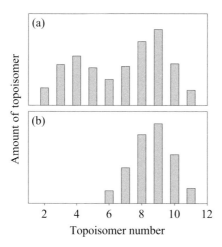

Figure 6.35 The effect of the B–Z transition on the distribution of topoisomers of plasmid DNA extracted from *E. coli* cells. (a) The B–Z transition in a fraction of the DNA molecules containing a d(CG)$_{12}$ · d(CG)$_{12}$ insert caused reduction of the supercoiling tension. The reduction was compensated by the decrease of Lk by DNA gyrase, which together with topoisomerase I maintains the supercoiling tension. The resulting group of topoisomers corresponds to the left half of the bimodal distribution of the plasmid topoisomers. The right half of the distribution represents the molecules where the insert remained in the B form. It matches the distribution pattern for the reference plasmid that did not have the insert (b). The figure is based on the data of Haniford and Pulleyblank (1983b).

This method of registering the formation of alternative structures in a cell has been used for evaluating DNA's torsional stress in vivo. Indeed, by finding the minimum length of inserts that assume alternative forms in the cells under specific conditions, it is possible to evaluate the torsional stress of DNA. This was done by using the method described above or by chemical modification of the alternative structures inside the cells (Zacharias *et al.* 1988, McClellan *et al.* 1990, Dayn *et al.* 1991). Such analysis, however, has to take into account that the superhelix density of the transition depends on the ionic conditions. The ionic conditions inside the cells correspond to 0.2 M NaCl, while the majority of the experiments detecting the alternative structures in vitro were run at a considerably lower concentration of sodium ions. This point was taken into account by Dayn *et al.*, who showed that in cells with suppressed protein synthesis the cruciforms evolved in practically all molecules with d(AT)$_n$ · d(AT)$_n$ inserts of 32 and 42 bp and in some of the inserts of 22 bp (Dayn *et al.* 1991). The authors concluded that the superhelix tension in these cells corresponded to a σ of −0.055. After accounting for the rise of the superhelix density caused by the suppression of protein synthesis, the authors concluded that supercoiling tension in growing cells corresponds to a σ of −0.045. Zacharias *et al.* managed to observe the B–Z transition in growing cells in long d(CG)$_n$ · d(CG)$_n$ inserts (Zacharias *et al.* 1988). The evaluation of supercoiling tension in their study did not take into account the dependence of the transition point on the ionic conditions. After the correction for this factor the supercoiling tension found in that study corresponded to a σ of −0.04.

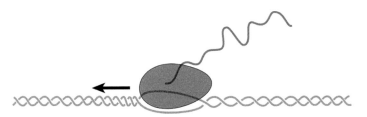

Figure 6.36 Diagram of DNA supercoiling induced by transcription. The local positive and negative supercoiling appears due to local rotation of the double helix caused by the moving RNA polymerase (shown in gray). This model assumes sufficiently high frictional resistance to the rotation of the enzyme and newly synthesized RNA (shown in dark gray) around DNA or DNA around the enzyme complex. For simplicity of the diagram the excess of both positive and negative supercoiling is shown as a change of DNA local twist, although actual partitioning between twist and writhe depends on the local conditions.

Thus, different studies showed that the supercoiling tension in prokaryotic cells corresponds to the range of σ from -0.025 to -0.05. It has become clear now that such a wide range may be due not only to inaccuracy of the specific methods, but also to the nonuniform distribution of the supercoiling tension inside the cells.

6.8.3   Transcription-induced DNA supercoiling

After the discovery of DNA gyrase, an enzyme capable of introducing negative supercoiling by using the energy of ATP hydrolysis, it seemed clear that the enzyme should play a key role in creating negative DNA supercoiling inside cells. In 1987, however, Liu and Wang came up with another hypothesis, which has been convincingly proved by later experiments. According to their idea an important role in the process of supercoiling belongs to DNA transcription (Liu & Wang 1987). In the course of this process, RNA polymerase synthesizes mRNA on DNA molecules. Since DNA has the structure of a double helix, in the course of this process either the RNA polymerase complex rotates around the axis of the double helix, or the double helix turns upon itself. The rotation of the RNA polymerase complex runs into considerable complications because of its very large size: RNA polymerase itself is one of the largest enzymes, then there is the newly synthesized RNA chain which can be bound with ribosomes, the sites of protein synthesis. DNA, on the other hand, can rotate fast around its axis, even if this axis is curved in space (similarly to a speedometer cable in a car) (Krasilnikov *et al.* 1999). However, this rotation is hindered by DNA-bound proteins forming the chromatin structure. Therefore, there is no a simple way for the cell to overcome this problem. This is why, as RNA polymerase complex moves along the DNA, it creates an area of positive supercoiling in front of itself and of negative supercoiling behind (Fig. 6.36). Such supercoiling may dissolve as a result of the diffusion of supercoils and their disappearance at the ends of the molecule, or mutual annihilation in the case of circular DNA.

The mutual annihilation of the supercoils is especially difficult when RNA synthesis occurs simultaneously in two regions of circular DNA in opposite directions. In this case, two regions arise in DNA, one with positive and the other with negative supercoiling, which are separated by RNA polymerase complexes. Another possible reason for restricting the diffusion of supercoils is the attachment of DNA at some points to large cellular structures. In such cases the transcription can continue only if the torsional stress arising with the movement of RNA polymerase is released by the action of DNA topoisomerases. In *E. coli* cells DNA gyrase causes the relaxation of positive supercoiling while topoisomerase I removes negative supercoiling. It was shown that suppression of the activity of one of these enzymes resulted in strong negative or positive supercoiling of plasmid DNAs isolated from the cells (Wu *et al.* 1988). No changes in supercoiling were observed in control experiments when the transcription was suppressed by adding rifampicin to the cells. It was also shown that the supercoiling of DNA isolated from the cells with a suppressed gyrase activity depended on the number and orientation of the transcription initiation sites (Wu *et al.* 1988).

These results showed, above all, that topoisomerases are absolutely essential for normal transcription. It was unclear, however, whether transcription in cells with normal topoisomerase activity can lead to the development of a local supercoiling that is substantially different from the mean value. In cells with normal topoisomerase activity the overall supercoiling of isolated DNA changed very insignificantly with changes in transcription intensity and in the number and intensity of promoters (Wu *et al.* 1988). It is possible, in principle, that topoisomerases very quickly remove the excess supercoiling tension, so that local supercoiling practically always corresponds to the mean value. On the other hand, the relaxation of local supercoiling tension could happen during DNA isolation, while in the course of transcription it might differ substantially from the mean value. Analysis of Z-DNA formation in $d(CG)_n \cdot d(CG)_n$ inserts (Rahmouni & Wells 1989) and cruciform formation in $d(AT)_n \cdot d(AT)_n$ inserts (Dayn *et al.* 1992) introduced into plasmids upstream from the active promoters clearly pointed to the latter possibility. Both studies detected substantial increase of the supercoiling tension behind the moving polymerases, although there were quantitative differences in the results. Rahmouni and Wells detected a relatively short wave of negative supercoiling behind the enzyme while Dayn *et al.* found that the effect of transcription spreads to about 1 kb upstream of the active promoter. Dayn *et al.* estimated that in the vicinity of the promoter the supercoiling tension is close to -0.05. Qualitatively similar results were observed in eukaryotic cells by Rich and co-workers (Wittig *et al.* 1992, Wolfl *et al.* 1996).

Another approach to measure dynamic supercoiling, based on enhanced trimethylpsoralen binding with negatively supercoiled DNA, allowed local supercoiling tension to be estimated in specific segments of eukaryotic genome (Kramer *et al.* 1999). The approach was based on blot hybridization or high-density oligonucleotide microarrays to detect possible change in the supercoiling tension in chosen DNA segments. It was found by this method that transcription-induced dynamic supercoiling spreads upstream of the start sites of active genes over distances of about 1 kb (Ljungman & Hanawalt 1992, Kramer & Sinden 1997, Kouzine *et al.* 2013, Naughton *et al.* 2013). The approach also confirmed the earlier data that showed that the eukaryotic genome is organized into supercoiling domains (see Naughton *et al.* (2013) and references therein). Still, it should be noted that the effect of supercoiling is relatively small, so the methods of supercoiling detection using trimethylpsoralen binding have limited accuracy.

Thus, it is generally accepted that DNA transcription can change local supercoiling tension in cells with normal activity of DNA topoisomerases, although specific details of this supercoiling depend on various surrounding factors. On the other hand, insufficient relaxation of the transcription-induced torsional stress can affect the transcription elongation. This seems to be the case for transcription of extremely long genes (longer than 67 kb) in neurons (King *et al.* 2013). It was found that inhibition of topoisomerases slows down the elongation of transcription of such genes, probably due to accumulation of positive supercoiling tension ahead of the transcription complex. Some of these genes are related to the autism spectrum disorders, and in accordance with this rare mutations in topoisomerases were found recently in patients with these disorders (Neale *et al.* 2012). Thus, transcription-induced supercoiling and insufficient activity of topoisomerases may be responsible for these and some other genetic disorders in which the genes are exceptionally long.

6.9   DNA topoisomerases

It does not require sophisticated knowledge of topology to understand that the helical structure of double-stranded circular DNA creates very serious problem during its replication. Indeed, the complementary strands of the molecule form a torus link of very high order, and this topology should be preserved during the replication when both strands of the double helix serve as templates for building the complementary strands. Therefore, the daughter dsDNA molecules have to form a torus link with very large Lk value. Clearly, these linked molecules cannot be segregated into different daughter cells if their backbones remain intact. This problem even prompted suggestions that maybe the structure of dsDNA is different from the one suggested by Watson and Crick, and is not a double helix. It turned out, however, that nature solved this and other related problems by creating DNA topoisomerases, the enzymes that can alter the topological state of circular DNA molecules.

Since a topological state of circular DNA cannot be altered without a break in the chain, the work of these enzymes comes down to creating temporary single-stranded or double-stranded breaks, allowing other segments to pass through the breaks and then resealing these breaks. These remarkable enzymes have been a subject of intensive biochemical, biophysical and genetic studies since 1971, when the first topoisomerase was discovered (Wang 1971). They are the subject of many comprehensive reviews (Wang 1996, Wang 1998, Champoux 2001, Schoeffler & Berger 2008), and even a monograph written by J. Wang, who made an outstanding personal contribution to this field (Wang 2009). Here we will give only a brief consideration of the basic properties of these enzymes and discuss, at the end, some thermodynamic aspects of their work.

There are four major types of topoisomerase, IA, IB, IIA and IIB. Roman figures I and II refer to the number of DNA strands the enzymes cut simultaneously during the catalytic act. Type IA and type IB topoisomerases are very different, and we consider them separately. On the other hand, the difference between type IIA and type IIB topoisomerases is not so dramatic, although their amino-acid sequences have very low similarity. From a biophysical point of view they are quite similar enzymes, and therefore we will not touch type IIB topoisomerases here. There are also two special topoisomerases, DNA gyrase and reverse gyrase, which belong to types IIA and IA, respectively. These are the only enzymes that can introduce supercoiling into circular DNA, negative in the case of DNA gyrase and positive in the case of reverse gyrase. We will consider these enzymes separately.

From a chemical point of view all topoisomerases catalyze the same type of reaction, called transesterification. The reaction consists of two coupled chemical events: breaking an O–P bond in a DNA strand and simultaneously creating an O–P bond between one of the formed DNA ends and the protein. At the end of the catalytic cycle the reaction goes in the reverse direction, resealing the break in the DNA strand. A remarkable feature of this reaction is that it does not require additional energy, which is needed for joining DNA ends per se (the reaction catalyzed by DNA ligase). The transesterification reaction was postulated by Wang, who discovered the first DNA topoisomerase, later labeled type IA, in 1971 (Wang 1971).

6.9.1   Type IA DNA topoisomerases

These enzymes bind to a ssDNA segment. Usually this is a strand of dsDNA. They introduce a single-stranded break into the bound strand, simultaneously forming a covalent bond between the tyrosyl residue of the protein and the 5′-end of the broken strand. The protein holds, by noncovalent interaction, the 3′-end of the broken DNA strand, so both ends of the DNA strand remain attached to the enzyme. At the same time, the enzyme creates a gap between the strand ends (Fig. 6.37). This gap serves as an entrance to the large cavity in the proteins, which can accommodate not only a ssDNA segment but even a segment of the double helix. When a segment passes through the break and appears in the cavity, the break is resealed and the protein dissociates from the strand. So the net result of the catalytic act is passing a DNA segment through the single strand. The described method of the enzyme action was established by numerous biochemical and structural studies (reviewed by Wang (1996), Champoux (2001) and Schoeffler and Berger (2008)).

Type IA topoisomerases can relax negative DNA supercoiling. The fact that the substrate DNA has to be negatively supercoiled is crucially important in the reaction pathway. First, negative supercoiling greatly facilitates formation of open regions, needed for the protein binding, while positive supercoiling suppresses formation of such regions. Second, negative supercoiling causes left-handed interwinding of the complementary strands in the open regions, a prerequisite of increasing Lk during the strand-passing reaction. It was shown that the enzymes can relax positively supercoiled DNA molecules as well, if they have a stable single-stranded loop, which is needed for the topoisomerase binding. Such loops were created by incorporating into DNA a short double-stranded segment with a mismatch of several unpaired nucleotides (Kirkegaard & Wang 1985,

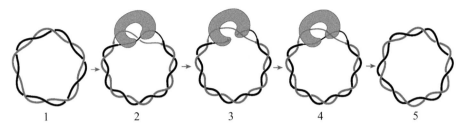

Figure 6.37 Diagram of supercoiling relaxation by type IA DNA topoisomerase. The enzyme is able to relax only negative supercoiling, and such a circular DNA is diagramed in panel 1. The enzyme binds only with a ssDNA segment (shown by thin lines), whose appearance is much more common in negatively supercoiled molecules, since base-pair opening accompanied by local unwinding reduces the torsional stress in the rest of the circular DNA (panel 2). An open region can accommodate either positive or negative interwinding of one strand around the other, but the left-handed interwinding is much more probable since it further reduces the torsional stress in the double-stranded region of negatively supercoiled DNA. After the binding the enzyme breaks the bound DNA strand, covalently attaches one end to itself, holds the other end of the broken strand, and creates a gap between the ends (panel 3). The segment of the complementary strand passes through the gap to a large cavity in the enzyme and the break is resealed after this (panel 4). The protein dissociates from the DNA, completing the catalytic cycle (panel 5). The reaction increases Lk of the complementary strands by 1, reducing negative supercoiling of the molecule. A black-and-white version of this figure will appear in some formats. For the color version, please refer to the plate section.

Dekker *et al.* 2002). Of course, in the latter case the handedness of interwinding in the open region corresponds to Lk reduction during the strand-passing reaction.

The reaction passway also explains how type IA topoisomerases can unlink or link double-stranded circular DNAs, if at least one of the participating molecules has a single-stranded nick (Tse & Wang 1980, Brown & Cozzarelli 1981). The nick strongly facilitates opening of the adjacent base pairs (Krueger *et al.* 2006), so a DNA segment in the opposite strand is a favorable spot for the enzyme to bind and create a transient break in the intact strand. A double-stranded segment can pass through the nick and the enzyme-created break into the large cavity of the enzyme. In such a way type IA topoisomerases can remove or create links between double-stranded circular DNA molecules and also create and remove knots (DiGate & Marians 1988, Hiasa & Marians 1994).

By now topoisomerases of type IA have been found in all classes of living organisms.

6.9.2   Type 1B DNA topoisomerases

These enzymes were discovered in 1972 by Dulbecco and Champoux (Champoux & Dulbecco 1972). Type IB topoisomerases bind dsDNA, forming a clamp around the DNA duplex. Similarly to the type IA enzymes, they make a break in one DNA strand and create a covalent bond with one of the created ends (3′-end in this case). However, they do not hold the other end of the strand firmly, allowing its rotation around the backbone of the intact strand (Fig. 6.38). Eventually they catch the rotating end in the position needed

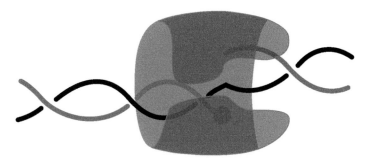

Figure 6.38 Diagram of type IB topoisomerase action. The enzyme binds a double-stranded segment and cuts one strand. One end of the cut strand is covalently attached to the protein while the other end is released for rotation around the intact strand. Eventually this end is caught by the enzyme and the broken strand is resealed. In this way the enzyme can relax both negative and positive supercoiling. A black-and-white version of this figure will appear in some formats. For the color version, please refer to the plate section.

to reseal the break (Stewart *et al.* 1998). Thus, the enzymes create a temporary swivel in the dsDNA. They can relax equally well both negatively and positively supercoiled DNA molecules, since they do not need a single-stranded segment for binding to DNA. It has been shown that the enzymes can remove several DNA supercoils during a single catalytic act (Koster *et al.* 2005). The latter property also differantiates type IB topoisomerases from type IA enzymes, which allow only a single strand-passing during the catalytic act (Dekker *et al.* 2002). The enzymes create the same equilibrium distribution of topoisomers in a solution of circular DNA molecules as does DNA ligase from nicked DNA circles (Depew & Wang 1975).

Type IB topoisomerases were found in all classes of eukaryotes and in many bacteria, although *E. coli* does not have one.

The reactions catalyzed by type IA and IB topoisomerases do not require any source of energy, such as ATP hydrolysis, coupled with the topological changes.

6.9.3 Type IIA DNA topoisomerases

The first topoisomerase of type II, named DNA gyrase, was discovered by Gellert and his colleagues in 1976 (Gellert *et al.* 1976). Today we know that the enzyme extracted from *E. coli* cells has all the features of type IIA topoisomerases, but in addition to this it is able to introduce negative supercoiling in closed circular DNA. We will discuss this property of DNA gyrase later in this section. Type IIA topoisomerases consist of two identical halves that can be monomers, dimers or even trimers (see Schoeffler and Berger (2008) for a review). In all cases the structure of the enzymes has twofold symmetry (Fig. 6.39). After the discovery of DNA gyrase, a few other topoisomerases of this type were soon found, and in 1979–80 it was established that the enzymes break two DNA strands simultaneously during their catalytic act (Brown & Cozzarelli 1979, Liu *et al.* 1979, Liu *et al.* 1980, Mizuuchi *et al.* 1980). The current model of the reaction act is diagramcd in Fig. 6.39.

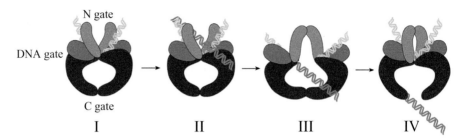

Figure 6.39 Diagram of type IIA topoisomerase action. In I, the enzyme in its open-clamp conformation binds a G segment. In II, a T segment enters the enzyme through the open clamp. The closure of the clamp upon ATP binding traps the T segment. The closure is coupled with breaking the G segment and opening the DNA gate. In III, the T segment reaches the central cavity of the enzyme, passing through the DNA gate. After this the DNA gate closes and the G segment is religated. In IV, closure of the DNA gate triggers the opening of the exit gate (C gate) and release of the T segment. A black-and-white version of this figure will appear in some formats. For the color version, please refer to the plate section.

First, the enzyme binds a DNA segment (G segment). Then it captures a second segment of the same or another DNA molecule (T segment), which enters through the N gate and triggers its closing. This trapping of the T segment causes breaking of the G segment and formation of the covalent links between the DNA ends and the halves of the enzyme. This event triggers opening of the DNA gate, allowing transport of the T segment through the break in the G segment. When the T segment appears in the large cavity of the enzyme, the DNA gate closes and the G segment is religated. Then the C gate opens and the T segment passes through it and leaves the enzyme. Thus, the T segment always enters the DNA-bound enzyme through the N gate and exits through the C gate (Roca & Wang 1994, Roca *et al.* 1996).

The first proof that type IIA topoisomerases cut both DNA strands came from the elegant experiments performed by Brown and Cozzarelli with DNA gyrase (Brown & Cozzarelli 1979). They used the fact that passing one double-stranded segment through another changes the value of Lk of the complementary strands by $+2$ or -2 (see Section 6.2.2). Thus, if the DNA sample consists of a single topoisomer (all molecules have the same value of Lk) the enzyme can only transform it into a mixture of topoisomers whose values of Lk differ from one another by an even number. This was precisely what the experiments demonstrated. In accordance with this mechanism of action, it was found soon that type IIA topoisomerases can link and unlink circular DNA molecules, create knotted circular DNAs and unknot them (Liu *et al.* 1980, Mizuuchi *et al.* 1980).

Type IIA topoisomerases can relax both negative and positive supercoiling. It was found, however, that the enzyme from *E. coli*, Topo IV, can relax positive supercoiling much more efficiently than negative supercoiling (Crisona *et al.* 2000, Charvin *et al.* 2003). This property is related to the requirement of a certain angle between the protein-bound G segment and a potential T segment entering the N gate (Stone *et al.* 2003). In negatively supercoiled DNA the angles between the opposite segments of the interwound superhelix are distributed around 120° while in positively supercoiled molecules they

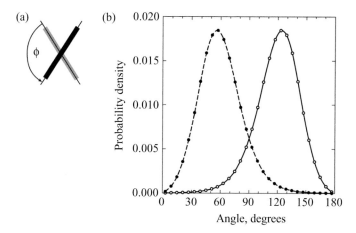

Figure 6.40 The distribution of angles between juxtaposed segments in supercoiled DNA. (a) The definition of the angle, φ, between two segments that have no directionality (we should assume here that the segments have no directionality since type IIA topoisomerases have no sequence specificity in their interaction with DNA). The angle is measured in the counterclockwise direction from the underlying segment to the overlying one in the plane that is parallel to both segments. Thus, the angle changes in the [0, 180°] interval. (b) The distribution of angles between juxtaposed segments in positively (dashed line) and negatively (solid line) supercoiled DNA. The distributions were obtained by computer simulation of the equilibrium set of DNA conformations (Vologodskii & Cozzarelli 1996). One can see from the plot that, if a type IIA topoisomerase binds a T segment when the angle with the enzyme-bound G segment is close to 60°, it has to relax, mainly, positive rather than negative supercoiling (Stone *et al.* 2003).

are distributed around 60° (Fig. 6.40). Clearly, Topo IV prefers angles between G and T segments that are close to 60°. This chiral discrimination has not been observed among eukaryotic type IIA topoisomerases: the enzyme from *Drosophila melanogaster* equally well relaxes negative and positive supercoiling (Charvin *et al.* 2003).

Nearly all reactions catalyzed by type II topoisomerases are coupled with hydrolysis of ATP. The only exception is slow relaxation of supercoiling by DNA gyrase, which occurs without ATP consumption.

6.9.4   DNA gyrase

DNA gyrase is the only topoisomerase that introduces negative supercoiling into circular DNA. It belongs to the family of type IIA DNA topoisomerases and has strong sequence and structural homology with other topoisomerases of this type. Type IIA topoisomerases can change Lk of the complementary strands by ± 2 (see above). DNA gyrase changes Lk by -2, until the superhelix density approaches a value close to -0.1 (Gellert *et al.* 1976).

Strand-passing reactions catalyzed by type IIA topoisomerases, which change DNA supercoiling, represent switching of the handedness of loops formed by the double helix (Fig. 6.41). In the case of all other type IIA enzymes, except DNA gyrase, the loops are much larger than the size of the proteins, and the directionality of the reaction shown in

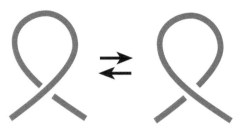

Figure 6.41 Two major topological intramolecular reactions catalyzed by type IIA topoisomerases. This reaction, depending on its direction, changes the value of Lk by $+2$ or -2. Only a small fraction of the intramolecular strand-passing reactions involve more complex loops; such reactions can create various knots and unknot double-stranded circular DNA.

Figure 6.42 Diagram of the DNA pathway in the DNA–gyrase complex. Wrapping DNA around either the left or right C-terminal domain (shown in gray) creates a left-handed loop, so the strand-passing will reduce the value of Lk of the complementary strands. A black-and-white version of this figure will appear in some formats. For the color version, please refer to the plate section.

Fig. 6.41 cannot be controlled by the enzymes. In general, the directionality is random, although DNA supercoiling can influence it. To introduce negative supercoiling the reaction has to go from right to left in Fig. 6.41, that is, to convert left-handed loops into right-handed ones. The only way to provide this condition is to control the conformation of the loop, and this is exactly what DNA gyrase does (Kampranis *et al.* 1999). The model of its action is shown in Fig. 6.42.

Still, with a small probability DNA gyrase is able to bind a T segment that is not wrapped around a C-terminal domain and well separated from the bound G segment along the DNA contour. In this way the enzyme catalyzes linking/unlinking of circular DNA molecules as well as their knotting/unknotting (Kreuzer & Cozzarelli 1980, Mizuuchi *et al.* 1980).

DNA gyrase has been found in the majority of bacteria, but eukaryotes do not have this enzyme.

6.9.5   Reverse gyrase

The enzyme was discovered by Kikuchi and Asai in hyperthermophilic bacteria (Kikuchi & Asai 1984). It creates positive supercoiling in circular DNA by consuming the energy of ATP hydrolysis coupled with the strand-passing reaction. The enzyme was classified as type IA topoisomerase since its catalytic part has structural homology with other enzymes of this type. Reverse gyrase cuts a single DNA strand during the enzymatic act. Although the 3D structure of the enzyme is known (Rodriguez & Stock 2002, Rudolph *et al.* 2012), there is only limited understanding of how the reverse gyrase introduces positive supercoiling. In general, it has to follow the path diagramed in Fig. 6.37, showing how type IA topoisomerases increase the Lk of circular DNA by 1, exactly what the reverse gyrase does during the catalytic act. However, the latter enzyme has to overcome two obstacles there. First, spontaneous formation of open regions is extremely rare in positively supercoiled DNA. Therefore, the enzyme creates an open region itself, by using its helicase domain (Rodriguez & Stock 2002, Rudolph *et al.* 2012). The presence of an extended AT sequence in the circular DNA facilitates this task (Slesarev 1988). The needed free energy comes from ATP hydrolysis incorporated in the catalytic act. Second, in positively supercoiled DNA spontaneous interwinding in the open region has to be right handed rather than left handed which is needed to increase the value of Lk. The only solution for the enzyme is a forceful creation of the left-handed interwinding in the open region by interacting with the intact strand.

So far the enzyme has been found only in hyperthermophilic bacteria.

6.9.6   Thermodynamics of the strand-passing reaction and simplification of DNA topology

Some of the topoisomerases consume the energy of ATP hydrolysis during their catalytic acts, while others do not. It is interesting to analyze this issue in some detail. Clearly, DNA gyrase and reverse gyrase must consume energy since they introduce supercoiling in circular DNA, increasing the free energy of the molecules. Therefore, we exclude these enzymes from the following discussion.

Type IA and type IB topoisomerases do not consume energy during their catalytic acts. This seems reasonable since they only reduce the free energy of DNA molecules and the chemical transformation accompanying the strand-passing, transesterification, does not require additional energy. This reasoning, however, does not work in the case of type IIA topoisomerases. These enzymes do not introduce supercoiling either and could work similarly to type IA enzymes, without energy consumption. The difference, from a thermodynamic point of view, is that, while type IA topoisomerases use the same gate for entrance and exit of the second segment, type IIA topoisomerases use the two-gate mechanism. A T segment enters the enzyme through the N gate and exits through the C gate (Fig. 6.39). Thus, type IIA topoisomerases catalyze a unidirectional transport of DNA segments relative to their bodies. A unidirectional transport requires energy since, in principle, it can be used to shift a system away from the thermodynamic equilibrium. When Roca and Wang established the two gate mechanism of the enzyme action

(Roca & Wang 1994, Roca *et al.* 1996), it was not clear why nature used this mechanism rather then one with a single gate that does not require energy consumption. An unexpected discovery shed light on this issue.

It is natural to assume that type IIA topoisomerases have to drive the distribution of DNA topological states in a solution to its equilibrium form. Indeed, the equilibrium distribution of topological states would be achieved if DNA segments freely passed through one another, and the enzymes nearly provide this condition. Contrary to this expectation, Rybenkov and his colleagues found that type IIA DNA topoisomerases dramatically reduce fractions of knotted and linked circular DNA molecules compared to the corresponding fractions at thermodynamic equilibrium (Rybenkov *et al.* 1997a). This property of the enzymes makes clear biological sense, since removing links between DNA molecules is one of the major functions of the enzymes. It explains why nature chose the energy-consuming two-gate reaction mechanism for type IIA DNA topoisomerases. There is no way to shift a system away from thermodynamic equilibrium without energy consumption, so in one way or another type IIA topoisomerases have to consume energy.

It still remains not completely clear how the enzymes accomplish the topology simplification. Indeed, topology is a global property of circular DNA molecules. These molecules are random coils in solution, with a coil size many times larger than the size of proteins. Thus, the topology of DNA molecules cannot be deduced by enzymes that only interact with DNA locally. Correspondingly, there is no way for the enzymes to uniquely determine the topological consequences of a particular strand-passing event that can either unlink or link two circular DNA molecules. The only way for a small enzyme to achieve topology simplification is to take advantage of the fact that the probability of some specific local conformations of DNA segments depends on DNA topology. If the enzymes only catalyze strand-passing for a local conformation that appears more often in knotted/linked molecules, they will reduce the steady-state fractions of knots and links. A few models have been suggested to explain the phenomenon, and explicitly or implicitly this idea was behind all of them (Yan *et al.* 1999, Vologodskii *et al.* 2001, Buck & Zechiedrich 2004, Vologodskii 2009).

The most promising model suggests that the enzymes create a sharp bend in the G segment and catalyze unidirectional passage of a T segment from inside to outside the hairpin formed by the protein-bound G segment (Vologodskii *et al.* 2001). This directionality of the strand passage is only local, since the hairpin can have any orientation relative to the DNA chain. Quantitative analysis shows, however, that this mechanism can provide a large decrease of the steady-state fraction of knots and catenanes relative to the corresponding equilibrium fractions (Vologodskii *et al.* 2001). The mechanism of topology simplification in this model is attributed to the entropic effect of a knot/link localization (Fig. 6.43).

The model received very strong support when the X-ray structure of a complex between a truncated yeast Topo II and a G segment was resolved (Dong & Berger 2007). The DNA segment was bent at about 150° in the complex. The second requirement of the model was established earlier, when Roca *et al.* showed in their very elegant experiments that yeast Topo II catalyzes the passage of a T segment in a specific direction relative to

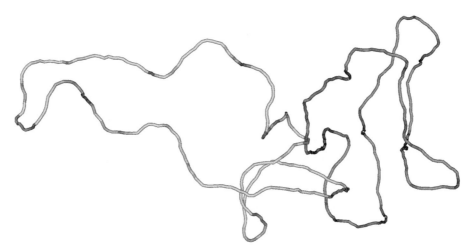

Figure 6.43 Mechanism of topology simplification in the model of a hairpin-like G segment (Vologodskii *et al.* 2001). In this simulated conformation of linked DNA molecules a potential T segment is located inside the hairpin made by the protein-bound G segment (the segment is shown in red, without bound protein). Such conformations make the link between the chains nearly local and increase conformational freedom for the rest of the linked chains. Therefore, the hairpin-like conformation of the G segment increases the probability that T and G segments will be properly juxtaposed in linked DNA molecules, accelerating unlinking. A black-and-white version of this figure will appear in some formats. For the color version, please refer to the plate section.

the enzyme body (Roca *et al.* 1996). Thus, at the qualitative level both assumptions of the model have been validated experimentally. Still, the hairpin-like G segment model fails to explain the phenomenon quantitatively. More studies of the issue are needed to reach complete understanding of the phenomenon.

References

Aboul-ela, F., Bowater, R. P. & Lilley, D. M. J. (1992). Competing B–Z and helix–coil conformational transitions in supercoiled plasmid DNA. *J. Biol. Chem.* 267, 1776–85.

Adams, C. C. (1994). *The Knot Book.* New York: Freeman.

Adrian, M., ten Heggeler-Bordier, B., Wahli, W., Stasiak, A. Z., Stasiak, A. & Dubochet, J. (1990). Direct visualization of supercoiled DNA molecules in solution. *EMBO J.* 9, 4551–4.

Alexander, J. W. (1928). Topological invariants of knots and links. *Trans. Am. Math. Soc.* 30, 275–306.

Bauer, W. & Vinograd, J. (1968). The interaction of closed circular DNA with intercalative dyes. I. The superhelix density of SV40 DNA in the presence and absence of dye. *J. Mol. Biol.* 33, 141–71.

Bauer, W. & Vinograd, J. (1970). The interaction of closed circular DNA with intercalative dyes. II. The free energy of superhelix formation in SV40 DNA. *J. Mol. Biol.* 47, 419–35.

Bauer, W. R. (1978). Structure and reactions of closed duplex DNA. *Annu. Rev. Biophys. Bioeng.* 7, 287–313.

Bauer, W. R. & Benham, C. J. (1993). The free energy, enthalpy and entropy of native and of partially denatured closed circular DNA. *J. Mol. Biol.* 234, 1184–96.

Bauer, W. R., Crick, F. H. C. & White, J. H. (1980). Supercoiled DNA. *Sci. Am.* 243, 100–13.

Beard, P., Morrow, J. F. & Berg, P. (1973). Cleavage of circular, superhelical simian virus 40 DNA to a linear duplex by S1 nuclease. *J. Virol.* 12, 1303–13.

Bednar, J., Furrer, P., Stasiak, A., Dubochet, J., Egelman, E. H. & Bates, A. D. (1994). The twist, writhe and overall shape of supercoiled DNA change during counterion-induced transition from a loosely to a tightly interwound superhelix. *J. Mol. Biol.* 235, 825–47.

Benham, C. J. (1978). The statistics of superhelicity. *J. Mol. Biol.* 123, 361–70.

Benham, C. J. (1979). Torsional stress and local denaturation in supercoiled DNA. *Proc. Natl. Acad. Sci. U. S. A.* 76, 3870–4.

Benham, C. J. (1983). Statistical mechanical analysis of competing conformational transitions in superhelical DNA. *Cold Spring Harbor Symp. Quant. Biol.* 47, 219–27.

Bliska, J. B. & Cozzarelli, N. R. (1987). Use of site-specific recombination as a probe of DNA structure and metabolism in vivo. *J. Mol. Biol.* 194, 205–18.

Boles, T. C., White, J. H. & Cozzarelli, N. R. (1990). Structure of plectonemically supercoiled DNA. *J. Mol. Biol.* 213, 931–51.

Borst, P., Overdulve, J. P., Weijers, P. J., Fase-Fowler, F. & Van den Berg, M. (1984). DNA circles with cruciforms from *Isospora (Toxoplasma) gondii. Biochim. Biophys. Acta* 781, 100–11.

Brahms, S., Nakasu, S., Kikuchi, A. & Brahms, J. G. (1989). Structural changes in positively and negatively supercoiled DNA. *Eur. J. Biochem.* 184, 297–303.

Brahms, S., Vergne, J., Brahms, J. G., Di Capua, E., Bucher, P. & Koller, T. (1982). Natural DNA sequences can form left-handed helices in low salt solution under conditions of topological constraint. *J. Mol. Biol.* 162, 473–93.

Brown, P. O. & Cozzarelli, N. R. (1979). A sign inversion mechanism for enzymatic supercoiling of DNA. *Science* 206, 1081–3.

Brown, P. O. & Cozzarelli, N. R. (1981). Catenation and knotting of duplex DNA by type 1 topoisomerases: a mechanistic parallel with type 2 topoisomerases. *Proc. Natl. Acad. Sci. U. S. A.* 78, 843–7.

Browning, D. F., Grainger, D. C. & Busby, S. J. W. (2010). Effects of nucleoid-associated proteins on bacterial chromosome structure and gene expression. *Curr. Opin. Microbiol.* 13, 773–80.

Buck, G. R. & Zechiedrich, E. L. (2004). DNA disentangling by type-2 topoisomerases. *J. Mol. Biol.* 340, 933–9.

Calugareanu, G. (1961). Sur las classes d'isotopie des noeuds tridimensionnels et leurs invariants. *Czech. Math. J.* 11, 588–625.

Champoux, J. J. (2001). DNA topoisomerases: structure, function, and mechanism. *Annu. Rev. Biochem.* 70, 369–413.

Champoux, J. J. & Dulbecco, R. (1972). An activity from mammalian cells that untwists super-helical DNA – a possible swivel for DNA replication. *Proc. Natl. Acad. Sci. U. S. A.* 69, 143–6.

Charvin, G., Bensimon, D. & Croquette, V. (2003). Single-molecule study of DNA unlinking by eukaryotic and prokaryotic type-II topoisomerases. *Proc. Natl. Acad. Sci. U. S. A.* 100, 9820–5.

Courey, A. J. & Wang, J. C. (1983). Cruciform formation in a negatively supercoiled DNA may be kinetically forbidden under physiological conditions. *Cell* 33, 817–29.

Crisona, N. J., Kanaar, R., Gonzalez, T. N., Zechiedrich, E. L., Klippel, A. & Cozzarelli, N. R. (1994). Processive recombination by wild-type Gin and an enhancer-independent mutant:

Insight into the mechanisms of recombination selectivity and strand exchange. *J. Mol. Biol.* 243, 437–57.

Crisona, N. J., Strick, T. R., Bensimon, D., Croquette, V. & Cozzarelli, N. R. (2000). Preferential relaxation of positively supercoiled DNA by *E. coli* topoisomerase IV in single-molecule and ensemble measurements. *Genes Dev.* 14, 2881–92.

Dayn, A., Malkhosyan, S., Duzhy, D., Lyamichev, V., Panchenko, Y. & Mirkin, S. (1991). Formation of (dA-dT)n cruciform in E. coli cells under different environmental conditions. *J. Bacteriol.* 173, 2658–64.

Dayn, A., Malkhosyan, S. & Mirkin, S. M. (1992). Transcriptionally driven cruciform formation in vivo. *Nucleic Acids Res.* 20, 5991–7.

Dedon, P. C., Dederich, D. A. & Barth, M. C. (2009). An improved method for large-scale preparation of negatively and positively supercoiled plasmid DNA. *BioTechniques* 47, 633–5.

Dekker, N. H., Rybenkov, V. V., Duguet, M., Crisona, N. J., Cozzarelli, N. R., Bensimon, D. & Croquette, V. (2002). The mechanism of type IA topoisomerases. *Proc. Natl. Acad. Sci. U. S. A.* 99, 12126–31.

Depew, R. E. & Wang, J. C. (1975). Conformational fluctuations of DNA helix. *Proc. Natl. Acad. Sci. U. S. A.* 72, 4275–9.

DiGate, R. J. & Marians, K. J. (1988). Identification of a potent decatenating enzyme from *Escherichia coli*. *J. Biol. Chem.* 263, 13366–73.

Dong, K. C. & Berger, J. M. (2007). Structural basis for gate-DNA recognition and bending by type IIA topoisomerases. *Nature* 450, 1201–5.

Drlica, K. (1992). Control of bacterial DNA supercoiling. *Mol. Microbiol.* 6, 425–33.

Du, Q., Livshits, A., Kwiatek, A., Jayaram, M. & Vologodskii, A. (2007). Protein-induced local DNA bends regulate global topology of recombination products. *J. Mol. Biol.* 368, 170–82.

Dulbecco, R. & Vogt, M. (1963). Evidence for a ring structure of polyoma virus DNA. *Proc. Natl. Acad. Sci. U. S. A.* 50, 236–43.

Ellison, M. J., Fenton, M. J., Ho, P. S. & Rich, A. (1987). Long-range interactions of multiple DNA structural transitions within a common topological domain. *EMBO J.* 6, 1513–22.

Ellison, M. J., Kelleher, R. J., 3rd, Wang, A. H., Habener, J. F. & Rich, A. (1985). Sequence-dependent energetics of the B-Z transition in supercoiled DNA containing nonalternating purine-pyrimidine sequences. *Proc. Natl. Acad. Sci. U. S. A.* 82, 8320–4.

Frank-Kamenetskii, M. D., Lukashin, A. V., Anshelevich, V. V. & Vologodskii, A. V. (1985). Torsional and bending rigidity of the double helix from data on small DNA rings. *J. Biomol. Struct. Dyn.* 2, 1005–12.

Frank-Kamenetskii, M. D. & Vologodskii, A. V. (1981). Topological aspects of the physics of polymers: the theory and its biophysical applications. *Sov. Phys. Usp.* 24, 679–96.

Frank-Kamenetskii, M. D. & Vologodskii, A. V. (1984). Thermodynamics of the B–Z transition in superhelical DNA. *Nature* 307, 481–2.

Fuller, F. B. (1971). The writhing number of a space curve. *Proc. Natl. Acad. Sci. U. S. A.* 68, 815–19.

Gagua, A. V., Belintsev, B. N. & Lyubchenko, Y. L. (1981). Effect of base-pair stability on the melting of superhelical DNA. *Nature* 294, 662–3.

Gebe, J. A., Delrow, J. J., Heath, P. J., Fujimoto, B. S., Stewart, D. W. & Schurr, J. M. (1996). Effects of Na^+ and Mg^{2+} on the structures of supercoiled DNAs: comparison of simulations with experiments. *J. Mol. Biol.* 262, 105–28.

Gebe, J. A. & Schurr, J. M. (1996). Thermodynamics of the first transition in writhe of a small circular DNA by Monte Carlo simulation. *Biopolymers* 38, 493–503.

Geggier, S., Kotlyar, A. & Vologodskii, A. (2011). Temperature dependence of DNA persistence length. *Nucleic Acids Res.* 39, 1419–26.

Geggier, S. & Vologodskii, A. (2010). Sequence dependence of DNA bending rigidity. *Proc. Natl. Acad. Sci. U. S. A.* 107, 15421–6.

Gellert, M., Mizuuchi, K., O'Dea, M. H. & Nash, H. A. (1976). DNA gyrase: an enzyme that introduces superhelical turns into DNA. *Proc. Natl. Acad. Sci. U. S. A.* 73, 3872–6.

Gellert, M., Mizuuchi, K., O'Dea, M. H., Ohmori, H. & Tomizawa, J. (1979). DNA gyrase and DNA supercoiling. *Cold Spring Harbor Symp. Quant. Biol.* 43, 35–40.

Gellert, M., O'Dea, M. H. & Mizuuchi, K. (1983). Slow cruciform transitions in palindromic DNA. *Proc. Natl. Acad. Sci. U. S. A.* 80, 5545–9.

Gray, H. B., Jr., Ostrander, D. A., Hodnett, J. L., Legerski, R. J. & Robberson, D. L. (1975). Extracellular nucleases of Pseudomonas BAL 31. I. Characterization of single strand-specific deoxyriboendonuclease and double-strand deoxyriboexonuclease activities. *Nucleic Acids Res.* 2, 1459–92.

Greaves, D. R., Patient, R. K. & Lilley, D. M. (1985). Facile cruciform formation by an $(A-T)_{34}$ sequence from a *Xenopus* globin gene. *J. Mol. Biol.* 185, 461–78.

Halford, S. E., Welsh, A. J. & Szczelkun, M. D. (2004). Enzyme-mediated DNA looping. *Annu. Rev. Biophys. Biomol. Struct.* 33, 1–24.

Hammermann, M., Brun, N., Klenin, K. V., May, R., Toth, K. & Langowski, J. (1998). Salt-dependent DNA superhelix diameter studied by small angle neutron scattering measurements and Monte Carlo simulations. *Biophys. J.* 75, 3057–63.

Haniford, D. B. & Pulleyblank, D. E. (1983a). Facile transition of poly[d(TG) · d(CA)] into a left-handed helix in physiological conditions. *Nature* 302, 632–4.

Haniford, D. B. & Pulleyblank, D. E. (1983b). The in-vivo occurrence of Z DNA. *J. Biomol. Struct. Dyn.* 1, 593–609.

Haniford, D. B. & Pulleyblank, D. E. (1985). Transition of a cloned $d(AT)_n · d(AT)_n$ tract to a cruciform in vivo. *Nucleic Acids Res.* 13, 4343–63.

Hentschel, C. C. (1982). Homocopolymer sequences in the spacer of a sea urchin histone gene repeat are sensitive to S1 nuclease. *Nature* 295, 714–6.

Hiasa, H. & Marians, K. J. (1994). Topoisomerase III, but not topoisomerase I, can support nascent chain elongation during theta-type DNA replication. *J. Biol. Chem.* 269, 32655–9.

Higgins, N. P. & Vologodskii, A. V. (2004). Topological behavior of plasmid DNA. In *Plasmid Biology*, eds. G. Phillips & B. Funnell, 181–201. Washington, DC: ASM Press.

Ho, P. S., Ellison, M. J., Quigley, G. J. & Rich, A. (1986). A computer aided thermodynamic approach for predicting the formation of Z-DNA in naturally occurring sequences. *EMBO. J.* 5, 2737–44.

Hochschild, A. (1990). Protein–protein interactions and DNA loop formation. In *DNA Topology and its Biological Effects*, eds. N. R. Cozzarelli & J. C. Wang, 107–38. Cold Spring Harbor, NY: Cold Spring Habor Laboratory Press.

Hochschild, A. & Ptashne, M. (1986). Cooperative binding of lambda repressors to sites separated by integral turns of the DNA helix. *Cell* 44, 681–7.

Horowitz, D. S. & Wang, J. C. (1984). Torsional rigidity of DNA and length dependence of the free energy of DNA supercoiling. *J. Mol. Biol.* 173, 75–91.

Hsieh, T. S. & Wang, J. C. (1975). Thermodynamic properties of superhelical DNAs. *Biochemistry* 14, 527–35.

Htun, H., Lund, E. & Dahlberg, J. E. (1984). Human U1 RNA genes contain an unusually sensitive nuclease S1 cleavage site within the conserved 3′ flanking region. *Proc. Natl. Acad. Sci. U. S. A.* 81, 7288–92.

Kampranis, S. C., Bates, A. D. & Maxwell, A. (1999). A model for the mechanism of strand passage by DNA gyrase. *Proc. Natl. Acad. Sci. U. S. A.* 96, 8414–9.

Kelleher, R. J., 3rd, Ellison, M. J., Ho, P. S. & Rich, A. (1986). Competitive behavior of multiple, discrete B-Z transitions in supercoiled DNA. *Proc. Natl. Acad. Sci. U. S. A.* 83, 6342–6.

Keller, W. (1975). Determination of the number of superhelical turns in simian virus 40 DNA by gel electrophoresis. *Proc. Natl. Acad. Sci. U. S. A.* 72, 4876–80.

Kikuchi, A. & Asai, K. (1984). Reverse gyrase – a topoisomerase which introduces positive superhelical turns into DNA. *Nature* 309, 677–81.

King, I. F., Yandava, C. N., Mabb, A. M., Hsiao, J. S., Huang, H. S., Pearson, B. L., Calabrese, J. M., Starmer, J., Parker, J. S., Magnuson, T., Chamberlain, S. J., Philpot, B. D. & Zylka, M. J. (2013). Topoisomerases facilitate transcription of long genes linked to autism. *Nature* 501, 58–62.

Kirkegaard, K. & Wang, J. C. (1985). Bacterial DNA topoisomerase I can relax positively super-coiled DNA containing a single-stranded loop. *J. Mol. Biol.* 185, 625–37.

Klenin, K. V., Vologodskii, A. V., Anshelevich, V. V., Dykhne, A. M. & Frank-Kamenetskii, M. D. (1988). Effect of excluded volume on topological properties of circular DNA. *J. Biomol. Struct. Dyn.* 5, 1173–85.

Klenin, K. V., Vologodskii, A. V., Anshelevich, V. V., Klisko, V. Y., Dykhne, A. M. & Frank-Kamenetskii, M. D. (1989). Variance of writhe for wormlike DNA rings with excluded volume. *J. Biomol. Struct. Dyn.* 6, 707–14.

Klug, A. & Lutter, L. C. (1981). The helical periodicity of DNA on the nucleosome. *Nucleic Acids Res.* 17, 4267–83.

Koster, D. A., Croquette, V., Dekker, C., Shuman, S. & Dekker, N. H. (2005). Friction and torque govern the relaxation of DNA supercoils by eukaryotic topoisomerase IB. *Nature* 434, 671–4.

Kouzine, F., Gupta, A., Baranello, L., Wojtowicz, D., Ben-Aissa, K., Liu, J., Przytycka, T. M. & Levens, D. (2013). Transcription-dependent dynamic supercoiling is a short-range genomic force. *Nat. Struct. Mol. Biol.* 20, 396–403.

Kramer, H., Amouyal, M., Nordheim, A. & Muller-Hill, B. (1988). DNA supercoiling changes the spacing requirement of two lac operators for DNA loop formation with lac repressor. *EMBO J.* 7, 547–56.

Kramer, P. R., Bat, O. & Sinden, R. R. (1999). Measurement of localized DNA supercoiling and topological domain size in eukaryotic cells. *Methods Enzymol.* 304, 639–50.

Kramer, P. R. & Sinden, R. R. (1997). Measurement of unrestrained negative supercoiling and topological domain size in living human cells. *Biochemistry* 36, 3151–8.

Krasilnikov, A. S., Podtelezhnikov, A., Vologodskii, A. & Mirkin, S. M. (1999). Large-scale effects of transcriptional DNA supercoiling in vivo. *J. Mol. Biol.* 292, 1149–60.

Krasnow, M. A., Stasiak, A., Spengler, S. J., Dean, F., Koller, T. & Cozzarelli, N. R. (1983). Determination of the absolute handedness of knots and catenanes of DNA. *Nature* 304, 559–60.

Kreuzer, K. N. & Cozzarelli, N. R. (1980). Formation and resolution of DNA catenanes by DNA gyrase. *Cell* 20, 245–54.

Krueger, A., Protozanova, E. & Frank-Kamenetskii, M. D. (2006). Sequence-dependent base pair opening in DNA double helix. *Biophys. J.* 90, 3091–9.

Krylov, D. Y., Makarov, V. L. & Ivanov, V. I. (1990). The B–A transition in superhelical DNA. *Nucleic Acids Res.* 18, 759–61.

LaMarr, W. A., Sandman, K. M., Reeve, J. N. & Dedon, P. C. (1997). Large scale preparation of positively supercoiled DNA using the archaeal histone HMf. *Nucleic Acids Res.* 25, 1660–1.

Larsen, A. & Weintraub, H. (1982). An altered DNA conformation detected by S1 nuclease occurs at specific regions in active chick globin chromatin. *Cell* 29, 609–22.

Laundon, C. H. & Griffith, J. D. (1988). Curved helix segments can uniquely orient the topology of supertwisted DNA. *Cell* 52, 545–9.

Le Bret, M. (1979). Catostrophic variation of twist and writhing of circular DNAs with constraint? *Biopolymers* 18, 1709–25.

Le Bret, M. (1980). Monte Carlo computation of supercoiling energy, the sedimentation constant, and the radius of gyration of unknotted and knotted circular DNA. *Biopolymers* 19, 619–37.

Lee, C. H., Mizusawa, H. & Kakefuda, T. (1981). Unwinding of double-stranded DNA helix by dehydration. *Proc. Natl. Acad. Sci. U. S. A.* 78, 2838–42.

Lee, D. W. & Schleif, R. F. (1989). In vivo DNA loops in araCBAD: Size limits and helical repeat. *Proc. Natl. Acad. Sci. U. S. A.* 86, 476–80.

Lilley, D. M. (1980). The inverted repeat as a recognizable structural feature in supercoiled DNA molecules. *Proc. Natl. Acad. Sci. U. S. A.* 77, 6468–72.

Liu, L. F., Liu, C.-C. & Alberts, B. M. (1980). Type II DNA topoisomerases: enzymes that can unknot a topologically knotted DNA molecule via a reversible double-strand break. *Cell* 19, 697–707.

Liu, L. F., Liu, C. C. & Alberts, B. M. (1979). T4 DNA topoisomerase: a new ATP-dependent enzyme essential for initiation of T4 bacteriophage DNA replication. *Nature* 281, 456–61.

Liu, L. F. & Wang, J. C. (1987). Supercoiling of the DNA template during transcription. *Proc. Natl. Acad. Sci. U. S. A.* 84, 7024–7.

Ljungman, M. & Hanawalt, P. C. (1992). Localized torsional tension in the DNA of human cells. *Proc. Natl. Acad. Sci. U. S. A.* 89, 6055–9.

Lyamichev, V., Panyutin, I. & Mirkin, S. (1984). The absence of cruciform structures from pAO3 plasmid DNA in vivo. *J. Biomol. Struct. Dyn.* 2, 291–301.

Lyamichev, V. I., Mirkin, S. M. & Frank-Kamenetskii, M. D. (1985). A pH-dependent structural transition in the homopurine–homopyrimidine tract in superhelical DNA. *J. Biomol. Struct. Dyn.* 3, 327–38.

Lyamichev, V. I., Mirkin, S. M. & Frank-Kamenetskii, M. D. (1986). Structures of homopurine–homopyrimidine tract in superhelical DNA. *J. Biomol. Struct. Dyn.* 3, 667–9.

Lyamichev, V. I., Mirkin, S. M. & Frank-Kamenetskii, M. D. (1987). Structure of (dG)n.(dC)n under superhelical stress and acid pH. *J. Biomol. Struct. Dyn.* 5, 275–82.

Lyamichev, V. I., Mirkin, S. M., Kumarev, V. P., Baranova, L. V., Vologodskii, A. V. & Frank-Kamenetskii, M. D. (1989). Energetics of the B–H transition in supercoiled DNA carrying d(CT)$_x$ · d(AG)$_x$ and d(C)$_n$ · d(G)$_n$ inserts. *Nucleic Acids Res.* 17, 9417–23.

Lyamichev, V. I., Panyutin, I. G. & Frank-Kamenetskii, M. D. (1983). Evidence of cruciform structures in superhelical DNA provided by two-dimensional gel electrophoresis. *FEBS Lett.* 153, 298–302.

Lyubchenko, Y. L. & Shlyakhtenko, L. S. (1997). Visualization of supercoiled DNA with atomic force microscopy in situ. *Proc. Natl. Acad. Sci. U. S. A.* 94, 496–501.

McClellan, J., Boublikova, P., Palecek, E. & Lilley, D. (1990). Superhelical torsion in cellular DNA responds directly to enironmental and genetic factors. *Proc. Natl. Acad. Sci. U. S. A.* 87, 8373–7.

Menzel, R. & Gellert, M. (1987). Regulation of the genes for E. coli DNA gyrase: homeostatic control of DNA supercoiling. *Cell* 34, 105–13.

Mirkin, S. M. & Frank-Kamenetskii, M. D. (1994). H-DNA and related structures. *Annu. Rev. Biophys. Biomol. Struct.* 23, 541–76.

Mirkin, S. M., Lyamichev, V. I., Drushlyak, K. N., Dobrynin, V. N., Filippov, S. A. & Frank-Kamenetskii, M. D. (1987a). DNA H form requires a homopurine–homopyrimidine mirror repeat. *Nature* 330, 495–7.

Mirkin, S. M., Lyamichev, V. I., Kumarev, V. P., Kobzev, V. F., Nosikov, V. V. & Vologodskii, A. V. (1987b). The energetics of the B–Z transition in DNA. *J. Biomol. Struct. Dyn.* 5, 79–88.

Mizuuchi, K., Fisher, L. M., O'Dea, M. H. & Gellert, M. (1980). DNA gyrase action involves the introduction of transient double-strand breaks into DNA. *Proc. Natl. Acad. Sci. U. S. A.* 77, 1847–51.

Naughton, C., Avlonitis, N., Corless, S., Prendergast, J. G., Mati, I. K., Eijk, P. P., Cockroft, S. L., Bradley, M., Ylstra, B. & Gilbert, N. (2013). Transcription forms and remodels supercoiling domains unfolding large-scale chromatin structures. *Nat. Struct. Mol. Biol.* 20, 387–95.

Naylor, L. H., Lilley, D. M. & van de Sande, J. H. (1986). Stress-induced cruciform formation in a cloned d(CATG)$_{10}$ sequence. *EMBO J.* 5, 2407–13.

Neale, B. M., Kou, Y., Liu, L., Ma'ayan, A., Samocha, K. E., Sabo, A., Lin, C. F., Stevens, C., Wang, L. S., Makarov, V., Polak, P., Yoon, S., Maguire, J., Crawford, E. L., Campbell, N. G., Geller, E. T., Valladares, O., Schafer, C., Liu, H., Zhao, T., Cai, G., Lihm, J., Dannenfelser, R., Jabado, O., Peralta, Z., Nagaswamy, U., Muzny, D., Reid, J. G., Newsham, I., Wu, Y., Lewis, L., Han, Y., Voight, B. F., Lim, E., Rossin, E., Kirby, A., Flannick, J., Fromer, M., Shakir, K., Fennell, T., Garimella, K., Banks, E., Poplin, R., Gabriel, S., DePristo, M., Wimbish, J. R., Boone, B. E., Levy, S. E., Betancur, C., Sunyaev, S., Boerwinkle, E., Buxbaum, J. D., Cook, E. H., Jr., Devlin, B., Gibbs, R. A., Roeder, K., Schellenberg, G. D., Sutcliffe, J. S. & Daly, M. J. (2012). Patterns and rates of exonic de novo mutations in autism spectrum disorders. *Nature* 485, 242–5.

Panayotatos, N. & Wells, R. D. (1981). Cruciform structures in supercoiled DNA. *Nature* 289, 466–70.

Panyutin, I., Klishko, V. & Lyamichev, V. (1984). Kinetics of cruciform formation and stability of cruciform structure in superhelical DNA. *J. Biomol. Struct. Dyn.* 1, 1311–24.

Panyutin, I., Lyamichev, V. & Mirkin, S. (1985). A structural transition in d(AT)$_n$ · d(AT)$_n$ inserts within superhelical DNA. *J. Biomol. Struct. Dyn.* 2, 1221–34.

Peck, L. J., Wang, J. C., Nordheim, A. & Rich, A. (1986). Rate of B to Z structural transition of supercoiled DNA. *J. Mol. Biol.* 190, 125–7.

Pettijohn, D. & Phenninger, O. (1980). Supercoils in prokaryotic DNA restrained in vivo. *Proc. Natl. Acad. Sci.* 77, 1331–5.

Pohl, F. M. (1986). Dynamics of the B-to-Z transition in supercoiled DNA. *Proc. Natl. Acad. Sci. U. S. A.* 83, 4783–7.

Pohl, F. M., Thomae, R. & Di Capua, E. (1982). Antibodies to Z-DNA interact with form V DNA. *Nature* 300, 545–6.

Pulleyblank, D. E., Shure, M., Tang, D., Vinograd, J. & Vosberg, H. P. (1975). Action of nicking–closing enzyme on supercoiled and nonsupercoiled closed circular DNA: formation of a Boltzmann distribution of topological isomers. *Proc. Natl. Acad. Sci. U. S. A.* 72, 4280–4.

Rahmouni, A. R. & Wells, R. D. (1989). Stabilization of Z DNA in vivo by localized supercoiling. *Science* 246, 358–63.

Roca, J., Berger, J. M., Harrison, S. C. & Wang, J. C. (1996). DNA transport by a type II topoisomerase – direct evidence for a two-gate mechanism. *Proc. Natl. Acad. Sci. U. S. A.* 93, 4057–62.

Roca, J. & Wang, J. (1994). DNA transport by a type II DNA topoisomerase – evidence in favor of a two-gate mechanism. *Cell* 77, 609–16.

Rodriguez, A. C. & Stock, D. (2002). Crystal structure of reverse gyrase: insights into the positive supercoiling of DNA. *EMBO J.* 21, 418–26.

Rolfsen, D. (1976). *Knots and Links*. Berkeley, CA: Publish or Perish.

Rudolph, M. G., del Toro Duany, Y., Jungblut, S. P., Ganguly, A. & Klostermeier, D. (2012). Crystal structures of Thermotoga maritima reverse gyrase: inferences for the mechanism of positive DNA supercoiling. *Nucleic Acids Res.* DOI: 10.1093/nar/gks1073

Rybenkov, V. V., Cozzarelli, N. R. & Vologodskii, A. V. (1993). Probability of DNA knotting and the effective diameter of the DNA double helix. *Proc. Natl. Acad. Sci. U. S. A.* 90, 5307–11.

Rybenkov, V. V., Ullsperger, C., Vologodskii, A. V. & Cozzarelli, N. R. (1997a). Simplification of DNA topology below equilibrium values by type II topoisomerases. *Science* 277, 690–3.

Rybenkov, V. V., Vologodskii, A. V. & Cozzarelli, N. R. (1997b). The effect of ionic conditions on DNA helical repeat, effective diameter, and free energy of supercoiling. *Nucleic Acids Res.* 25, 1412–18.

Rybenkov, V. V., Vologodskii, A. V. & Cozzarelli, N. R. (1997c). The effect of ionic conditions on the conformations of supercoiled DNA. I. Sedimentation analysis. *J. Mol. Biol.* 267, 299–311.

Rybenkov, V. V., Vologodskii, A. V. & Cozzarelli, N. R. (1997d). The effect of ionic conditions on the conformations of supercoiled DNA. II. Equilibrium catenation. *J. Mol. Biol.* 267, 312–23.

Schoeffler, A. J. & Berger, J. M. (2008). DNA topoisomerases: harnessing and constraining energy to govern chromosome topology. *Q. Rev. Biophys.* 41, 41–101.

Shaw, S. Y. & Wang, J. C. (1993). Knotting of a DNA chain during ring closure. *Science* 260, 533–6.

Shimada, J. & Yamakawa, H. (1988). Moments for DNA topoisomers: the helical wormlike chain. *Biopolymers* 27, 657–73.

Sinden, R. R., Broyles, S. S. & Pettijohn, D. E. (1983). Perfect palindromic lac operator DNA sequence exists as a stable cruciform structure in supercoiled DNA in vitro but not in vivo. *Proc. Natl. Acad. Sci. U. S. A.* 80, 1797–801.

Sinden, R. R., Carlson, J. O. & Pettijohn, D. E. (1980). Torsional tension in the DNA double helix measured with trimethylpsoralen in living E. coli cells: analogous measurements in insect and human cells. *Cell* 21, 773–83.

Singleton, C. K., Klysik, J., Stirdivant, S. M. & Wells, R. D. (1982). Left-handed Z-DNA is induced by supercoiling in physiological ionic conditions. *Nature* 299, 312–6.

Slesarev, A. I. (1988). Positive supercoiling catalysed *in vitro* by ATP-dependent topoisomerase from *Desulfurococcus amylolyticus*. *Eur. J. Biochem.* 173, 395–9.

Stark, W. M. & Boocock, M. R. (1995). Topological selectivity in site-specific recombination. In *Mobile Genetic Elements*, ed. D. J. Sherratt, 101–29. Oxford: Oxford University Press.

Stewart, L., Redinbo, M. R., Qiu, X., Hol, W. G. & Champoux, J. J. (1998). A model for the mechanism of human topoisomerase I. *Science* 279, 1534–41.

Stone, M. D., Bryant, Z., Crisona, N. J., Smith, S. B., Vologodskii, A., Bustamante, C. & Cozzarelli, N. R. (2003). Chirality sensing by *Escherichia coli* topoisomerase IV and the mechanism of type II topoisomerases. *Proc. Natl. Acad. Sci. U. S. A.* 100, 8654–9.

Tse, Y. & Wang, J. C. (1980). E. coli and M. luteus DNA topoisomerase I can catalyze catenation of decatenation of double-stranded DNA rings. *Cell* 22, 269–76.

Ullsperger, C. J., Vologodskii, A. V. & Cozzarelli, A. V. (1995). Unlinking of DNA by topoisomerases during DNA replication. *Nucl. Acids Mol. Biol.* 9, 115–42.

Vinograd, J., Lebowitz, J., Radloff, R., Watson, R. & Laipis, P. (1965). The twisted circular form of polyoma viral DNA. *Proc. Natl. Acad. Sci. U. S. A.* 53, 1104–11.

Vinograd, J., Lebowitz, J. & Watson, R. (1968). Early and late helix–coil transitions in closed circular DNA. The number of superhelical turns in polyoma DNA. *J. Mol. Biol.* 33, 173–97.

Vologodskaia, M. Y. & Vologodskii, A. V. (1999). Effect of magnesium on cruciform extrusion in supercoiled DNA. *J. Mol. Biol.* 289, 851–9.

Vologodskii, A. 2007. Monte Carlo simulation of DNA topological properties. In *Topology in Molecular Biology*, ed. M. Monastryrsky, 23–41. Berlin: Springer.

Vologodskii, A. (2009). Theoretical models of DNA topology simplification by type IIA DNA topoisomerases. *Nucleic Acids Res.* 37, 3125–33.

Vologodskii, A. V., Anshelevich, V. V., Lukashin, A. V. & Frank-Kamenetskii, M. D. (1979a). Statistical mechanics of supercoils and the torsional stiffness of the DNA. *Nature* 280, 294–298.

Vologodskii, A. V. & Cozzarelli, N. R. (1994). Conformational and thermodynamic properties of supercoiled DNA. *Annu. Rev. Biophys. Biomol. Struct.* 23, 609–43.

(1996). Effect of supercoiling on the juxtaposition and relative orientation of DNA sites. *Biophys. J.* 70, 2548–56.

Vologodskii, A. V., Crisona, N. J., Laurie, B., Pieranski, P., Katritch, V., Dubochet, J. & Stasiak, A. (1998). Sedimentation and electrophoretic migration of DNA knots and catenanes. *J. Mol. Biol.* 278, 1–3.

Vologodskii, A. V. & Frank-Kamenetskii, M. D. (1983). The relaxation time for a cruciform structure in superhelical DNA. *FEBS Lett.* 160, 173–6.

Vologodskii, A. V., Levene, S. D., Klenin, K. V., Frank-Kamenetskii, M. & Cozzarelli, N. R. (1992). Conformational and thermodynamic properties of supercoiled DNA. *J. Mol. Biol.* 227, 1224–43.

Vologodskii, A. V., Lukashin, A. V., Anshelevich, V. V. & Frank-Kamenetskii, M. D. (1979b). Fluctuations in superhelical DNA. *Nucleic Acids Res.* 6, 967–82.

Vologodskii, A. V., Zhang, W., Rybenkov, V. V., Podtelezhnikov, A. A., Subramanian, D., Griffith, J. D. & Cozzarelli, N. R. (2001). Mechanism of topology simplification by type II DNA topoisomerases. *Proc. Natl. Acad. Sci. U. S. A.* 98, 3045–9.

Wang, A. H., Gessner, R. V., van der Marel, G. A., van Boom, J. H. & Rich, A. (1985). Crystal structure of Z-DNA without an alternating purine–pyrimidine sequence. *Proc. Natl. Acad. Sci. U. S. A.* 82, 3611–5.

Wang, J. C. (1971). Interaction between DNA and an *Escherichia coli* protein ω. *J. Mol. Biol.* 55, 523–33.

Wang, J. C. (1974a). The degree of unwinding of the DNA helix by ethidium. I. Titration of twisted PM2 DNA molecules in alkaline cesium chloride density gradients. *J. Mol. Biol.* 89, 783–801.

Wang, J. C. (1974b). Interactions between twisted DNAs and enzymes: the effects of superhelical turns. *J. Mol. Biol.* 87, 797–816.

Wang, J. C. (1996). DNA topoisomerases. *Annu. Rev. Biochem.* 65, 635–95.

Wang, J. C. (1998). Moving one DNA double helix through another by a type II DNA topoisomerase: the story of a simple molecular machine. *Q. Rev. Biophys.* 31, 107–44.

Wang, J. C. (2009). *Untangling the Double Helix: DNA Entanglement and the Action of the DNA Topoisomerases.* Cold Spring Harbor, NY: Cold Spring Harbor Laboratory Press.

Wang, J. C., Peck, L. J. & Becherer, K. (1983). DNA supercoiling and its effects on DNA structure and function. *Cold Spring Harbor Symp. Quant. Biol.* 47, 85–91.

Wang, J. C. & Schwartz, H. (1967). Noncomplementarity in base sequences between the cohesive ends of coliphages 186 and lambda and the formation of interlocked rings between the two DNAs. *Biopolymers* 5, 953–66.

Wasserman, S. A. & Cozzarelli, N. R. (1986). Biochemical topology: applications to DNA recombination and replication. *Science* 232, 951–60.

Wasserman, S. A., White, J. H. & Cozzarelli, N. R. (1988). The helical repeat of double-stranded DNA varies as a function of catenation and supercoiling. *Nature* 334, 448–50.

Weil, R. & Vinograd, J. (1963). The cyclic helix and cyclic coil forms of polyoma viral DNA. *Proc. Natl. Acad. Sci. U. S. A.* 50, 730–8.

White, J. H. (1969). Self-linking and the Gauss integral in higher dimensions. *Am. J. Math.* 91, 693–728.

White, J. H., Cozzarelli, N. R. & Bauer, W. R. (1988). Helical repeat and linking number of surface-wrapped DNA. *Science* 241, 323–7.

Wittig, B., Wolfl, S., Dorbic, T., Vahrson, W. & Rich, A. (1992). Transcription of human c-myc in permeabilized nuclei is associated with formation of Z-DNA in three discrete regions of the gene. *EMBO J.* 11, 4653–63.

Wolfl, S., Martinez, C., Rich, A. & Majzoub, J. A. (1996). Transcription of the human corticotropin-releasing hormone gene in NPLC cells is correlated with Z-DNA formation. *Proc. Natl. Acad. Sci. U. S. A.* 93, 3664–8.

Wu, H.-Y., Shyy, S., Wang, J. C. & Liu, L. F. (1988). Transcription generates positively and negatively supercoiled domains in the template. *Cell* 53, 433–40.

Yan, J., Magnasco, M. O. & Marko, J. F. (1999). A kinetic proofreading mechanism for disentanglement of DNA by topoisomerases. *Nature* 401, 932–5.

Zacharias, W., Jaworski, A., Larson, J. E. & Wells, R. D. (1988). The B- to Z-DNA equilibrium in vivo is perturbed by biological processes. *Proc. Natl. Acad. Sci.* 85, 7069.

Index

Printed in the United States
by Baker & Taylor Publisher Services